D0849331

PRINCIPLES OF
INORGANIC MATERIALS
DESIGN

PRINCIPLES OF INORGANIC MATERIALS DESIGN

John. N. Lalena

Consultant
Formerly Senior Research Scientist
Honeywell Electronic Materials

David. A. Cleary

Chairman, Department of Chemistry
Gonzaga University

⊛WILEY-INTERSCIENCE

A JOHN WILEY & SONS, INC., PUBLICATION

Copyright © 2005 by John Wiley & Sons, Inc. All rights reserved.

Published by John Wiley & Sons, Inc., Hoboken, New Jersey.
Published simultaneously in Canada.

No part of this publication may be reproduced, stored in a retrieval system, or transmitted in any form or
by any means, electronic, mechanical, photocopying, recording, scanning, or otherwise, except as
permitted under Section 107 or 108 of the 1976 United States Copyright Act, without either the prior
written permission of the Publisher, or authorization through payment of the appropriate per-copy fee to
the Copyright Clearance Center, Inc., 222 Rosewood Drive, Danvers, MA 01923, 978-750-8400,
fax 978-646-8600, or on the web at www.copyright.com. Requests to the Publisher for permission
should be addressed to the Permissions Department, John Wiley & Sons, Inc., 111 River Street,
Hoboken, NJ 07030, (201) 748-6011, fax (201) 748-6008.

Limit of Liability/Disclaimer of Warranty: While the publisher and author have used their best efforts
in preparing this book, they make no representations or warranties with respect to the accuracy or
completeness of the contents of this book and specifically disclaim any implied warranties of
merchantability or fitness for a particular purpose. No warranty may be created or extended by sales
representatives or written sales materials. The advice and strategies contained herein may not be suitable
for your situation. You should consult with a professional where appropriate. Neither the publisher
nor author shall be liable for any loss of profit or any other commercial damages, including but not
limited to special, incidental, consequential, or other damages.

For general information on our other products and services please contact our Customer Care Department
within the U.S. at 877-762-2974, outside the U.S. at 317-572-3993 or fax 317-572-4002.

Wiley also publishes its books in a variety of electronic formats. Some content that appears in print,
however, may not be available in electronic format.

Library of Congress Cataloging-in-Publication Data is Available

0-471-43418-3

Printed in the United States of America

10 9 8 7 6 5 4 3 2 1

To our families

CONTENTS

FOREWORD

Whereas solid-state physics is concerned with the mathematical description of the varied physical phenomena that solids exhibit and the solid-state chemist is interested in probing the relationships between structural chemistry and physical phenomena, the materials scientist has the task of using these descriptions and relationships to design materials that will perform specified engineering functions. However, the physicist and the chemist are often called on to act as material designers, and the practice of materials design commonly requires the exploration of novel chemistry that may lead to the discovery of physical phenomena of fundamental importance for the body of solid-state physics. I cite three illustrations where an engineering need has led to new physics and chemistry in the course of materials design.

In 1952, I joined a group at the M.I.T. Lincoln Laboratory that had been charged with the task of developing a square $B–H$ hysteresis loop in a ceramic ferrospinel that could have its magnetization reversed in less than 1 μs by an applied magnetic-field strength less than twice the coercive field strength. At that time, the phenomenon of a square $B–H$ loop had been obtained in a few iron alloys by rolling them into tapes so as to align the grains, and hence the easy magnetization directions, along the axis of the tape. The observation of a square $B–H$ loop led Jay Forrester, an electrical engineer, to invent the coincident-current, random-access magnetic memory for the digital computer since, at that time, the only memory available was a 16×16 byte electrostatic storage tube. Unfortunately, the alloy tapes gave too slow a switching speed. As an electrical engineer, Jay Forrester assumed the problem was eddy-current losses in the tapes, so he had turned to the ferrimagnetic ferrospinels that were known to be magnetic insulators. However, the polycrystalline ferrospinels are ceramics that cannot be rolled! Nevertheless, the U.S. Air Force had financed the M.I.T. Lincoln Laboratory to develop an Air Defense System, of which the digital computer was to be a key component. Therefore, Jay Forrester was able to put together an interdisciplinary team of electrical engineers, ceramists, and physicists to realize his random-access magnetic memory with ceramic ferrospinels.

The magnetic memory was achieved by a combination of systematic empiricism, careful materials characterization, theoretical analysis, and the emergence of an unanticipated phenomenon that proved to be a stroke of good fortune. A systematic mapping of the structural, magnetic, and switching properties of the Mg–Mn–Fe ferrospinels as a function of their heat treatments revealed that the spinels in one part of the phase diagram were tetragonal rather than cubic and that compositions just on the cubic side of the cubic-tetragonal phase boundary yield sufficiently square B–H loops if given a carefully controlled heat treatment. This observation led me to propose that the tetragonal distortion was due to a cooperative orbital ordering on the Mn^{3+} ions that would lift the cubic-field orbital degeneracy; cooperativity of the site distortions minimizes the cost in elastic energy and leads to a distortion of the entire structure. This phenomenon is now known as a cooperative Jahn–Teller distortion since Jahn and Teller had earlier pointed out that a molecule or molecular complex having an orbital degeneracy would lower its energy by deforming its configuration to a lower symmetry that removed the degeneracy. Armed with this concept, I was able almost immediately to apply it to interpret the structure and the anisotropic magnetic interactions that had been found in the manganese-oxide perovskites, since the orbital order revealed the basis for specifying the rules for the sign of a magnetic interaction in terms of the occupancies of the overlapping orbitals responsible for the interatomic interactions. These rules are now known as the Goodenough–Kanamori rules for the sign of a superexchange interaction. Thus an engineering problem prompted the discovery and description of two fundamental phenomena in solids that have ever since been used by chemists and physicists to interpret structural and magnetic phenomena in transition-metal compounds and to design new magnetic materials. Moreover, the discovery of cooperative orbital ordering fed back to an understanding of our empirical solution to the engineering problem. By annealing at the optimum temperature for a specified time, the Mn^{3+} ions of a cubic spinel would migrate to form Mn-rich regions where their energy is lowered through cooperative, dynamic orbital ordering. The resulting chemical inhomogeneities acted as nucleating centers for domains of reverse magnetization that, once nucleated, grew away from the nucleating center. We also showed that eddy currents were not responsible for the slow switching of the tapes, but a small coercive field strength and an intrinsic damping factor for spin rotation.

In the early 1970s, an oil shortage focused worldwide attention on the need to develop alternative energy sources, and it soon became apparent that these sources would benefit from energy storage. Moreover, replacing the internal combustion engine with electric-powered vehicles, or at least the introduction of hybrid vehicles, would improve the air quality, particularly in big cities. Therefore, a proposal by the Ford Motor Company to develop a sodium–sulfur battery operating at 300°C with molten electrodes and a ceramic Na^+-ion electrolyte stimulated interest in the design of fast alkali-ion conductors. More significant was interest in a battery in which Li^+ rather than H^+ is the working ion, since the energy density that can be achieved with an aqueous electrolyte is lower than what, in principle, can be obtained with a nonaqueous Li^+-ion electrolyte. However, realization of a Li^+-ion rechargeable battery would require identification of a cathode material into/from which Li^+ ions can be inserted/extracted reversibly. Brian Steele of Imperial College, London, first suggested use of TiS_2, which contains TiS_2 layers held together only by Vander Waals S^{2-}—S^{2-} bonding; lithium can be inserted reversibly between the TiS_2 layers. M. Stanley Whittingham's demonstration was the first to reduce this suggestion to practice while he was at the Exxon Corporation. Whittingham's demonstration of a rechargeable $Li–TiS_2$ battery was commercially nonviable because the lithium anode proved unsafe. Nevertheless, his demonstration focused attention on the work of the chemists Jean Rouxel of Nantes, France, and R. Schöllhorn of Berlin on insertion compounds that provide a convenient means of changing continuously the mixed valency of a fixed transition-metal array across a redox couple. Although work at Exxon was halted, their demonstration had shown that if an insertion compound such as graphite were used as the anode, a viable lithium battery could be achieved, but use of a less electropositive anode would require an alternative insertion-compound cathode material that provided a higher voltage vs. a lithium anode than TiS_2. I was able to deduce that no sulfide would give a significantly higher voltage than that obtained with TiS_2 and therefore that it would be necessary to go to a transition-metal oxide. Although oxides other than V_2O_5 and MoO_3, which contain vandyl or molybdyl ions, do not form layered structures analogous to TiS_2, I knew that $LiMO_2$ compounds exist that have a layered structure similar to that of $LiTiS_2$. It was only necessary to choose the correct M^{3+} cation and to determine how much Li could be extracted before the structure collapsed. That was how the $Li_{1-x}CoO_2$ cathode material was developed that now powers the cell telephones and laptop computers. The choice of M = Co, Ni, or $Ni_{0.5+\delta}Mn_{0.5-\delta}$ was dictated by the position of the redox energies and an octahedral site-preference energy strong enough to inhibit migration of the M atom to the Li layers on removal of Li. Electrochemical studies of these cathode materials, and particularly of $Li_{1-x}Ni_{0.5+\delta}Mn_{0.5-\delta}O_2$, have provided a demonstration of the pinning of a redox couple at the top of the valence band, a concept of singular importance for interpretation of metallic oxides having only M–O–M interactions, of the reason for oxygen evolution at critical Co(IV)/Co(III) or Ni(IV)/Ni(III) ratios in $Li_{1-x}MO_2$ studies, and of why Cu(III) in an oxide has a low-spin configuration. Moreover, exploration of other oxide structures that can act as hosts for insertion of Li as a guest species have provided a means of quantitatively determining the influence of a countercation on the energy of a transition-metal redox couple. This determination allows tuning of the energy of a redox couple, which may prove important for the design of heterogenous catalysts.

As a third example, I turn to the discovery of high-temperature superconductivity in the copper oxides first announced by Bednorz and Müller of IBM Zürich in the summer of 1986. Karl A. Müller, the physicist of the pair, had been thinking that a dynamic Jahn–Teller ordering might provide an enhanced electron–phonon coupling that would raise the superconductive critical temperature T_C. He turned to his chemist colleague Bednorz to make a mixed-valent Cu^{3+}/Cu^{2+} compound, since Cu^{2+} has an orbital degeneracy in an octahedral site. This speculation led to the discovery of the family of high-T_C copper oxides; however, the enhanced electron–phonon coupling is not due to a conventional dynamic Jahn–Teller orbital ordering, but rather to the first-order character of the transition from localized to itinerant electronic behavior of σ-bonding Cu:3d electrons of $(x^2 - y^2)$ symmetry in CuO_2 planes. In this case, the search for an improved engineering material has led to a demonstration that the celebrated Mott–Hubbard transition is generally not smooth as originally assumed, and it has introduced an unanticipated new physics associated with bond-length fluctuations and vibronic electronic properties. It has challenged the theorist to develop new theories of the crossover regime that can describe the mechanism of superconductive pair formation in the copper oxides, quantum critical-point behavior at low temperatures, and an anomalous temperature dependence of the resistivity at higher temperatures as a result of strong electron–phonon interactions.

These examples show how the challenge of materials design from the engineer may lead to new physics as well as to new chemistry. Sorting out the physical and chemical origins of the new phenomena feed back to the range of concepts available to the designer of new engineering materials. In recognition of the critical role in materials design of interdisciplinary cooperation between physicists, chemists, ceramists, metallergists, and engineers that is practiced in industry and government research laboratories, John N. Lalena and David A. Cleary have initiated with their book what should prove to be a growing trend toward greater interdisciplinarity in the education of those who will be engaged in the design and characterization of tomorrow's engineering materials.

JOHN B. GOODENOUGH

Virginia H. Cockrell Centennial Chair in Engineering
The University of Texas at Austin

■■■■■■ PREFACE

Inorganic solid-state chemistry has matured into its own distinct subdiscipline. The reader may wonder why we have decided to add another textbook to the plethora of books already published. Our response is that we see a need for a single-source presentation that recognizes the interdisciplinary nature of the field. Solid-state chemists typically receive a small amount of training in condensed-matter physics, but none in materials science or engineering, and yet all of these traditional fields are inextricable components of inorganic solid-state chemistry.

Materials scientists and engineers have traditionally been primarily concerned with the fabrication and utilization of materials already synthesized by the chemist and identified by the physicist as having the appropriate intrinsic properties for a particular engineering function. Although the demarcation between the three disciplines remains in an academic sense, the separate job distinctions for those working in the field are fading. This is especially obvious in the private sector, where one must ensure that materials used in real commercial devices not only perform their primary function, but also meet a variety of secondary requirements.

Individuals involved with such multidisciplinary projects must be prepared to work independently or to collaborate with other specialists in facing design challenges. In the latter case, communication is enhanced, if each individual is able to speak the "language" of the other. Therefore, in this book we introduce a number of concepts that are not usually covered in standard solid-state chemistry textbooks. When this occurs, we try to follow the introduction of the concept with an appropriate worked example to demonstrate its use. Two areas that have lacked thorough coverage in most solid-state chemistry texts in the past, namely, microstructure and mechanical properties, are treated extensively in this book.

We have kept the mathematics to a minimum—but adequate—level, suitable for a descriptive treatment. Appropriate citations are included for those needing the quantitative details. It is assumed that the reader has sufficient knowledge of calculus and elementary linear algebra, particularly matrix manipulations, and some prior exposure to thermodynamics, quantum theory, and group theory. The book should be satisfactory for senior-level undergraduate or beginning graduate students in chemistry. One will recognize from the Table of Contents that entire textbooks have been devoted to each of the chapters in this book, which indicates the necessary limits on the depth of coverage. Along with their chemistry colleagues, physics and engineering students should also find the book to be informative and useful.

Every attempt has been made to extensively cite all the original and pertinent research in a fashion similar to that found in a review article. Students are encouraged to seek out this work. We have also included biographies of several individuals who have made significant *fundamental* contributions to inorganic materials science in the twentieth century. Limiting these to the small number we have room for was, of course, difficult. The reader should be warned that some topics have been left out. In this book, we only cover nonmolecular inorganic materials. Polymers and other molecular substances are not discussed. Also omitted are coverages of surface science, self-assembly, and composite materials.

We are grateful to Prof. John B. Wiley, Dr. Nancy F. Dean, Dr. Martin W. Weiser, Dr. Everett E. Carpenter, and Dr. Thomas K. Kodenkandath for reviewing various chapters in this book. We are grateful to Prof. John F. Nye, Prof. John B. Goodenough, Dr. Frans Spaepen, Dr. Larry Kaufman, and Dr. Bert Chamberland for providing biographical information. We also thank Prof. Philip Andersn, Prof. Mats H. Hillert, Prof. Nye, Dr. Kaufman, Dr. Terrell Vanderah, Dr. Barbara Sewall, and Mrs. Jennifer Moss for allowing us to use photographs from their personal collections. Finally, we acknowledge the inevitable neglect our families must have felt during the period taken to write this book. We are grateful for their understanding and tolerance.

CHAPTER 1

The Mesoscale

The prefix *meso-* comes from the Greek *mesos*, meaning "intermediate" or "in the middle." Materials scientists and engineers describe the structure of a substance at four different length scales: macroscopic > mesocopic > microscopic > molecular/nano level. Sometimes the labels for the two intermediate levels are interchanged. To avoid confusion in this textbook, we group these two levels into one "meso" length scale between the nano- and macroscopic levels.

Before the advent of X-ray diffractometry, mineralogists could only visually examine crystals. An entire classification scheme was developed and still in use today for describing a single crystal's external morphology, or macroscopic appearance. One or more of 47 possible *forms* are usually apparent in the morphology. A form is a collection of symmetry-equivalent faces. The crystal *habit*, which depends on the relative sizes of the faces of the various forms present, may be described as cubic, octahedral, fibrous, acicular, prismatic, dendritic (tree-like), platy, or blade-like, among others. If a crystal is grown in a symmetrical environment, for example, freely suspended in a liquid, its morphological symmetry is exactly that of the point group *isogonal* (same angular relation) with its space group. It will depart from true point group symmetry under nonsymmetrical growth conditions.

With conventionally processed polycrystals, the smallest particles that are discernible with a high-quality optical microscope are the individual crystallites, or grains, that make up the sample. The term *microstructure* refers to the grain morphology, or grain size, shape, and orientation. Different techniques may be used to examine specific structural features. For example, high-resolution imaging with a scanning electron microscope (SEM) enables observation of dislocations. Information about preferred orientation can be obtained with an X-ray diffractometer equipped with a texture goniometer or by electron backscattered diffraction (EBSD).

Given the penetration depths in Table 1.1, it is obvious that electron diffraction and microscopy only probe the surfaces of solids (the topmost atomic layers), whereas neutron and X-ray diffraction provide information about the bulk. It is well known that the surface crystalline structure of a solid may differ from that of the bulk. The surfaces of most samples, however, are usually subjected to some sort of

Principles of Inorganic Materials Design   By John N. Lalena and David A. Cleary
ISBN 0-471-43418-3   Copyright © 2005 John Wiley & Sons, Inc.

TABLE 1.1 Some Probes Used in Materials Characterization

Source	Wavelength[a] (Å)	Penetration Depth
Light	4×10^3–7×10^3	0
Neutrons	1–2.5	cm–dm
X-rays	0.1–10	μm–mm
Electrons	0.04	nm

[a] For elementary particles, $\lambda = hc/\sqrt{(2mc^2E)}$; for light and X-rays, $\lambda = hc/E$.

chemical–mechanical polishing prior to microstructural analysis to ensure that bulk grain morphology is apparent.

Inorganic materials are commonly grouped into one of two structure categories: crystalline or amorphous (glassy). Amorphous materials possess no long-range structural order, or periodicity. By contrast, crystalline solids are composed of arrays of atoms or molecules, whose positions may be referenced to a translationally invariant lattice. All crystals possess one or more of the basic symmetry elements. Some authors also classify fractals as a distinct structural class. In this case, the structure is self-similar, or scale-invariant, looking identical at all length scales (e.g., cauliflowers and silica aerogels). Crystal structure, however, is the topic of the next chapter. In this chapter, we focus on the microstructures of polycrystalline solids. The majority of solid materials of technological interest are used in polycrystalline form.

Microstructure is determined by the conditions used during the material processing. Hence, our objective is to clarify that a major goal of inorganic materials engineering is the systematic generation of specific grain morphologies in order to vary and adapt the properties of polycrystalline materials to given applications. We focus on describing the microstructures of solidification products (metals), formed powder aggregates (ceramics), and thin films. Microstructure/property correlation is also discussed. Mechanical, chemical, and transport properties are markedly influenced by microstructure.

1.1 INTERFACES IN POLYCRYSTALS

The regions separating different grains, or crystallites, within a polycrystalline solid are called *grain boundaries*. Although grain boundaries are often regarded as regions of structural disorder, it is now well established that many have a periodic structure. True incoherency, in which there is little correlation between atomic positions across the boundary, only sets in when the mismatch between adjacent crystals is very high (Bhadeshia, 1987). This is primarily determined by the relative orientations of the adjoining grains. In a polycrystalline sample, both the grain orientation distribution, or *texture*, and the structure of the grain boundary itself can be crucial to the bulk materials properties. Therefore, it is appropriate to begin with orientation relationships.

1.1.1 Orientation Relationships in Bicrystals

The orientation relationship between a pair of grains of the same substance (the only kind we will consider here), a *bicrystal*, is often expressed by an axis-angle description, since one crystal always can be generated from the other by a rigid-body rotation about a suitable axis. More precisely, the lattices can be made to coincide by turning one of the crystals about a suitable rotation axis. Rotation axes are commonly denoted as unit vectors, in terms of three indices of direction written in square brackets, [*uvw*], while the misorientation angle is expressed in degrees about this axis. The [*uvw*] indices are obtained by taking the projections of the vector on the *x*, *y*, and *z*-axes, respectively, of a Cartesian coordinate system and dividing these three numbers by their highest common denominator.

It is always possible to describe the orientation relationship between a pair of grains in terms of more than one axis-angle pair. Consider a pair of adjacent identical cubic crystals of different orientation, *A* and *B*. Suppose further that *B* can be generated from *A* by a right-hand rotation of $60°$ counterclockwise about the *A* crystal's body-diagonal axis, or the $[111]_A$ direction. This particular orientation relationship is called a *twin*, since the two domains are related by a symmetry element (a twin operation) that is not part of the space group symmetry of a single crystal of the material. The extra symmetry element may be a reflection plane (twin plane) or a rotation axis (twin axis). The high symmetry of the cubic lattice allows us to find numerous equivalent axis-angle pairs for any orientation relationship. Using this twin boundary as an example, we now show how other axis-angle pairs, which are equivalent to a $60°$ right-hand rotation about the $[111]_A$ axis, can be obtained.

Indices are convenient for describing directions (vectors or axes) in crystals. However, direction cosines are much more useful for calculations. Therefore, one must first convert the direction indices, [*uvw*], designating the rotation axis into direction cosines. In our present example, the body diagonal of a cube of unit length has direction indices [111]. This is seen by using a Cartesian coordinate system, where the origin of the cube is taken to be one of its corners and which is designated as $(x_1, y_1, z_1) = (0, 0, 0)$. The body diagonal is obtained by drawing a line segment of length $|r|$ from the origin and terminating at the coordinates $(x_2, y_2, z_2) = (1, 1, 1)$. The direction cosines are given by the equations:

$$\cos \alpha = r_1 = (x_2 - x_1)/ |r|$$
$$\cos \beta = r_2 = (y_2 - y_1)/ |r| \qquad (1.1)$$
$$\cos \gamma = r_3 = (z_2 - z_1)/ |r|$$

where $|r|$ is given by $[r_1^2 + r_2^2 + r_3^2]^{1/2} = [(x_2 - x_1)^2 + (y_2 - y_1)^2 + (z_2 - z_1)^2]^{1/2}$. Hence, in the cubic crystal, we get $\cos \alpha = \cos \beta = \cos \gamma = 0.5773$, satisfying the requirement that $\cos^2 \alpha + \cos^2 \beta + \cos^2 \gamma = 1$.

A (3×3) square rotation matrix, $\mathbf{R}$, may now be obtained, which has the following elements:

$$\begin{pmatrix} r_1 r_1 (1-\cos\theta)+\cos\theta & r_1 r_2 (1-\cos\theta)+r_3\sin\theta & r_1 r_3 (1-\cos\theta)-r_2\sin\theta \\ r_1 r_2 (1-\cos\theta)-r_3\sin\theta & r_2 r_2 (1-\cos\theta)+\cos\theta & r_2 r_3 (1-\cos\theta)+r_1\sin\theta \\ r_1 r_3 (1-\cos\theta)+r_2\sin\theta & r_2 r_3 (1-\cos\theta)-r_1\sin\theta & r_3 r_3 (1-\cos\theta)+\cos\theta \end{pmatrix}$$

$$(1.2)$$

In this book, we follow the standard convention for all matrices, that the elements $a_{i1}, a_{i2}, \ldots, a_{in}$ are the elements of the ith row, and the elements $a_{1j}, a_{2j}, \ldots, a_{mj}$ are the elements of the jth column. That is, the first subscript for an element denotes the column and the second subscript gives the row. Equation 1.2 transforms the components of a vector referred to one basis to those referred to the other basis as:

$$a_1 = R_{11}b_1 + R_{21}b_2 + R_{31}b_3$$
$$a_2 = R_{12}b_1 + R_{22}b_2 + R_{32}b_3 \qquad (1.3)$$
$$a_3 = R_{13}b_1 + R_{23}b_2 + R_{33}b_3$$

In Eq. 1.3, R_{23}, for example, is the second element in the third row (or, equivalently, the third element of the second column) of Eq. 1.2. For $r_1 = r_2 = r_3 = 0.5773$ and $\theta = 60°$, Eq. 1.2 gives:

$$\mathbf{R} = \begin{pmatrix} 0.6667 & 0.6667 & 0.3333 \\ -0.3333 & 0.6667 & 0.6667 \\ 0.6667 & 0.3333 & 0.6667 \end{pmatrix} \qquad (1.4)$$

In order to obtain the equivalent axis-angle pairs, $\mathbf{R}$ must be multiplied by the matrices representing the 24 rotation operations of the cubic lattice. The rotational degeneracy of all crystal lattices can be obtained from the character tables for their respective point groups: cubic, O_h (24), hexagonal, D_{6h} (12), hexagonal close packed, D_{3d} (6), tetragonal, D_{4h} (8), trigonal, D_{3d} (6), orthorhombic, D_{2h} (4), monoclinic, C_{2h} (2), and triclinic, C_i (1).

Continuing with the present example, we can operate on Eq. 1.4 with the (3×3) square matrix representing, say, a 90° right-hand rotation about [100], which is obtained from Eq. 1.2 with $r_1 = 1$, $r_2 = r_3 = 0$ and $\theta = 90°$. The result is a product matrix, which we call $\mathbf{J}$:

$$\mathbf{J} = \begin{pmatrix} 1 & 0 & 0 \\ 0 & 0 & 1 \\ 0 & -1 & 0 \end{pmatrix} \begin{pmatrix} 0.6667 & 0.6667 & -0.3333 \\ -0.3333 & 0.6667 & 0.6667 \\ 0.6667 & -0.3333 & 0.6667 \end{pmatrix}$$

$$= \begin{pmatrix} 0.6667 & 0.6667 & -0.3333 \\ 0.6667 & -0.3333 & 0.6667 \\ 0.3333 & -0.6667 & -0.6667 \end{pmatrix} \qquad (1.5)$$

Note that $\mathbf{J}$ is *not* the product of two symmetry operations, because the first rotation took crystal A into crystal B, rather than back into itself. The A and B orientations are distinguishable. We can now use $\mathbf{J}$ to extract an equivalent axis-angle pair. The new rotation angle, θ, is given by

$$j_{11} + j_{22} + j_{33} = 1 + 2\cos\theta \qquad (1.6)$$

where the terms on the left-hand side are the diagonal elements of $\mathbf{J}$. The equivalent rotation axis for $\theta \neq \pi$ or, for nonsymmetric matrices (when ${}^{t}\mathbf{J} \neq \mathbf{J}$), is obtained from the relations:

$$r_1 = [j_{23} - j_{32}]/2\sin\theta, \qquad r_2 = [j_{31} - j_{13}]/2\sin\theta, \qquad r_3 = [j_{12} - j_{21}]/2\sin\theta \qquad (1.7)$$

where $r_1^2 + r_2^2 + r_3^2 = 1$. When the product matrix is symmetric (${}^{t}\mathbf{J} = \mathbf{J}$), for example, if $\theta = 180°$, Eq. 1.6 does not apply. In this case, the following equation is needed to determine the rotation matrix:

$$
\begin{aligned}
j_{11} &= 1 - 2(r_2^2 + r_3^2) & j_{12} &= 2\,r_1 r_2 & j_{13} &= 2\,r_1 r_3 \\
j_{21} &= 2\,r_1 r_2 & j_{22} &= 1 - 2(r_1^2 + r_3^2) & j_{23} &= 2\,r_2 r_3 \\
j_{31} &= 2\,r_1 r_3 & j_{32} &= 2\,r_2 r_3 & j_{33} &= 1 - 2(r_1^2 + r_2^2)
\end{aligned}
\qquad (1.8)
$$

where $r_1^2 + r_2^2 + r_3^2 = 1$. When using Eq. 1.8, the idea is to extract the maximum component from the diagonal elements of the matrix. If j_{11} is of maximum magnitude, compute:

$$r_1 = [j_{11} - j_{22} - j_{33} + 1]^{1/2}/2 \qquad r_2 = j_{13}/2r_1 \qquad r_3 = j_{13}/2r_1$$

If j_{22} is the maximum, compute:

$$r_2 = [j_{22} - j_{11} - j_{33} + 1]^{1/2}/2 \qquad r_1 = j_{12}/2u_2 \qquad r_3 = j_{23}/2r_2$$

If j_{33} is the maximum, compute:

$$r_3 = [j_{33} - j_{11} - j_{22} + 1]^{1/2}/2 \qquad r_1 = j_{13}/2r_3 \qquad r_2 = j_{23}/2r_3$$

Example 1.1   Calculate the axis-angle pair from the product matrix in Eq. 1.5 that is equivalent to a $60°$ rotation about $[111]_A$.

Solution   The rotation angle, using Eq. 1.6, is

$$\cos^{-1}([0.6667 - 0.3333 - 0.6667 - 1]/2) = \theta = 2.300\,\text{rad}$$
$$2.3000 \times 180/\pi = 131.8°$$

TABLE 1.2 Axis-Angle Pairs Equivalent to a
60° Rotation about $\langle 111 \rangle_A$ in a Cubic Bicrystal

Axis	Angle
$\langle 111 \rangle_A$	180°
$\langle 012 \rangle_A$	131.8°
$\langle 112 \rangle_A$	180°
$\langle 113 \rangle_A$	146.4°
$\langle 11\bar{3} \rangle_A$	146.4°
$\langle 011 \rangle_A$	70.5°
$\langle 011 \rangle_A$	109.5°

We see by inspection that $\mathbf{J}$ is a nonsymmetric matrix ($j_{ij} \neq j_{ji}$). Therefore, we can use Eq. 1.7 to compute the components of the rotation axis:

$$r_1 = [0.6667 - (-0.6667)]/2 \sin(131.8) = 0.8943$$
$$r_2 = [0.3333 - (-0.3333)]/2 \sin(131.8) = 0.4470$$
$$r_3 = [0.6667 - 0.6667]/2 \sin(131.8) = 0$$

From vector algebra, we know that any ordered set of three numbers that can be obtained from $[r_1\ r_2\ r_3]$ by multiplying all of them by the same positive constant k is also a set of direction numbers for the vector $\mathbf{r}$, in that they define the direction of the vector. Hence choosing k to be $(1/0.4470)$ gives: [0.8943/0.4470, 0.4470/0.4470, 0] or [210]. Therefore, the equivalent axis-angle pair is rotated by 131.8° about $[210]_A$.

Using the procedure just outlined with other symmetry operations of the cubic lattice, we can calculate other axis-angle pairs that, for the purposes of expressing the orientation relationship in a cubic bicrystal, are equivalent to a 60° rotation about $\langle 111 \rangle_A$. The results are given in Table 1.2

Obviously, a completely unambiguous description of the relative orientation between two identical crystals must contain the axis-angle pair (rather than an angle alone). As we have just seen, however, a rotation matrix can also be used to specify the orientation relation within a bicrystal. We have just gone to great lengths to show how these matrix elements are computed. The advantage of expressing the orientation relationship in this manner will be apparent in Section 1.1.3 where we quantify the "goodness of fit" at the interface between grains.

1.1.2 Grain Boundary Orientations

We have been discussing orientation relationships between pairs of grains. This is *not* the same as the orientation of the grain boundary. For example, Figure 1.1 shows a twinned bicrystal like that discussed earlier. As illustrated in the figure, the grain boundary plane between two crystals with this orientation relationship need not coincide with the twin plane. The orientation relationship between the grains

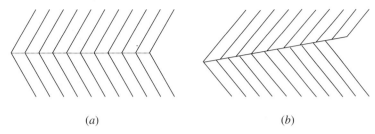

Figure 1.1   (a) The twin plane coincides with the boundary plane. (b) The twin plane and boundary plane do not coincide.

does provide us with three of the five degrees of freedom needed to specify the grain boundary orientation, however. One of these degrees of freedom, we have seen, is a rotation angle. The rotation is carried out about a rotation axis, which we have been denoting by three indices of direction, $[uvw]$. Because an axis is a polar vector in spherical coordinates, it can also be specified by a polar angle and an azimuthal angle relative to the grain boundary plane. Thus, three of our five degrees of freedom are Euler angles that, taken together, describe the orientation relationship between the grains: $0 \leq \phi_E \leq 2\pi$; $0 \leq \theta_E \leq \pi$; and $0 \leq \psi_E \leq 2\pi$ (the subscript E simply denotes that these are Euler angles). The remaining two degrees of freedom define the boundary plane in the coordinate system of the reference grain. They are spherical angles that specify the boundary plane inclination: $0 \leq \theta_S \leq 2\pi$; $0 \leq \phi_S \leq \pi$, where the subscript S denotes spherical angles.

One might naturally ask: How many different grain boundary orientations are observable? The number of distinguishable orientations, N, depends on the precision with which the various angular measurements are made, and the number of symmetry operators for the crystal class. For example, for a cubic bicrystal the boundary normal can be selected in two directions, the crystals can be exchanged, and one can apply 24 rotation operations to either crystal. There are thus $2 \cdot 2 \cdot 24^2$ combinations of the five angular parameters that lead to identical bicrystals. To generalize, if we represent the number of symmetry operations for the crystal class by η, the precision of the angular measurements by Δ, and the number of degrees of freedom by n, we have the following formula for the number of distinguishable orientations (Saylor et al., 2000).

$$
\begin{aligned}
N &= 1/(4\eta^2)\prod_{n,\Delta}(n/\Delta) \\
&= [(2\pi)(\pi)(2\pi)(2\pi)(\pi)]/(4\eta^2\Delta^5) \qquad (1.9) \\
&= 8\pi^5/(4\eta^2\Delta^5)
\end{aligned}
$$

where Δ is in radians. The $8\pi^5$ factor is the product of the full ranges for each angular parameter. For a cubic system, if $\Delta = 0.087$ (5°), Eq. 1.9 predicts 2.1×10^5 distinct boundaries. The number of distinguishable boundaries obviously increases with increases in the angular precision.

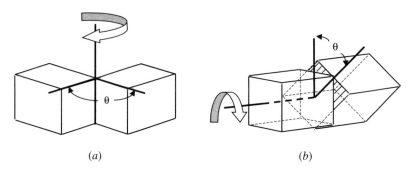

Figure 1.2 (a) A tilt boundary. (b) A twist boundary.

Despite such a large number of possible orientations, it has been observed experimentally that grain orientation relationships do not occur in a random manner. For example, low-energy grain boundaries like the twin boundary are very commonly observed in cubic systems. It is hard to say whether this is a result of thermodynamic or kinetic control. Interfacial energy minimization could be responsible, or the activation energies for nucleation and grain growth in certain orientations could be lower, or possibly both factors could be at work.

The Dislocation Model of Low Angle Grain Boundaries   A general grain boundary has a mixture of *tilt* and *twist* character. We can think of a pure tilt boundary as consisting of an axis of rotation that is in the grain boundary plane (Figure 1.2a). In contrast, twist boundaries contain an axis of rotation that is perpendicular to the grain boundary plane (Figure 1.2b). A useful way to picture the symmetrical tilt boundary (a boundary in which the boundary plane contains the rotation axis and bisects the rotation angle) is to consider it as a straight array of *edge dislocations*, as in Figure 1.3. In a single-crystal metal, edge dislocations consist of extra half-planes of atoms. In ionic or covalent crystals, edge dislocations involve extra half planes of unit cells. As long as the misorientation angle is low (i.e., small-angle grain boundaries), tilt boundaries may be regarded as the coalescence of these line defects into a dislocation network. The spacing between the dislocations, D, is

$$D = \boldsymbol{b}/\sin\theta \qquad (1.10)$$

where $\boldsymbol{b}$ is the *Burgers vector*, perpendicular to the line of the dislocation, and θ is the misorientation angle.

If the dislocation density is low (the value of D is large), a semicoherent interface results, in which regions of good fit are separated by the individually recognizable interface dislocations. Note how the extra half-planes in Figure 1.3 all have a single Burgers vector. In an unsymmetrical low-angle tilt boundary, different Burgers vectors are required to accommodate the mismatch. The dislocation model is really only valid for low-angle grain boundaries. In the cubic crystal class, for

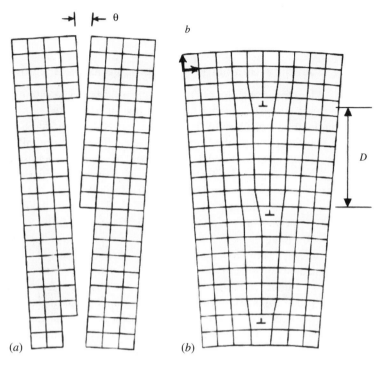

Figure 1.3 (a) A low-angle tilt boundary. (b) Representation as an array of parallel edge dislocations.

values $\theta > \sim 15°$, D can get so small, corresponding to a high dislocation density, that dislocations become indistinguishable (Read and Shockley, 1950). The symmetrical low-angle twist boundary can similarly be represented by a screw dislocation (Figure 1.4). Screw dislocations have been likened to multistoried parking garages, the atomic planes spiraling around the dislocation line in the same

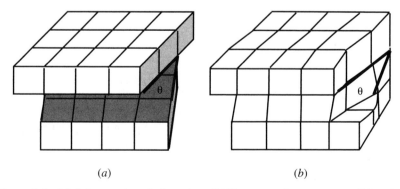

Figure 1.4 (a) A low-angle twist boundary. (b) Representation as a screw dislocation.

manner as a parking garage floor spirals around a central pole of the garage (Weertman and Weertman, 1992).

1.1.3 Grain Boundary Geometry: The Coincidence Site Lattice

The evolution of our modern picture of crystalline interfaces can be summarized as follows. The earliest geometric models of crystalline interfaces were the "amorphous" high-angle grain boundary by Hargreaves and Hill (Hargreaves and Hill, 1929), and the twin interface by the French mining engineer and crystallographer Georges Friedel (1865–1933) (Friedel, 1926), son of organic chemist Charles Friedel. (1832–1899). N. F. Mott first suggested that grain boundaries should contain regions of fit and misfit (Mott, 1948). Kronberg and Wilson then pointed out the importance of the coincidence of atom positions across grain boundaries in influencing metal properties such as diffusion coefficients and mobilities (Kronberg and Wilson, 1949). Ranganthan presented a general procedure for obtaining coincidence relationships between lattices about rotation axes (Ranganathan, 1966).

The modern method for quantifying the goodness of fit between two adjacent grains examines the number of lattice points (*not* atomic positions) from each grain that coincide. In special cases, for example when the grain boundary plane is a twin plane, the lattice sites for each of the adjacent crystals coincide *in* the boundary. These are called *coherent* boundaries. It has long since been experimentally verified that coherent grain boundaries possess special properties. For example, coherent boundaries migrate faster than random boundaries during recrystallization (Aust and Rutter, 1959).

Consider a pair of adjacent crystals. We mentally expand the two neighboring crystal lattices until they interpenetrate and fill all space. Without loss of generality, it is assumed that the two lattices possess a common origin. If we now hold one crystal fixed and rotate the other, it is found that a number of lattice sites for each crystal, in addition to the origin, coincide with certain relative orientations. The set of coinciding points form a *coincidence site lattice*, or CSL, which is a sublattice for both the individual crystals.

In order to quantify the lattice coincidence between the two grains, A and B, the symbol Σ customarily designates the reciprocal of the fraction of A (or B) lattice sites that are common to both A and B.

$$\Sigma = \text{Number of crystal lattice sites/Number of coincidence lattice sites} \quad (1.11)$$

For example, if one-third of the A (or B) crystal lattice sites are coincidence points belonging to both the A and B lattices, then, $\Sigma = 1/(1/3) = 3$. The value of Σ also gives the ratio between the areas enclosed by the CSL unit cell and crystal unit cell. The value of Σ is a function of the lattice types and grain misorientation. The two grains need not have the same crystal structure or unit cell parameters. Hence, they need not be related by a rigid-body rotation. The boundary plane intersects the CSL and will have the same periodicity as that portion of the CSL along which the intersection occurs.

The simple CSL model is directly applicable to the cubic crystal class. The lower symmetry of the other crystal classes necessitates the more sophisticated formalism known as the *constrained coincidence site lattice*, or CCSL (Chen and King, 1988). In this book, we only treat cubic systems. Interestingly, whenever an *even* value is obtained for Σ in a cubic system, it will always be found that an additional lattice point lies in the center of the CSL unit cell. The true area ratio is then half the apparent value. This operation can always be applied in succession, until an odd value is obtained—thus Σ is always *odd* in the cubic system. A rigorous mathematical proof of this would require that we invoke what is known as O-lattice theory (Bollman, 1967) The O-lattice takes into account all equivalence points between two neighboring crystal lattices. It includes as a subset not only coinciding lattice points (the CSL) but also all nonlattice sites of identical internal coordinates. However, to expand on that topic would be well beyond the scope of this textbook. The interested reader is referred to Bhadeshia (1987) or Bollman (1970).

Single crystals and bicrystals with no misorientation (i.e., $\theta = 0$), by convention, are denoted $\Sigma 1$. In practice, small- or low-angle grain boundaries with a misorientation angle less than $10°$–$15°$ are also included under the $\Sigma 1$ term. Since Σ is always odd, the coincidence orientation for high-angle boundaries with the largest fraction of coinciding lattice points is $\Sigma 3$ (signifying that 1/3 of the lattice sites coincide). Next in line would be $\Sigma 5$, then $\Sigma 7$, and so on.

Figure 1.5 shows a tilt boundary between two cubic crystals. The grain boundary plane is perpendicular to the plane of the page. In the figure, we are looking down one of the $\langle 100 \rangle$ directions, and the [100] axis about which grain B is rotated is also perpendicular to the page and passes through the origin. At the precise misorientation angle of $36.9°$, one-fifth of the B crystal lattice sites are coincidence points, which also belong to the expanded lattice of crystal A; this is a $\Sigma 5$ CSL misorientation. The set of coincidence points forms the CSL, the unit cell of which is outlined. Note that the area enclosed by the CSL unit cell is five times that enclosed by the crystal unit cell.

Fortunately, there is an easy, although tedious, way to determine the true value for Σ. If an integer, N, can be found such that all the elements of the rotation matrix become integers when multiplied by N, then that integer will be the Σ value. The value of N is found simply by multiplying all the matrix elements by integers, in increments of one beginning with the number 1, until the matrix elements are all integers. If the value of Σ turns out to be even using this procedure, then the true value is obtained by successively dividing N by two until the result is an odd integer. This method can be used to compute the value of Σ for any general rotation matrix. For example, factoring out 1/3 from **R** in Eq. 1.4 gives a matrix with integral elements, in which Σ is equal to three:

$$\mathbf{R} = 1/3 \begin{pmatrix} 2 & 2 & -1 \\ -1 & 2 & 2 \\ 2 & -1 & 2 \end{pmatrix}$$

Hence, the $60°$ $\langle 111 \rangle$ twin boundary has a $\Sigma 3$ CSL misorientation. It is also a coherent boundary because of the large number of coincidence points along the

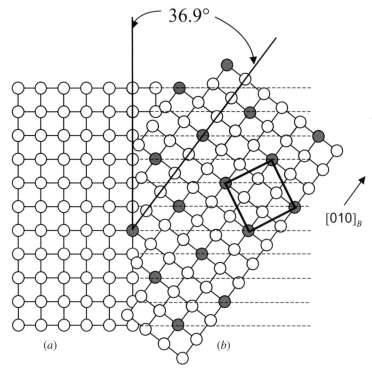

Figure 1.5   A view down the [001] direction of a tilt boundary between two crystals (*A*, *B*) with a misorientation angle of 36.9° about [001]. The grain boundary is perpendicular to the plane of the page. Every fifth atom in the [010] direction in *B* is a coincidence point (shaded). The area enclosed by the CSL unit cell (bold lines) is five times that of the crystal unit cell, so $\Sigma = 5$.

twin plane itself, as shown in Figure 1.6. In this figure, we are looking down *on* the $\langle 111 \rangle$ rotation axis. The lattice sites shown are in rows off the $\langle 111 \rangle$ axis, along the set of (111) planes, as illustrated in the bottom left corner. It will be recalled that the boundary plane intersects the CSL and will have the same periodicity as that portion of the CSL along which the intersection occurs. Thus, not all $\Sigma 3$ boundaries will be coherent. For example, although rotations of 70.5° and 109.5° about $\langle 011 \rangle_A$ are also equivalent $\Sigma 3$ misorientations, only the 70.5° rotation will result in a high degree of coincidence along the $\langle 011 \rangle$.

Example 1.2   The matrix corresponding to a rotation of 50.5° about [110] is given in Bhadeshia's monograph *Worked Examples in the Geometry of Crystals* as:

$$\mathbf{R} = \begin{pmatrix} 0.545621 & -0.545621 & 0.636079 \\ 0.181961 & 0.818039 & 0.545621 \\ 0.818039 & 0.181961 & -0.545621 \end{pmatrix}$$

Calculate the value of Σ.

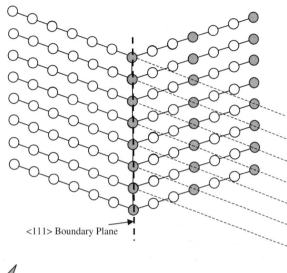

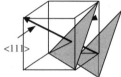

Figure 1.6 The twin boundary (perpendicular to the plane of the page) is a Σ3 CSL misorientation. Note that there is complete coincidence in the boundary plane itself.

Solution Multiplying each matrix element by an integer, starting with the number 1, and progressing in increments of 1 until the products are integers shows that when $N = 11$, the rotation matrix can be written as

$$\mathbf{R} = 1/11 \begin{pmatrix} 6 & -6 & 6 \\ 6 & 9 & 6 \\ 9 & 2 & -6 \end{pmatrix}$$

Hence, $\Sigma = 11$.

For tilt boundaries, the value of Σ can also be calculated if the plane of the boundary is specified in the coordinate systems for both adjoining grains. This method is called the *interface-plane scheme* (Wolfe and Lutsko, 1989). In a crystal, lattice planes are imaginary sets of planes that intersect the unit cell edges (Section 2.1.3). These planes are denoted by Miller indices, a group of integers that are the reciprocals of the fractional coordinates of the points where the planes intercept each of the unit cell edges. In cubic crystals, the (hkl) planes are orthogonal to the $[uvw]$ direction. The tilt and twist boundaries can be defined in terms of the Miller indices for each of the adjoining lattices and the twist angle, Φ, of both plane stacks normal to the boundary plane, as follows:

$$(h_1k_1l_1) = (h_2k_2l_2); \Phi = 0 \qquad \text{symmetric tilt boundary}$$
$$(h_1k_1l_1) \neq (h_2k_2l_2); \Phi = 0 \qquad \text{asymmetric tilt boundary}$$
$$(h_1k_1l_1) = (h_2k_2l_2); \Phi > 0 \qquad \text{low-angle twist boundary}$$
$$(h_1k_1l_1) \neq (h_2k_2l_2); \Phi > 0 \qquad \text{high-angle twist boundary}$$

Thus, the value of the CSL-Σ value is obtained for symmetric tilt boundaries between cubic crystals as follows:

$$\Sigma = h^2 + k^2 + l^2 \qquad \text{for } h^2 + k^2 + l^2 = \text{odd}$$
$$= (h^2 + k^2 + l^2)/2 \qquad \text{for } h^2 + k^2 + l^2 = \text{even} \qquad (1.12)$$

For asymmetric tilt boundaries between cubic crystals, Σ is calculated from (Randle, 1993):

$$\Sigma = [(h_1^2 + k_1^2 + l_1^2)/(h_2^2 + k_2^2 + l_2^2)]^{1/2} \qquad (1.13)$$

For example, if we mentally expand the lattices of both A and B in Figure 1.5, it will be seen that the grain boundary plane cuts the B unit cell at (340) in the B coordinate system and the A unit cell at (010) in the A coordinate system. Thus, Eq. 1.13 yields $\Sigma = (25/1)^{1/2} = 5$.

In polycrystals, misorientation angles rarely correspond to *exact* CSL configurations. There are ways of dealing with this deviation, which set criteria for the proximity to an exact CSL orientation that an interface must have in order to be classified as belonging to the class $\Sigma = n$. The Brandon criterion (Brandon et al., 1964) asserts that the maximum permitted deviation is $v_0\Sigma^{-1/2}$. For example, the maximum deviation that a $\Sigma 3$ CSL configuration with a misorientation angle of $15°$ is allowed to have and still be classified as $\Sigma 3$ is $15°(3)^{-1/2} = 8.7°$. The coarsest lattice characterizing the deviation from an exact CSL orientation, which contains the lattice points for each of the adjacent crystals, is referred to as the displacement shift complete lattice (DSL).

Despite the difficulties associated with characterizing inexact CSL orientations, the CSL concept is useful because grain boundary structure, which depends on the orientation relationship between the grains and, hence, the CSL, directly influences intragranular properties like chemical reactivity (e.g., corrosion resistance), segregation, and fracture resistance. *Grain boundary engineering* is a relatively new field that concentrates on controlling the intragranular structure, or CSL geometry, to improve these properties, in turn, improving bulk materials performance (Watanabe, 1984, 1993). For the most part, this means introducing a large fraction of low-Σ boundaries, particularly twin boundaries. It is believed, however, that optimal grain boundary properties may be restricted to narrow regions (small deviations) about exact CSL orientations.

1.1.4 Grain Boundary Energy

A finite number of point defects (e.g., vacancies, impurities) can be found in any crystalline material because the configurational entropy term, $-T\Delta S$, for a low point-defect concentration outweighs the positive formation enthalpy in the free-energy expression, $\Delta G = \Delta H - T\Delta S$. Thus, introduction of a small number of point defects into a perfect crystal gives rise to a free-energy minimum, as illustrated in Figure 1.7a. Further increases in the point-defect concentration,

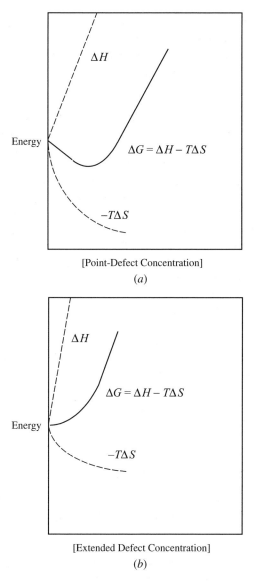

Figure 1.7   Energy changes associated with the incorporation of defects into a perfect crystal. (a) For point defects, the minimum in the free energy occurs at some finite concentration of defects. (b) For extended defects, the minimum in the free energy corresponds to the defect-free structure.

however, will raise the free energy of the system. Point defects in crystals are discussed in Sections 2.5.1 and 5.4.1.

On the other hand, the positive enthalpy of formation is so high for extended defects that the entropy gain is not sufficient to give rise to minima in the free energy (Figure 1.7b). Recall how a tilt boundary can be regarded as an array of edge dislocations. Edge dislocations are extended defects in which the formation energy must be proportional to the linear dimensions of the sample (Elliot, 1998). Hence, dislocations and grain boundaries are higher energy *metastable* configurations, introduced primarily from processing. On annealing, polycrystals tend to evolve toward single crystals through grain growth, and grains with low dislocation densities tend to grow by consuming grains with high dislocation densities.

Unfortunately, reliable grain boundary energies are hard to obtain. As one might imagine, measuring grain boundary energies is difficult and tedious. So is calculation from first principles, since this requires accurate atom positions—the determination of grain boundary structure and orientation requires careful sample preparation and high-resolution instruments. Nonetheless, some experimental work has been performed, and it is possible to make some generalized statements.

First, low Σ boundaries tend to have relatively lower grain boundary energy, on average. The entropy term is undoubtedly the dominant contribution to the free energy in these cases. With small misorientation angles ($\theta < 15°$), D in Eq. 1.10 is large. Since grain boundary energy is proportional to $1/D$, γ tends to be small for low-angle boundaries, and it has been found experimentally that as the angle exceeds 15°, the grain boundary energy typically begins to level off as it becomes independent of θ. However, one must be very cautious when attempting to correlate the three parameters, Σ, θ, and γ. Increases in Σ do often correspond to increases in θ, but not all high-angle boundaries are high Σ also. For example, the high-angle coherent twin boundary ($\theta = 60°$) is a "low sigma" $\Sigma 3$ structure. Furthermore, the atoms at the interface of a high-angle coherent twin boundary are coherent, which results in a very low-energy boundary. Likewise, although many low-Σ boundaries tend to have relatively low energies, the energy does not always show a simple relationship to Σ.

1.1.5 Special Types of Low-Energy Grain Boundaries

In addition to low-energy coherent twin boundaries, other low-energy grain boundaries exist that do not involve a grain misorientation ($\theta = 0$). In a cubic close-packed (ccp) crystal, for example, the stacking repeat sequence of the close-packed layers can be represented as $\ldots ABCABCABC \ldots$, where each letter represents a layer of hexagonally coordinated atoms with a particular displacement relative to its adjacent layers. *Stacking faults* occur when the layer sequence is interrupted, for example, $\ldots ABCABABC \ldots$. This type of defect commonly occurs in polytopic metals, in which the polytypes are different types of close packing. However, stacking faults can occur in non-close-packed structures as well. Figure 1.8 shows a (110) section through a diamond lattice (e.g., silicon) containing a stacking fault by removal of two adjacent layers. The diamond structure may be thought of as two interpenetrating face-centered cubic (fcc) lattices.

A second type of boundary, in which there is no misorientation between grains, is the *antiphase boundary*. This occurs when "wrong" atoms are next to each other on the boundary plane. For example, with hexagonal close-packed (hcp) crystals,

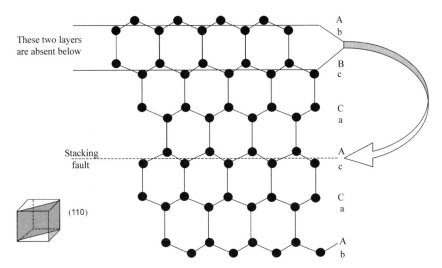

Figure 1.8   A (110) section through a diamond lattice showing a stacking fault by the absence of two adjacent atomic layers. The layer sequence along the ⟨111⟩ body-diagonal direction should be ... *A b B c C a A b*... (Adapted from Runyan and Bean, 1990, *Semiconductor Integrated Circuit Processing Technology*. Copyright © Addison-Wesley Publishing Company, Inc. Reproduced with permission.)

the sequence ...*ABABAB*... can be reversed at the boundary to *ABABA|ABABA*, where | represents the boundary plane. Antiphase boundaries and stacking faults are typically of very low energy, comparable to that of a coherent twin boundary.

1.1.6   Grain Boundary Dynamics

Thus far, we have confined our discussion to the "static" properties of grain boundaries. However, grain boundaries are metastable configurations and, as such, in response to external forces (e.g., thermal, mechanical) they exhibit dynamical behavior. We will briefly mention only two of the more important ones: grain boundary migration and sliding. *Grain boundary migration* is an example of when a heat-treated system attempts to minimize its free energy. Polycrystals tend to evolve toward single crystals through grain growth. In this phenomenon, atoms move from the side of the grain boundary with a high free energy to the low-energy side. The free energy of the system can thus be reduced as the low-energy crystals consume the high-energy crystals. For example, on annealing, polycrystalline grains with low dislocation densities will grow by consuming grains with high dislocation densities. Likewise, at curved boundaries atoms are more likely to diffuse from the convex side to the concave side in order to flatten the interface. In this way, the interfacial area and energy are reduced.

Grain boundary sliding is a process in which adjacent grains slide past each other along their common boundary. It is a deformation mechanism that contributes to plastic (nonrecoverable) flow and *superplasticity* in polycrystalline samples with very small grain sizes. Superplasticity has been observed in both metals and nonmetals. Superplasticity is an important property because it allows engineers to fabricate complex shapes out of a material, which might otherwise be unobtainable. Grain boundary sliding and migration modify the texture, or preferred orientation, of polycrystalline materials during recrystallization.

1.1.7 Orientation Distributions in Polycrystalline Aggregates

Methods such as high-resolution transmission electron microscopy (HRTEM) enable direct examination of orientation relationships between pairs of crystals. However, their use for the determination of preferred orientation, or *texture*, in bulk polycrystalline samples is not convenient due to the very large number of grains present. Rather, texture is normally determined from X-ray diffraction data.

In a polycrystal, the grains may all be oriented at random, exhibit some preferred orientation, or there may be multiple regions, called *domains*, possessing different preferred orientations. The most common way of illustrating texture is the use of *pole figures*. The inclination to the normal to a particular type of crystal plane [e.g., (100)], relative to some reference plane, is specified for a large number of grains. It is thus necessary to consider two coordinate systems, the *crystal coordinate system* and the *sample coordinate system*. If a sphere is imagined to enclose the polycrystalline sample, then each plane normal will intersect the sphere's surface at a point called the *pole*, which by its position on the surface represents the orientation to that crystal plane. Because it is difficult to draw a three-dimensional (3D) sphere on a 2D piece of paper, the orientation distribution is displayed with a stereographic projection of the sphere, called a pole figure. Several pole figures, one for each type of crystal plane examined, collectively describe the texture.

As an example, consider a single crystallite contained within a thin film or rolled specimen of cubic symmetry, with its (001) plane parallel to the substrate, as in Figure 1.9a. The normal to the (001) plane is pointing straight up, its $(00\bar{1})$ plane normal straight down, its (100) and $(\bar{1}00)$ plane normals pointing left and right and its (010) and $(0\bar{1}0)$ plane normals pointing front and back. Note from Figure 1.9a that the $\langle 100 \rangle$ plane normals are parallel to the sample's radial directions (RD), transverse directions (TD), and normal directions (ND). This particular arrangement is called *cube texture*.

The normal to each crystallite produces a pole by its intersection with the surface of the sphere. Hence, a large number of poles will produce a spot with a diameter that is dependent on the distance between the sample and the sphere surface. By convention, the surface of the sphere is taken to be at sufficient distance such that strong texture (a large number of parallel-plane normals) is manifested as small diameter spots. Thus, for a polycrystalline aggregate with all its crystals aligned as in Figure 1.9a the pole figure of the {100} poles would show sharp maxima in the ND (top), RD (sides), and TD (center) locations, as illustrated in Figure 1.9b. If there were only weak cube texture, the maxima would smear out into lighter spots rather than form four distinct dark spots. This is because the plane normals are now no longer all parallel and their intersections with the sphere surface produces wide-

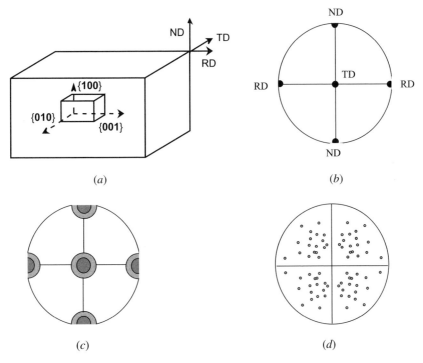

Figure 1.9 (a) Cube texture. (b) The (100) pole figure indicating sharp texture. (c) Weak texture in the (100) pole figure. (d) No preferred orientation in the (100) pole figure.

diameter less-densely populated spots, as illustrated in Figure 1.9c. If there were no preferred orientation, there would be a uniform distribution of poles in the pole figure. In this case, there is essentially one discernible pole corresponding to each crystallite whose orientation was measured, as illustrated in Figure 1.9d.

The implications of texture on the properties exhibited by a material are discussed in Sections 1.2.5, 1.3.1, and 1.3.2. We will see that the properties of polycrystalline aggregates with a sufficiently large number of grains possessing a completely random orientation distribution are macroscopically isotropic (independent of direction), even though the crystallites themselves may be anisotropic. Fabrication processes such as extrusion, rolling, and pressing, remove this isotropy, however.

1.2 SOLIDIFIED METALS AND ALLOYS

The most economical and, hence, most common method for fabricating metal pieces with a predefined size and shape is *casting*. In principle, solidification processes are equally applicable to nonmetals. However, most ceramics are very seldom prepared by this route, because of their high melting points. Typically, the molten material is poured into a thermally conducting mould to solidify. At first glance, solidification might appear to be a fundamentally simple process, but this would be deceptive. The quantitative mathematical relations governing this phenomenon are complicated moving interface diffusion equations. Fortunately, we need not present such mathematical expressions in any detail here.

In many respects, crystallization from the molten state is analogous to that from a solvent. For example, in both processes slow cooling tends to result in larger crystals, while faster cooling typically gives smaller crystals. Another important similarity involves the behavior of impurities. Impurity atoms usually do not "fit" into the crystal lattice of the solute crystallizing from a solvent. This allows crystallization from a solvent to be used as a purification technique. Likewise, impurity atoms in a polycrystalline metal tend to segregate at the grain boundaries during solidification because of their mismatch with the lattice of the metal atoms.

Conventional industrial casting processes usually involve directional heterogeneous solidification occurring in three stages: nucleus formation, crystal growth, and grain boundary formation. When enough heat is extracted, stable nuclei form in the liquid either on solid-phase impurities near the walls of the mould or on the mould itself, since this is the first region to cool sufficiently for crystals to form. Heterogeneous nucleation can also occur at the surface of the melt on solid-phase metal oxide particles. Oxides typically have much higher melting points than those of their parent metals. Other possible nucleation sites are inclusions and intentionally added grain refiners. At any rate, the solidification begins near the exterior edges and the solid–liquid interface subsequently moves inward toward the casting's center as heat is conducted through the freshly grown solid out through the mould. The nuclei consist of tiny aggregates of atoms arranged in the most favorable lattice under the process conditions. Crystals grow in all directions near the liquid–container interface. Hence, this region is called the *equiaxed zone*, as shown in Figure 1.10.

As solidification continues, an increasing number of atoms lose their kinetic energy, making the process exothermic. For a pure metal, the temperature of the melt will remain constant while the latent heat is given off (until freezing is complete). As the atoms coalesce, they may attach themselves to existing nuclei or form new nuclei. The process continues, with each crystal acquiring a random orientation and, as the gaps between crystals fill in, each grain acquires an irregular shape. The growth morphology is most likely under kinetic control, that is, the grain morphology that appears is the one with the maximum growth rate. Eventually, those grains that have a preferred growth direction will eliminate the others, resulting in the formation of a *columnar zone* where the crystals are elongated, or column-like. The growth direction is typically in the direction of heat flow. For alloys, an inner equiaxed zone can sometimes form in the casting's center, resulting from the growth of detached pieces of the columnar grains. This will be dependent on the degree of heat convection in that region.

Conventional casting procedures result in relatively coarse grains, with an average size in the range of several millimeters to several hundred micrometers. The interfaces between grains, formed by the last liquid to solidify, are the grain

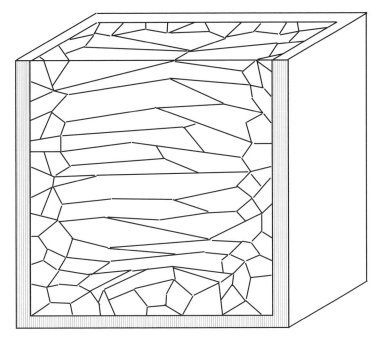

Figure 1.10   A schematic illustration of a section through an ingot showing the different solidification zones. The equiaxed zone forms near the mould walls. In the interior is the columnar zone, where crystal growth is in the direction of heat flow.

boundaries. A grain boundary is comprised of atoms that are not exactly aligned with the crystalline grains on either side of it. Hence, the grain boundaries have some degree of disorder and tend to contain a higher concentration of impurity atoms, which do not "fit" into the crystal lattices on either side of them (a melt is never *entirely* pure). The grain boundary has a slightly higher free energy due to the presence of defects. We will now investigate how grain morphology and composition, collectively termed *constitution,* is affected by the solidification rate. Of the many parameters affecting the development of the microstructure, none is more important than the cooling rate.

1.2.1   Grain Homogeneity

There are two limiting cases to consider. The first is equilibrium solidification, when the cooling rate is slow enough that solid-state diffusion can act to redistribute atoms and result in homogeneous crystals. In this case, complete diffusion occurs in both the liquid and solid. Under these conditions, the solid absorbs solute atoms from the liquid and solute atoms within the solid diffuse from the previously frozen material into subsequently deposited layers. The chemical compositions of the solid and liquid at any given temperature then follow the solidus and liquidus lines, respectively, of the equilibrium phase diagram.

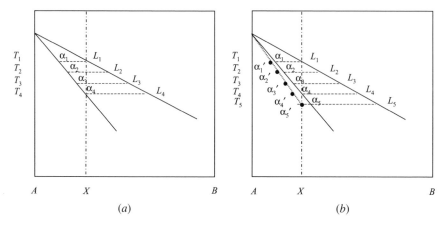

Figure 1.11 (a) A portion of a binary phase diagram illustrating equilibrium solidification. (b) Nonequilibrium (rapid) solidification, which results in a chemical composition gradient in the crystals, a condition known as coring.

Use of tie lines and the lever rule enable one to determine those compositions, as illustrated in Figure 1.11a for a binary system. The composition of the solid (C_s) as a function of the fraction solid transformed (f_s), assuming linear solidus and liquidus lines, is given by

$$C_s = kC_0/[f_s(k - 1) + 1] \qquad (1.14)$$

where k is the partition coefficient (the ratio of the solute concentration in the solid to that in the liquid) and C_0 is the composition of the original liquid alloy. The first crystals to freeze out have composition α_1. As the temperature is reduced to T_2, the liquid composition shifts to L_2. The compositions of the freezing solid and remaining liquid continuously shift to higher B contents and leaner A contents. The average solid composition follows the solidus line to T_4, where it equals the bulk composition of the alloy.

In order to qualify for equilibrium solidification, the solidification rate must be slower than the solute diffusivity in the solid:

$$D_s \gg L_x \nu \qquad (1.15)$$

where D_s is the solute diffusivity in the solid, L_x is the system length scale in one dimension, and ν is the solidification speed (Phanikumar and Chattopadhyay, 2001). Table 1.3 lists the self-diffusivities of pure metals at their freezing points (T_f) in the various structural classes. It can be deduced from Eq. 1.15 that, for a system length scale of 1 cm, the solidification rate (cm/s) must be lower than the numerical values given for the diffusivities (cm²/s) given in Table 1.3, which are very slow rates, indeed. In other words, equilibrium solidification occurs only when the melt is cooled extremely slowly!

TABLE 1.3   Self-diffusivities for Various Metals at the Melting Temperature

Class	D_s (cm^2/s)
fcc	5.5×10^{-9}
bcc[a] (Li, Na, K, Rb)	1.4×10^{-6}
bcc (Trans. elements)	2.9×10^{-8}
hcp (Mg, Cd, Zn)	1.6×10^{-8}
Alkali halides	3.2×10^{-9}

Source: Brown, A. M.; Ashby, M. F. 1980, *Acta Met, 28*, 1085.
[a] bcc = body-centered cubic.

The second limiting case approximates conventional industrial casting processes, in which the solidification rates are several orders of magnitude too fast to maintain equilibrium. The most widely used classic treatment of nonequilibrium solidification is by Scheil (Scheil, 1942). The model assumes negligible solute diffusion in the solid phase, complete diffusion in the liquid phase, and equilibrium at the solid–liquid interface. In this case, Eq. 1.14 can be rewritten as:

$$C_s = kC_0(1 - f_s)^{k-1} \qquad (1.16)$$

When the solid–liquid interface moves too fast to maintain equilibrium, it results in a chemical composition gradient within each grain (a condition known as *coring*). This is illustrated in Figure 1.11b. Without solid-state diffusion of the solute atoms in the material solidified at T_1 into the layers subsequently freezing out at T_2, the average composition of the crystals does not follow the solidus line, from α_1 to α_4, but rather follows the line α_1' to α_5', which is shifted to the left of the equilibrium solidus line. The faster the cooling rate, the greater the magnitude of the shift.

Note also that final freezing does not occur until a lower temperature, T_5 in Figure 1.11b, so that nonequilibrium solidification happens over a greater temperature range than equilibrium solidification. Because the time scale is too short for solid-state diffusion to homogenize the grains, their centers are enriched in the higher freezing component while the lowest freezing material gets segregated to the edges (recall how grain boundaries are formed from the last liquid to solidify). Grain boundary melting, *liquation*, can occur when subsequently heating such an alloy to temperatures below the equilibrium solidus line, which can have devastating consequences for metals used in structural applications.

1.2.2   Grain Morphology

In addition to controlling the compositional profile of the grains, the solidification velocity also determines the shape of the solidification front (solid–liquid interface), which, in turn, determines grain shape. The resulting structure arises from the competition between two effects. Undercooling of the liquid adjacent to the interface favors protrusions of the growing solid, which gives rise to *dendrites* with a characteristic treelike shape, while surface tension tends to restore the minimum surface configuration—a planar interface.

Consider the case of a molten pure metal cooling to its freezing point. When the temperature gradient across the interface is positive (the solid is below the freezing temperature, the interface is at the freezing temperature, and the liquid is above the freezing temperature), a planar solidification front is most stable. However, with only a very small number of impurities present in a pure melt on which nuclei can form, the bulk liquid becomes kinetically undercooled. Diffusion of the latent heat away from the solid–liquid interface via the liquid phase favors the formation of protrusions of the growing solid into the undercooled liquid (the undercooled liquid is a very effective medium for heat conduction). Ivantsov first mathematically modeled this for paraboloidal dendrites over half a century ago (Ivantsov, 1947). It is now known that this is true so long as the solidification velocity is not *too* fast. At the high velocities observed in some rapid quenching processes (e.g., >10 ms^{-1}), dendritic growth becomes unstable, as the perturbation wavelengths become small enough that surface tension can act to restore planarity (Hoglund et al., 1998; Mullins and Sekerka, 1963)

For a pure melt, dendritic growth is a function of the rate of latent heat removal from the interface. However, when we turn our attention to alloys, we see a slightly different situation. Here, in addition to heat flow, one must consider mass transport as well. In fact, the planar interface-destabilizing event primarily responsible for dendritic morphology in conventional alloy casting is termed *constitutional under-cooling*, to distinguish it from kinetic undercooling. The kinetic undercooling contribution can still be significant in some cases. In most models for two-component melts, it is assumed that the solid–liquid interface is in local equilibrium even under nonequilibrium solidification conditions based on the concept that interfaces will equilibrate much more rapidly than bulk phases. Solute atoms thus partition into a liquid boundary layer a few micrometers thick adjacent to the interface, slightly depressing the freezing point in that region. As in the case for pure meals, the positive temperature gradient criterion for planar interface stability still holds. However, although the bulk liquid is above the freezing point, once the boundary layer becomes undercooled, there is a large driving force for solidification *ahead* of the interface into the thin boundary layer.

The critical growth velocity, v, above which the planar interface in a two-component melt becomes unstable is related to the undercooling, ΔT_c, by an equation given by Tiller as:

$$G_L/v \geq \Delta T_c/D_L \qquad (1.17)$$

where $\Delta T_c = m_L C_0 (1 - k)/k$, and G_L is the thermal gradient in the liquid ahead of the interface, v is the solidification speed, m_L is the liquidus slope, C_0 is the initial liquid composition, k is the partition coefficient (previously defined), and D_L is the solute diffusivity in the liquid (Tiller et al., 1953).

Constitutional undercooling is difficult to avoid except with very slow growth rates. With moderate undercooling, a cellular structure, resembling arrays of parallel prisms, results. As the undercooling grows stronger, the interface breaks down completely as anisotropies in the surface tension and crystal structure leads to side branches at the growing tip of the cells along the "easy-growth" directions ($\langle 100 \rangle$ for fcc and body-centered cubic (bcc), $\langle 1010 \rangle$ for hcp), marking a transition from cellular to dendritic.

Since the mid-1950s, a large amount of work has gone into obtaining accurate mathematical descriptions of dendrite morphologies as functions of the solidification and materials parameters. Dendritic growth is well understood at a basic level. However, most solidification models fail to accurately predict *exact* dendrite morphology without taking into account effects like melt flow. In the presence of gravity, density gradients due to solute partitioning produce a convective stirring in the lower undercooling range corresponding to typical conditions encountered in the solidification of industrial alloys (Huang and Glicksman, 1981). Melt flow is a very effective heat transport mechanism during dendritic growth that may result in variations in the dendrite morphology, as well as spatially varying composition (macrosegragation).

1.2.3 Microstructural Features of Multiphase Alloys

The solidification of pure metals and single-phase solid solutions has been amply discussed. It would now be beneficial to briefly describe the microstructures of multiphase alloys. The simplest type is the binary eutectic system containing no intermetallic compounds or solid-phase miscibility. An example is the Ag–Si system, the phase diagram of which is shown in Figure 1.12. At the eutectic composition (e.g., 96.9 wt % Ag in Figure 1.12), both metals form nuclei and solidify simultaneously as two separate pure phases. Generally, high-volume fractions of both phases will tend to promote lamellar structures. If one phase is present in a small-volume fraction, that phase tends to solidify as fibers. However, some eutectic growths show no regularity in the distribution of the phases. Eutectic microstructures normally exhibit small interphase spacing, and the phases tend to grow with distinctly shaped particles of one phase in a matrix of the other phase, as shown in Figure 1.13. The microstructure will be affected by the cooling rate; it is possible for a eutectic alloy to contain some dendritic morphology, especially if it is rapidly cooled.

The microstructures of *hypo-* or *hyper*eutectic compositions normally consist of large particles of the primary phase (the component that begins to freeze first) surrounded by fine eutectic structure. Often times the primary particles will show a dendritic origin, but they can transform into *idiomorphic grains* (having their own characteristic shape) reflecting the phases' crystal structure.

Multiphasic alloys sometimes contain *intermediate phases* that are atomically ordered, but that do not always crystallize in accordance with normal valency rules. Intermediate phases that exist over small compositional ranges appear as line compounds on a phase diagram. These phases are known as *intermetallic compounds*,

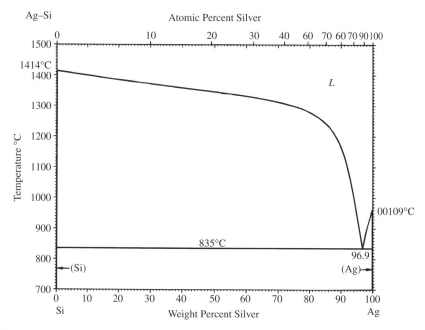

Figure 1.12 The Ag–Si phase diagram. This is the simplest type of binary eutectic system. There are no intermetallic compounds and no solid-phase miscibility. (After Baker, ed., 1992, *ASM Handbook*, Vol. 3: *Alloy Phase Diagrams*. Copyright © ASM International. Reproduced with permission.)

or IMCs (e.g., Cu_9Yb_2, Au_2Na). In contrast to the metallic bonds in a solid solution, the cohesive forces within an intermetallic may be partly covalent, ionic, or ordered metallic. Intermetallic phases are, in general, less "metallic" than metals or solid solutions. Their presence usually renders an alloy less ductile (but stronger) and lowers the thermal and electrical conductivities.

1.2.4 Metallic Glasses

It has been known for decades that in alloy systems with a deep eutectic or low-lying liquidus temperature, compared to the melting points of the pure metals, there is a strong tendency for metallic glass formation with cooling rates in the range 10^5–10^6 K s^{-1}. Such cooling rates are obtainable in industrial melt spinning and splat quenching techniques. Under these processing conditions, the highly disordered state of the supercooled liquid phase becomes "configurationally frozen" into a rigid amorphous (glassy) state. Glasses are monolithic materials, absent of grain boundaries or other crystalline defects. Some alloys can now be prepared as bulk metallic glasses (BMGs) in ribbons or rods with thickness of several centimeters, and at substantially lower cooling rates.

Glass formation can be compared to crystallization by referring to Figure 1.14, which is applicable to both metallic and nonmetallic systems. Crystallization

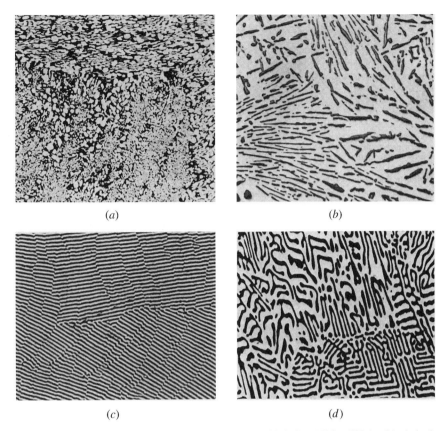

Figure 1.13 Various eutectic microstructures. (a) Globular (50Sn–50In). (b) Acicular (87Al–13Si). (c) Lamellar platelets (67Al–33Cu). (d) Lamellar "Chinese script" (63Mg–37Sn). (After Baker ed., 1992, *ASM Handbook*, Vol. 3: *Alloy Phase Diagrams*. Copyright © ASM International. Reproduced with permission.)

follows path *abcd*. As the temperature of a non-glass-forming melt is lowered, the molar volume of the alloy decreases continuously until it reaches the melting point, where it changes discontinuously, that is, where it experiences a *first-order* phase transition. The enthalpy and entropy behave similarly. By contrast, in glass formation the melt follows path *abef* with decreasing temperature. The liquid remains undercooled (it does not solidify) in the region *be*, below the melting point. The molar volume continuously decreases in the undercooled region and the viscosity increases rapidly. At the point T_g, called the *glass transition temperature*, the atomic arrangement becomes frozen into a rigid mass that is so viscous it behaves like a solid.

Turnbull's criterion for the ease of glass formation in supercooled melts predicts that a liquid will form a glass, if rapidly solidified, as the ratio of the glass transition temperature, T_g, to the liquidus temperature, T_l, becomes equal to or greater than 2/3 (Turnbull and Fisher, 1949; Turnbull, 1950). The T_g/T_l ratio is referred to as the reduced glass transition temperature, T_{rg}. Historically, most glass-forming alloys

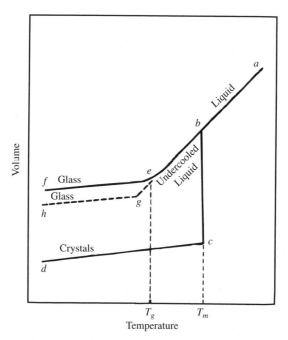

Figure 1.14 A comparison of glass formation (curve *abef*) and crystallization (curve *abcd*). The point T_g is the glass-transition temperature and T_m is the melting temperature. (After West, 1985, *Solid State Chemistry and Its Applications.* Copyright © John Wiley & Sons, Inc. Reproduced with permission.)

were metal–metalloid and metal–metal binary systems (where the metal is usually a transition element and the metalloid is B, Si, C, or P) with a T_g well above room temperature, in the range 300–700 K. With the exception of the group 12 elements (Zn, Cd, Hg) the transition metals have melting points exceeding 1200 K. Hence, those alloy systems containing very low melting eutectics (e.g., 636 K in the Au–Si system) tend to satisfy the Turnbull criterion.

Examples of binary metallic glasses include $Fe_{80}B_{20}$, $Ni_{60}Nb_{40}$, $Ni_{63}Zr27$, and $Ca_{65}Al_{35}$. The compositions of these glasses are near eutectic points (Turnbull, 1981). Turnbull's criterion has thus been validated in systems at cooling rates attainable by "conventional" casting procedures ($\sim 10^6$ K s^{-1}). Some alloy systems, such as $Cu_{60}Zr_{40}$, exhibit glass formation over composition ranges extending well beyond a eutectic point. By contrast, the T_{rg} of pure metals seem to be much smaller than 2/3. Furthermore, pure metallic liquids ($\sim 10^{-2}$ poise) have much lower viscosities than the glass-forming alloys. Therefore, glass formation from pure metal melts requires extremely high cooling rates, on the order of $\sim 10^{12}$ K s^{-1}.

There is both a kinetic and thermodynamic basis to Turnbull's criterion. The rate of homogeneous nucleation is dependent on the ease with which atomic rearrangement can occur (commonly taken as the atomic diffusion coefficient), which scales with fluidity or viscosity. Easy glass-forming substances form highly viscous melts (e.g., $>10^2$ poise), compared to non-glass-forming ones (e.g., water, with $\eta \sim 10^{-2}$ poise). In highly viscous melts, the atomic mobility is substantially reduced, which suppresses the homogeneous nucleation rate. Hence, the homogeneous nucleation rate is highly dependent on T_{rg}. The $T_{rg} > 2/3$ successfully predicts glass formation in metallic and nonmetallic liquids. Igor Evgenevich Tammann (1861–1938) pointed out, as early as 1904, that the higher the viscosity of a melt, the lower its crystallizability (Tammann, 1904). It must be noted, however, that heterogeneous nucleation (e.g., on "seed" particles) may prevent glass formation.

The preceding arguments are based on kinetics. It may also be shown on thermodynamic grounds that a high value for T_{rg} (and therefore the tendency to form a glass at lower cooling rates) is obtained for deep eutectic systems, that is, where the melting point is substantially lowered. These systems tend to be those with very little solid solubility between the components. When atoms do not "fit" together in the lattice (due to mismatches in size, valence, etc.), the tendency for crystallization diminishes. This is due to both a large negative heat of mixing and entropy of mixing for the liquid compared with the competing crystalline phase (Johnson, 2000).

Advances in metallic glasses have been made in the last couple of decades with the discovery of new families of multicomponent alloys with significantly improved glass-forming ability. Ternary glass formers include those systems in which the binary subsets exhibit limited mutual solid solubility, such as $Pd_{77.5}Cu_6Si_{6.5}$ and $Pd_{40}Ni_{40}P_{20}$. These systems have been found to form glasses at cooling rates as low as 10^3 K s^{-1} and 10 K s^{-1}, respectively. This is due to both an increased frustration of the homogeneous nucleation process and to the greater suppression of the liquidus temperature as the number of components is increased. Hence, the glass-forming ability appears to be even further enhanced in yet higher order systems, such as $Pd_{40}Cu_{30}Ni_{10}P_{20}$ (Inoue et al., 1997) and $Zr_{41.2}Ti_{13.8}Cu_{12.5}Ni_{10}Be_{22.5}$ (vitreloy 1) (Perker and Johnson, 1993). These alloys have T_g of about 582 K and 639 K, respectively, and critical cooling rates of just 1 K s^{-1}!

In the supercooled liquid state, BMGs have very high yield strength and a high elastic-strain limit (often exceeding 2%, compared with crystalline materials that are almost always less than 1%), which makes them very "springy." However, under tensile loads bulk metallic glasses lack any significant global plasticity, which limits applications as structural materials (Johnson, 1999). Recent efforts have focused on the development of engineering applications for metallic-glass containing composite materials. Such composites have been found to exhibit greatly enhanced ductility and impact resistance as compared to monolithic glasses.

Metallic glasses, like nonmetallic ones, are thermodynamically metastable states. However, metallic glasses appear to be more susceptible than nonmetallic glasses to *devitrification*, or crystallization at temperatures above T_g. They transform to more stable crystalline phases, typically around 300°–450°C. For example, nanocrystalline grains (grain size < 100 nm) can sometimes be obtained from a metallic glass when it is annealed at temperatures at which primary crystallization can occur. Nanocrystalline phases have been under increased study in recent years because they often have improved properties over their coarse-grained counterparts. Nanocrystalline alloys themselves, however, are also metastable phases, with a tendency toward grain growth.

David Turnbull (Courtesy of Materials Science Group, Division of Engineering & Applied Sciences, Harvard University © Harvard University. Reproduced with permission.)

DAVID TURNBULL

(b. 1915) earned his Ph.D. in physical chemistry from the University of Illinois at Urbana-Champaign in 1939 under T. E. Phipps. From 1939 to 1946, he was on the faculty of the Case Institute of Technology. Turnbull was a research scientist at General Electric from 1946 to 1962, as well as an adjunct professor at Rensselaer Polytechnic Institute from 1954 to 1962. He joined the faculty at Harvard University in 1962, where he was the Gordon McKay Professor of Applied Physics, becoming emeritus in 1985. Turnbull was a pioneer in the study of kinetic phenomena in condensed matter. He performed the critical experiment in several areas: nucleation and growth of crystals; diffusion in crystalline and amorphous materials; and viscous flow of amorphous materials. He formulated the classic theory for nucleation in condensed matter and, with Morrel Cohen, the free volume theory for the flow of liquids and glasses. He predicted that glass formation is universal, anticipated the discovery of metallic glasses, and demonstrated the first formation of metallic glasses in bulk form. The Turnbull criterion predicts the ease with which glass formation occurs. Turnbull was awarded the von Hippel Prize of the Materials Research Society, the Acta Metallurgica Gold Medal in 1979, and the Japan Prize in 1986. The Materials Research Society named an annual lecture after him. He was elected to the U.S. National Academy of Sciences in 1968.

(*Source*: B. Sewall and F. Spaepen, personal communication, February 25, 2004.)

1.2.5 Microstructure/Property Correlation and Control in Metals

Mechanical properties are studied in detail in Chapter 9. Here, we present a brief introduction appropriate for the present discussion. The microstructures of all metals and alloys greatly influence strength, hardness, malleability, and ductility. These properties describe a body's *plasticity*, or ability to withstand permanent deformation without rupture. Plastic deformation is due to the gliding motion, or *slip*, of planes of atoms. Slip most readily occurs on close-packed planes of high atomic density in the close-packed directions. However, much smaller stresses are required to move a dislocation through a crystal than a full plane of atoms. Therefore, dislocations actually govern the ability of a coarse-grained material to plastically deform. Dislocations have already been introduced. One type is the edge dislocation, which is an extra half-plane of atoms.

Polycrystalline materials are stronger than single crystals. It is not easy for dislocations to move across grain boundaries because of changes in the direction of the slip planes. Thus, the mechanical properties of polycrystalline metals are dependent on the average grain size and the orientation distribution, or *texture*. Deformation processes such as rolling and extrusion, which rely on plasticity, introduce texture into a sample. The *elasticity* of a polycrystal, or ability of the body to return to its original size and shape upon removal of a mechanical stress, is also a function of texture. Strictly speaking, this is true so long as the following conditions are met: (1) the number of grains in the sample is sufficiently large; (2) there is no preferred orientation; (3) the polycrystal is under homogenous strain and stress (i.e., all grains experience the same strain and stress); and (4) grain boundaries do not contribute to the body's elasticity.

Other physical properties of polycrystalline metals, such as electrical and thermal conduction, are also affected by microstructure. Point defects (vacancies, impurities) and extended defects scatter electrons and phonons, shortening their mean free paths. An approximation known as *Matthiessen's rule*, from the nineteenth-century physicist Ludwig Matthiessen (1830–1906), asserts that all the various scattering processes are independent (strictly true only so long as the scattering processes are isotropic). Hence, the contributions to the resistivity can simply be added up. The electrical resistivity of a polycrystalline metal is then the sum of the contributions due to electron scattering in the bulk crystal and at the interfaces, the latter of which is a function of the grain boundary structure. Furthermore, the resistivity of a fine-grained metal is higher than that of a coarse-grained sample because the former has a larger number of grain boundaries.

Transport properties are second-rank tensors, which are only isotropic for cubic crystals and polycrystalline aggregates with a random crystallite orientation. For all other cases, the conductivity will be dependent on direction. A noncubic polycrystalline metal or alloy will show a texture dependency. However, because most metals are in the cubic class, we postpone a discussion of this topic until Section 1.3.1 on the transport properties of ceramics, where the effect is more marked.

In addition to specifying the texture or orientation distribution to the grains (Section 1.1.7), we can specify the fraction of a particular CSL boundary type

(Section 1.1.3). This approach is also useful because the grain boundary structure often correlates with certain materials properties, particularly conductivity, creep (time-dependent deformation at constant load), and corrosion. The incorporation of a high percentage of a specific grain boundary type in a polycrystal is referred to as *grain boundary engineering*. For example, coherent twin boundaries are able to block dislocation motion and strengthen a metal. They also allow for a much more efficient transfer of current than do conventional grain boundaries. Hence, metals with a high percentage of coherent twin boundaries are strengthened without a loss to the electrical conductivity. Likewise, it has been shown that Ni–Cr–Fe alloys with a high fraction of special CSL boundaries possess higher creep resistances (lower strain rates) than those with general boundaries (Thaveepringsriporn and Was, 1997; Was et al., 1998).

Grain boundary structure also contributes to the chemical properties exhibited by a metal. *Inter*granular atomic diffusion processes occur more rapidly than *intra*granular ones, since these regions are usually not as dense as the grains. Remember also that grain boundaries have a higher free energy than the grains themselves. Because of these aspects, a metal will usually oxidize or corrode more quickly at the grain boundaries, a condition known as intergranular corrosion. The oxidation rate may be very dependent on the grain boundary structure. Again, this gives rise to the possibility for grain boundary engineering. For example, it has been shown that Ni–Fe (Yamaura et al., 2000) alloys and Pb electrodes (Palumbo et al., 1998) deformation processed or thermally processed to have a high fraction of low-Σ boundaries are much more resistant to intergranular corrosion, while those with random high-angle boundaries, which are probably higher in energy, more easily oxidize. For conventionally grained solids, however, the grain boundary volume is a small fraction of the total volume of the sample. Hence, intragranular diffusion is usually the dominant mass-transport process, except for very small grain sizes or at low temperatures. Nonetheless, because impurity atoms tend to segregate at grain boundaries, intergranular chemical reactions that may not occur intragranularly are possible.

Materials processing is generally aimed at achieving a texture and/or grain boundary structure that will maximize a specific property of interest. Approaches with polycrystalline metals and alloys usually involve deformation processing and annealing, or a combination thermomechanical method. During fabrication processes, such as cold rolling, for example, polycrystalline metals deform by mechanisms involving slip and, where slip is restricted, *rotation* of the individual grains. Both processes, of course, must satisfy the condition that the interfaces along which the grains are connected remain intact during deformation. As the extent of deformation increases, larger grains may break up into smaller grains of different orientations, giving rise to an orientation spread. Similar changes accompany both hot working and annealing. In these procedures, the free energy of a metal decreases due to rearrangement of dislocations into lower energy configurations of decreased dislocation density. This is termed *recovery*. However, in competition with this is *recrystallization*, in which the resulting grain structure and texture depend on the spatial distribution and orientation of the recrystallization nuclei. The competition between these two processes and their effect on texture and grain boundary structure is a hotly debated subject in metallurgy.

1.3 CERAMIC POWDER AGGREGATES

Traditionally, ceramics have been defined rather broadly as all man-made inorganic nonmolecular materials, excluding metals and semiconductors. Solidification processes, like those discussed in the previous section, are very seldom used to prepare ceramics. This is because ceramics normally have exceedingly high melting points, decompose, or react with most crucible materials at their melting temperatures. In the past, ceramics were prepared only by high-temperature solid-state reactions, which came to be known as the *ceramic* method. This restriction was lifted to a certain degree with the advent of lower temperature (up to a few hundred degrees) soft chemical routes, such as ion exchange, intercalation, and topochemical reactions, where the reactivity is controlled by the crystal structure of the starting material.

The products of most ceramic syntheses are powders. Fabricating a bulk part from a powder requires a forming process, usually compaction (either compression molding or injection molding) followed by sintering. The latter step involves heating the phase at a temperature below its solidus (melting point), but high enough that grain growth occurs via solid-state diffusion. In the early stages of sintering, volatile species can be removed (e.g., moisture content) and, at later stages, homogenization improved (e.g., the removal of macrosegragation), and densification increased as the particles bond together, eliminating pores. The resultant products have higher strength and reliability.

1.3.1 Transport Properties of Polycrystalline Ceramics

Although polycrystals are mechanically superior to single crystals, they have inferior transport properties. Quite a large number of ceramics are actually electrical conductors, that is, they exhibit a metallic-like conductivity. However, a polycrystal has a lower thermal and electrical conductivity than a single crystal because of phonon and electron scattering at grain boundaries. Generally, the smaller the grain size, the larger the total interfacial volume and the lower the conductivity. We will see a specific example of this grain-size dependence later in Section 7.4.2 of Chapter 7. In addition to this effect, it is also important to understand the effect of texture on transport properties. The conductivities of all crystal classes other than the cubic class are anisotropic, or dependent on direction. Some ceramics are even low-dimensional transport systems, in which conduction is much weaker or completely absent along one or more of the principal crystallographic axes. If the transport properties of the individual crystallites in a polycrystalline aggregate are anisotropic, such as in a low-dimensional system, sample texture will influence the anisotropy to the conductivity of the polycrystal.

When a polycrystal is free of preferred orientation, its transport properties will be, like a cubic crystal, macroscopically isotropic regardless of the anisotropy to a

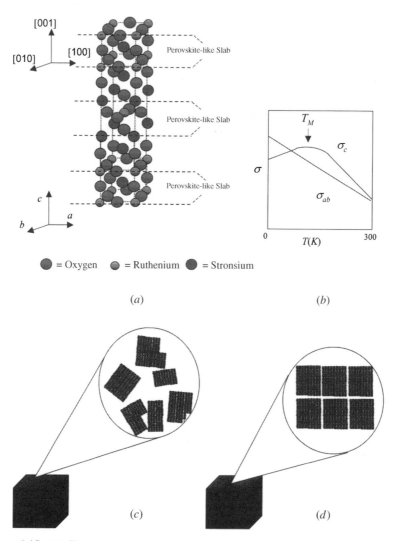

Figure 1.15 (a) The crystal structure of Sr_2RuO_4. (b) Above T_M, conduction in the *ab* planes within the perovskite layers is metallic, while the *c*-axis transport is semiconducting. (c) Random orientation in a powder aggregate with different crystallite sizes. Percolation paths exist for both σ_{ab} and σ_c in all directions, resulting in an isotropic conductivity at any given temperature. (d) If identical crystallites are all aligned with their *c* axes parallel, the polycrystal will have the same anisotropic conductivity as a single crystal.

single crystallite. For example, consider the low-dimensional metal Sr_2RuO_4, with the tetragonal crystal structure shown in Figure 1.15a. At high temperatures, metallic conduction occurs in the *ab* planes within the perovskite layers, parallel to the two principal axes $\langle 010 \rangle$ and $\langle 100 \rangle$. Transport along the *c*-axis, perpendicular to the *ab* plane, is semiconducting at high temperatures.

For polycrystals, two extreme orientation distribution functions, which give the volume fraction $f(x)$ of crystals in orientation x, can be envisioned. These are zero preferred orientation or perfect alignment. The absence of preferred orientation statistically *averages* out the parallel and perpendicular conduction effects in Sr_2RuO_4. The bidimensional metallic conductivity, which is clearly observed in a single crystal, is completely inhibited in free powders composed of millions or billions of randomly oriented crystallites. This is shown in Figure 1.15c. By contrast, we can expect the application of several thousand pounds of uniaxial pressure in forming a pellet (i.e., a cylindrical shaped specimen) to induce a high degree of texture. This is especially true for elongated platelet crystals, where the c axes of the crystallites align parallel to the pressure axis. We cannot assume that all the crystals are identical in size and shape. Thus, with the application of uniaxial pressure alone many crystallites will remain misoriented. Higher degrees of texture may sometimes be induced by magnetic fields because of the anisotropy to the magnetic susceptibility (Section 1.3.4).

If the Sr_2RuO_4 crystallites were to be identical and perfectly aligned with their c axes parallel (somewhat like stacking children's building blocks), the pellet's resistivity would exhibit the same anisotropy as a tetragonal single crystal of Sr_2RuO_4. The resistivity along the ab axis of the pellet would follow the ρ_{ab} curve in Figure 1.15b, while the resistivity along the c-axis would follow the ρ_c curve. This is illustrated in Figure 1.15d. Nonetheless, the current density of the pellet would be lower than that of a single crystal of the same dimensions due to the grain boundary resistance.

We can gain further enlightenment on electrical conduction in polycrystals by likening the phenomenon to a random resister network with the *bond percolation model*. In this model, we have a lattice, where each site corresponds to an individual crystallite of the polycrystalline aggregate. We make the simplifying assumption for now that all sites, or crystallites, are equivalent electrical conductors. The grain boundaries are the "bonds" connecting the individual lattice sites. Each grain boundary has an electrical resistance, R_{GB}, that is dependent on the orientation relationships of the grains. For now, let us also assume a bimodal R_{GB}, that is, that 0 (superconducting) or ∞ (nonconducting) are the only two possible values for R_{GB}. Each bond is either "open" with probability p, or "closed" with probability $(1-p)$. The open bonds allow passage of current and the closed bonds do not.

The main concept of the bond percolation model is the existence of a percolation threshold, p_c, corresponding to the point at which a cluster of open bonds (a conducting path) first extends across the entire sample. Such a cluster is called a *spanning cluster*. For all $p < p_c$, no current can flow due to the lack of a complete current path, but there may be nonspanning clusters, connecting a finite number of points, which exist for any nonzero p. At p_c, the system abruptly transitions to the electrically conductive state, and for all $p > p_c$, the sample is electrically conducting because of the presence of spanning clusters.

For $p > p_c$, the conductivity, σ, is proportional to a power of $(p - p_c)$:

$$\sigma \propto (p - p_c)^\mu \tag{1.18}$$

The exponent μ is called a scaling exponent. It depends on the dimensionality of the system. For 2D transport, $\mu = 1.30$ and for 3D transport, $\mu = 2.0$ (Stauffer and Aharony, 1994).

The numerical value of the bond percolation threshold is dependent on the geometry of the grain boundary network. For example, p_c is equal to 0.500 for a simple 2D square lattice and 0.388 for a 3D diamond lattice. For the simple cubic, fcc, and bcc lattices, p_c is equal to 0.2488, 0.1803, and 0.119, respectively. With nontextured polycrystals, the geometries (grain orientations and/or angles) are random, and hence, the exact value for p_c may not be known. Furthermore, our original assumption that the grain boundaries are either superconducting or insulating is obviously a drastic one. In reality, the grain boundary resistance is not bimodal. It can have values other than zero or infinity, which are dependent on the grain orientations and/or angles, as can be inferred from Figure 1.15c. In fact, a broad distribution of grain boundary resistances may be observed. For sufficiently broad distributions, however, the resistance of the bulk sample is often close to the resistance of the grain boundary cluster with exactly the percolation threshold concentration (Stauffer and Aharony, 1994). For a derivation of Eq. 1.18, as well as a detailed treatment of percolation theory, the interested reader is referred to the book by Stauffer and Aharony (1994).

Example 1.3 The oxide $Bi_4V_2O_{11}$ is similar in structure to the layered Aurivillius phase Bi_2MoO_6 (tetragonal crystal class). The aliovalent exchange of Mo^{6+} by V^{5+} results in oxygen vacancies in the BiO_6 octahedral layers, which are separated along the c-axis by edge-sharing BiO_4 pyramids. The oxygen-ion conduction in $Bi_4V_2O_{11}$ is highly bidimensional, being much stronger in the ab planes of the octahedral layers than in the perpendicular out-of-plane direction along the c-axis. Speculate on what the (001) pole figures for each sample given in Table 1.4 would look like.

Solution The (001) pole figure for the single crystal must have a single spot in the middle of the circle since all (001) planes in a single crystal are parallel. The anisotropy in the conductivity would be expected to be the greatest. The (001) pole figure for the pressed polycrystal should show the greatest spread, or orientation distribution, to the grains. This is because the anisotropy to the conductivity of the

TABLE 1.4 Sample of (001) Pole Figures[a]

	$\sigma_{\parallel}$	σ_{-}
Single crystal	73.2	2.6
Pressed polycrystal	40.2	20.2
Magnetically aligned polycrystal	59.1	13.8

Source: The data are from Muller, C.; Chateigner, D.; Anne, M.; Bacmann, M.; Fouletier, J.; de Rango, P. 1996, *J. Phys. D, 29.*
[a] Conductivities ($\times 10^3$ S cm^{-1}).

polycrystal was observed to be the smallest. Its properties are closest to that of a randomly textured sample. The (001) pole figure for the magnetically aligned polycrystal should have an intermediate spread since its anisotropy to the conductivity was intermediate.

1.3.2   Magnetic and Dielectric Properties of Polycrystalline Ceramics

Just as crystals can have an easy direction for the transport of charge carriers, they will likewise magnetize along easy axes of magnetization. The interactions between subgrain domains of ferromagnetic and ferroelectric polycrystals further complicate behavior. Subgrain domains form because the energy density associated with the magnetic flux density exerted outside a sample is decreased if domains with opposing magnetizations are created (Elliot, 1998). In the absence of an external field, each domain exhibits spontaneous magnetization, or polarization in the case of ferroelectrics, which can be represented by the resultant net vector for that domain. However, a macroscopic crystal generally will exhibit very little magnetization in zero fields. This is because the individual domains have their vectors pointing in different directions, giving rise to a zero net magnetization or polarization. In the presence of a sufficiently strong applied field, saturation is reached, where the dipoles in all the domains within the crystal align.

Polycrystalline materials possess a texture, or orientation distribution to the grains, and each grain may contain several domains (see Figure 1.16). We might therefore expect that neighboring grains, and the domains within them, would interact magnetically in very different ways, giving rise to a response that will be very dependent on the grain size and texture. Such behavior is expected in the plots of magnetization versus applied field, so-called *hysteresis* loops, that follow one

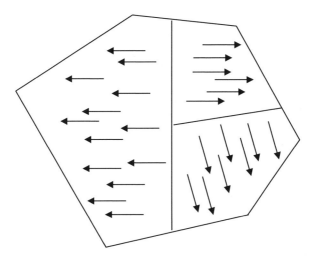

Figure 1.16   Schematic drawing of subgrain magnetic domains. Each of the three domains shown has a different net magnetic moment.

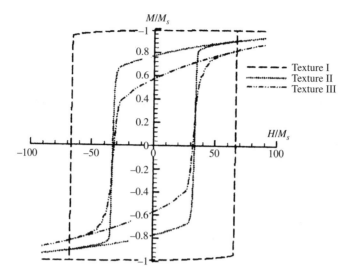

Figure 1.17 Simulated hysteresis loops for polycrystalline films with different textures. (After Jin et al. 2002, *J. App. Phys.*, *92*, 6172. Copyright © American Institute of Physics. Reproduced with permission.)

complete cycle of magnetization and demagnetization in an alternating field. Simulated hysteresis loops for polycrystalline films with different textures are shown in Figure 1.17. The greater the misalignment among grains, the less the magnetization that remains after the field is removed (*remanence*), and the lower the magnitude of the reverse field required for demagnetization (*coercivity*) (Jin et al., 2002). If these parameters can be controlled by manipulating the micro-structure, different magnetic properties can be designed. For example, if ultrasmall (i.e., nanosized) grains of a highly magnetocrystalline–anisotropic thin-film mate-rial can be deposited with all their easy magnetization axes normal to the film, magnetic storage media with much greater recording densities can be obtained.

1.3.3 Mechanical Properties of Ceramics

The mechanical properties of ceramics differ greatly from those of metals. Ceramics are typically brittle, very strong, hard, and resistant to deformation. It is found that dislocation motion is virtually impossible except at high temperatures. In ionic solids, slip is constrained because it requires bringing ions with like charges in contact. Furthermore, although many ionic solids are "almost" close packed, slip is very difficult on anything but truly close-packed planes. The strong and directional bonding present in covalent solids similarly impedes dislocation motion since it requires bond breaking and distortion.

The absence of a lattice-based mechanism, such as slip planes, does not neces-sarily preclude all deformation. Plastic flow can proceed in other modes. For example, at temperatures of about 40–50% of their melting points, grain boundary sliding can become important. Grain boundary sliding is believed to be the major contributor to the superplasticty observed in some polycrystalline ceramics.

1.3.4 Texture Control in Ceramics

We have already mentioned that one method of producing a textured ceramic involves the application of uniaxial pressure to form a cylindrical-shaped specimen. However, in this method, there usually remain a large number of misoriented grains. Furthermore, this simple approach is not applicable for every desired shape. There are other methods that have been used to produce textured ceramics. One is the templated grain growth (TGG) method, in which anisotropic seed particles regulate crystal growth in specific growth directions. This process involves the addition of a small amount of the template particles (usually less than 10 vol %) that is, seed particles of the same substance, to a powder matrix. The template particles are next oriented and the matrix is then sintered, causing the template particles to grow by consuming the randomly oriented matrix. This produces a highly oriented microstructure.

When the unit cell of a crystalline ceramic substance is anisotropic, this gives rise to the possibility that the accompanying anisotropy in its physical properties may aid the texture control of powder samples. For example, a crystal with an anisotropic magnetic susceptibility will rotate to an angle, minimizing the system energy when placed in a magnetic field. The reduction in magnetic energy is the driving force for magnetic alignment. However, alignment may be difficult in nondispersed powders because of strong particle–particle interactions that prevent the particles from moving. Therefore, dispersion of the powder in a suspension is usually necessary for effective utilization of a magnetic field. This can be readily accomplished in *slip casting*, where a suspension, or slurry, called the *slip*, is poured into a porous mold of the desired shape that absorbs the fluid. Magnetic alignment, in conjunction with slip casting, has been used to produce textured microstructures in a variety of substances, including some with only small anisotropic diamagnetic susceptibility, such as ZnO and TiO_2, if the magnetic field is high enough (e.g., several tesla).

1.4 THIN-FILM MICROSTRUCTURE

The cornerstone of the entire semiconductor microelectronics industry is thin-film technology. There are many methods available for growing thin films, including physical vapor deposition (PVD) (e.g., sputtering, evaporation, laser ablation); chemical vapor deposition (CVD); plating; and so-called soft chemical techniques (e.g., sol-gel coating). Most of these processes are well established in the microelectronics industry, but many also have important applications in the areas of advanced coatings and structural materials. An enormous variety of films can be prepared. The interested reader is encouraged to refer to any of numerous texts on thin films for details regarding the deposition processes.

An exciting new area of materials research that has begun to evolve in recent years is the application of combinatorial chemistry to the creation of thin-film libraries. By using masks (grids with separate squares), thousands of distinct combinations of materials can, in principle, be deposited onto a single substrate in order to greatly accelerate the screening of the resultant compounds for certain properties. This is part of a broad approach being sought for the rapid discovery of new materials known as *combinatorial materials science*, inspired by the success of combinatorial chemistry in revolutionizing the pharmaceutical industry.

1.4.1 Epitaxy

Thin films can be polycrystalline or single crystalline. The majority of thin films are polycrystalline. Single crystal films can be grown with a particular crystallographic orientation, relative to a single-crystal substrate that has a similar crystal structure, if the deposition conditions and substrate are very stringently selected. *Homoepitaxy* refers to the growth of a thin film of the same material as the substrate (e.g., silicon on silicon). In this case, the crystallographic orientation of the film will be identical to that of the substrate. In *heteroepitaxy*, such as silicon on sapphire, the layer's orientation may be different from that of the substrate. Both types of epitaxy often proceed by an island nucleation and growth mechanism referred to as Volmer–Weber growth.

Island growth also occurs with polycrystalline films, but in epitaxy, the islands combine to form a continuous single-crystal film, that is, one with no grain boundaries. In reality, nucleation is much more complex in the case of heteroepitaxy. Nucleation errors may result in relatively large areas, or domains, with different crystallographic orientations. The interfaces between domains are regions of structural mismatch called *subgrain* boundaries and will be visible in the microstructure.

Epitaxy has been used to stabilize films with crystal structures that are metastable in the bulk phases. Kinetic stabilization is obtained when the growth is performed under conditions of high surface diffusion, but low bulk diffusion. In this way, crystallographically oriented film growth occurs, while phase transformations are prohibited. The circumstances under which thermodynamic stabilization can be achieved have also been enumerated. Namely, by minimizing: (1) the lattice mismatch, or structural incoherency, and the free-energy difference between the growing film and the substrate; (2) the film thickness; and (3) the shear and elastic moduli of the film. Additionally, the growing film should be able to form a periodic multiple-domain structure (Gorbenko et al., 2002).

The high-temperature bulk fcc phase $\delta\text{-Bi}_2\text{O}_3$ is observed on heating $\alpha\text{-Bi}_2\text{O}_3$ (monoclinic structure) above $717°C$, but heteroepitaxial thin films of $\delta\text{-Bi}_2\text{O}_3$ have been deposited on an fcc gold substrate at temperatures as low $65°C$. Similarly, $YMnO_3$ crystallizes in the hexagonal structure under atmospheric pressure at high temperatures, yet was grown by pulsed laser deposition (PLD) as a stable perovskite film on an $NdGaO_3$ substrate (Salvador et al., 1998). It has also recently been shown how metal–organic chemical vapor deposition (MOCVD) results in the formation of metastable phases of GaS and GaTe, irrespective of the structure of the substrate (Keys et al., 1999; Gillan and Barron, 1997). Crystalline GaS was even grown on an amorphous substrate. It appears in these cases that the precursor's structure or its decomposition mechanism completely controls the structure of the thin film. Likewise, polycrystalline CVD diamond is commercially grown on a variety of substrates, including titanium, tungsten, molybdenum, SiO_2, and Si_3N_4. Each of these materials is capable of forming a carbide layer upon which an adherent diamond film can nucleate, although it is not clear whether carbide formation is essential.

1.4.2 Polycrystalline PVD Thin Films

The commonest morphology exhibited by PVD thin films is the highly oriented columnar (fiber) grain growth shown in Figure 1.18. This is a consequence of the fact that PVD processes generally deposit thin films atom by atom in a "line of site" fashion wherein the sputtered or evaporant *adatoms* travel from the source to substrate on a straight path. The resulting grains are thus aligned with their long axes perpendicular to the surface when the incident beam arrives at a normal angle of incidence. If the depositing atoms arrive at an angle away from normal incidence, the columns tilt into the oncoming beam. This gives rise to shadowing effects, which result in the columns separated from one another by voids. In either case, the grains can completely traverse the thickness of the film. Sputtered films generally tend to be denser, more amorphous in nature, and more adherent than evaporated films because of the higher energy of the arriving adatoms.

Slight variations can occur with columnar morphology, however. For example, Movchan and Thornton used structure zone models to illustrate how temperature influences the morphology of metal films (Movchan and Demchishin, 1969; Thornton, 1977). Grain growth often begins with island formation (nucleation sites). When the homologous temperature, T_h (for thin films this is the ratio of the substrate temperature to the melting point of the thin film), is <0.3, the surface mobility of the deposited atoms, or adatoms, is low. As the homologous

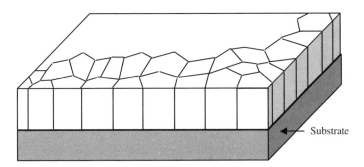

Figure 1.18 Highly oriented columnar (fiber) thin film. The grains are oriented with a preferred crystal direction normal to the substrate along the growth axis, but with no orientation relationship about the growth axis.

temperature increases $(0.3 < T_h < 0.5)$, the surface mobility increases. Thus, the islands may initially evolve three-dimensionally into V-shaped columns. Once the V-shaped columns impinge on one another, the grain boundaries become parallel and a dense structure is obtained. Finally, when $T_h > 0.5$, bulk diffusion increases substantially and allows for equiaxed grains. A major disadvantage of PVD thin films is its inability for conformal coating, that is, it produces poor step coverage. Residual stresses in PVD thin films are generally compressive.

As with solidification microstructures, kinetic factors are as important as thermodynamics. For example, the (200) surface has the lowest energy in TiN. But the preferred orientation of TiN thin films can vary between (111) and (200), depending on deposition conditions. This means that the texture depends on kinetic factors as well as the energy minimization.

1.4.3 Polycrystalline CVD Thin Films

Chemical-vapor deposition is a process in which a volatile species is transported in the vapor phase to the substrate to be coated. Typically, a chemical reduction then occurs at the substrate to deposit the desired film. For example, volatile $ReCl_5$ will deposit rhenium on a substrate heated to $1200°C$. Although adatom energy exerts the strongest influence on film morphology in PVD processes, CVD is dominated by chemical reaction kinetics. The islands grown in CVD processes tend to be larger in size and fewer in number. There also appears to be marked temperature dependence to the growth rate. Kinetic Monte-Carlo (MC) simulations have shown that grain morphology in CVD thin films is primarily due to an autocatalytic process, in which precursor molecules dissociate preferentially at existing nuclei sites (Mayer et al., 1994). Another difference between PVD and CVD films is that, in PVD, the residual stress in the deposited film is normally compressive, whereas in CVD, it is generally tensile. Nevertheless, as with PVD, the most common microstructure produced from CVD is the highly oriented columnar structure.

REFERENCES

Aust, K. T.; Rutter, J. W. 1959, *Trans. AIME, 215*, 119.

Bhadeshia, H. K. D. H. 1987, *Worked Examples in the Geometry of Crystals*, The Institute of Metals, London, p. 70.

Bollman, W. 1967, *Phil. Mag., 16*, 363.

Bollman, W. 1970, *Crystal Defects and Crystalline Interfaces*, Springer-Verlag, Berlin.

Brandon, D. G.; Ralph, B.; Ranganathan, S.; Wald, M. S. 1964, *Acta Metall., 12*, 813.

Chen, F.-R.; King, A. H. 1988, *Acta Crystallogr. B, 43*, 416.

Elliot, S. R. 1998, *The Physics and Chemistry of Solids*, John Wiley & Sons, Chichester, UK.

Friedel, G. 1926, *Lecons de Cristallographie*, Berger-Levrault, Paris.

Gillan, E. G.; Barron, A. R. 1997, *Chem. Mater., 9*, 3037.

Gorbenko, O. Y.; Samoilenkov, S. V.; Graboy, I. E.; Kaul, A. R. 2002, *Chem. Mater., 14*, 4026.

Hargreaves, F. Hill, R. T. 1929, *J. Inst. Metals, 41*, 237.

Hoglund, D. E.; Thompson, M. O.; Aziz, M. J. 1998, *Phys. Rev. B, 58*, 189.

Huang, S. C; Glicksman, M. E. 1981, *Acta Metall., 29*, 71.

Inoue, A.; Nishiyama, N.; Kimura, H. 1997, *Mater. Trans. JIM, 38*, 179.

Ivantsov, G. P. 1947, *Dokl. Akad. Nauk, 58*, 56.

Jin, Y. M.; Wang, Y. U.; Kazaryan, A.; Wang, Y.; Laughlin, D. E.; Khachaturyan, A. G. 2002, *J. App. Phys., 92*, 6172.

Keys, A.; Bott, S. G.; Barron, A. R. 1999, *Chem. Mater., 11*, 3578.

Johnson, W. L. 2000. In: Turchi, P. E. A., Shull, R. D., editors, *The Science of Alloys for the 21st Century: A Hume-Rothery Symposium Celebration*, The Minerals, Metals, and Materials Society, Warrendale, PA.

Johnson, W. L. 1999, *MRS Bull. 24*, 42.

Kronberg, M. L.; Wilson, F. H. 1949, *Trans. Met. Soc. AIME, 185*, 501.

Mayer, T. M.; Adams, D. P.; Swartzentruber, 1994, "Nucleation and Evolution of Film Structure in Chemical Vapor Deposition Processes," *Physical and Chemical Sciences Center Research Briefs*, Vol. 1-96, Sandia National Laboratories.

Mott, N. F. 1948, *Proc. Phys. Soc. London., 60*, 391.

Movchan, B. A.; Demchishin, A. V. 1969, *Phys. Met. Metallogr., 28*, 83.

Muller, C.; Chateigner, D.; Anne, M.; Bacmann, M.; Fouletier, J.; de Rango, P. 1996, *J. Phys. D, 29*, 3106.

Mullins, W. W.; Sekerka, R. F. 1963, *J. Appl. Phys., 34*, 323.

Palumbo, G.; Lehockey, E. M.; Lin, P. 1998, *JOM, 50*(2), 40.

Perker, A.; Johnson, W. L. 1993, *Appl. Phys. Lett., 63*, 2342.

Phanikumar, G.; Chattopadhyay. 2001, *Sadhana, 26*, 25.

Randle, V. 1993, *The Measurement of Grain Boundary Geometry*, IOP Publishing, Ltd., London.

Ranganathan, S. 1966, *Acta Crystallogr., 21*, 197.

Read, W. T.; Shockley. W. 1950, *Phys. Rev., 78*, 275.

Salvador, P. A.; Doan, T. D.; Mercey, B.; Raveau, B. 1998, *Chem. Mater, 10*, 2592.

Saylor, D. M.; Morawiec, A.; Adams, B. L.; Rohrer, G. S. 2000, *Interface Sci., 8*[2/3], 131.

Scheil, E. Z. 1942, *Metallkunde, 34*, 70.

Stauffer, D.; Aharony, A. 1994, *Introduction to Percolation Theory,* rev. 2nd ed., Taylor & Francis, London, p. 52.

Tammann, G. Z. 1904, *Elektrochem., 10*, 532.

Thaveepringsriporn, V.; Was, G. S. 1997, *Metall. Trans. A, 28*, 2101.

Thornton, J. A. 1977, *Ann. Rev. Mater. Sci., 7*, 239.

Tiller, W. A.; Jackson, K. A.; Rutter, R. W.; Chalmers, B., 1953, *Acta Metall. 1*, 50.

Turnbull, D.; Fisher, J. 1949, *J. Chem. Phys., 17*, 71.

Turnbull, D. 1950, *J. Chem. Phys., 18*, 198.

Turnbull, D. 1981, *Met. Trans. A, 12*, 695.

Was, G. S., Thaveepringsriporn, V.; Crawford, D. C. 1998, *JOM, 50*, 44.

Watanabe, T. 1984, *Res. Mech. 11*, 47.

Watanabe, T. 1993, In: Erb, U., Palumbo, G., editors, *Grain Boundary Engineering*, CIM, Montreal.

Weertman, J.; Weertman, J. R. 1992, *Elementary Dislocation Theory*, Oxford University Press, New York, p. 6.

Wolf, D.; Lutsko, J. F. 1989, *Z. Kristall.*, *189*, 239.

Yamaura, S.; Igarashi, Y.; Tsurekawa, S, Watanabe, T. 2000. In: Meike, A., Gonis, A., Turchi, P. E. A., Rajan, K., editors, *Properties of Complex Inorganic Solids 2*, Kluwer Academic/ Plenum Publishers, New York, pp. 27–37.

CHAPTER 2

Crystal Structure and Bonding

In Chapter 1, we discussed the structure of polycrystalline materials on a microscope scale, in which the smallest discernable particles are the crystallites, or grains. Microstructural analysis provides information on average grain size, shape, and orientation, but none regarding the atomic arrangement within the crystals. Crystal structures are normally determined from X-ray, electron, or neutron diffraction data. In the present chapter, we "zoom in" and explore crystal structure. Various ways of describing crystal structures are discussed, together with the different types of cohesive forces. An introduction to some of the most common structure types that are adopted is also presented.

2.1 STRUCTURE DESCRIPTION METHODS

There are primarily three methods employed to describe crystal structures: close packing, space-filling polyhedron, and the unit cell. The most concise, yet complete, description of a crystalline substance is the one that gives the unit cell, or repeating unit. This is the only one of the three methods that is applicable to all structure types. However, it is often advantageous to use the other two simpler approaches, in order to gain a different perspective.

2.1.1 Close Packing

The structures of many metals can be described rather simply if one assumes that the constituent atoms are close-packed spheres. In this arrangement, the maximum possible density is achieved. Close packing of spheres dates back to the seventeenth century when Sir Walter Raleigh asked the mathematician Thomas Harriot to study the stacking of cannon balls. Harriot debated this topic with Johannes Kepler for some time. In 1611, Kepler hypothesized from geometrical considerations that the densest possible arrangement of spheres is obtained with cubic close packing— 74.05%. This came to be known as the *Kepler conjecture*, and the problem of proving it was the *Kepler problem*. A rigorous mathematical proof eluded

Principles of Inorganic Materials Design By John N. Lalena and David A. Cleary
ISBN 0-471-43418-3 Copyright © 2005 John Wiley & Sons, Inc.

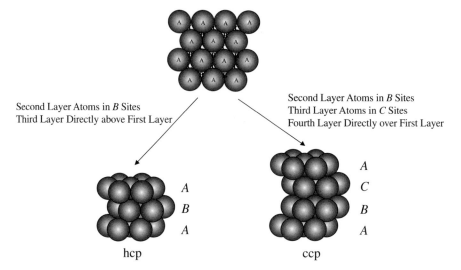

Second Layer Atoms in *B* Sites
Third Layer Directly above First Layer

Second Layer Atoms in *B* Sites
Third Layer Atoms in *C* Sites
Fourth Layer Directly over First Layer

A
B
A

hcp

A
C
B
A

ccp

Figure 2.1 Close-packing arrangements in three dimensions.

mathematicians until only recently. It was not until 1998 that the proof was accomplished by Professor Thomas Hales of M. I. T.!

It is well known that in two dimensions close-packed spheres have a hexagonal arrangement in which each sphere is tangent to six others in the plane. In three dimensions, there are two possible ways to achieve close packing. They have to do with the way closed-packed planes are stacked together. In both cases, however, each sphere is tangent to, or coordinated by, 12 others.

Figure 2.1 shows each type of arrangement. In the first case, each sphere in the upper layer, of the set of three layers, is directly above one sphere in the lower layer. The spheres of the middle layer rest in the hollows between three spheres in each of the adjacent layers. The staggered close-packed layers are sometimes represented as (...*ABABAB*...), where each letter corresponds to a 2D closed-packed layer, and in which the sequence required to achieve 3D close packing is clear. This is called *hexagonal close packed*, and is abbreviated hcp.

The second case corresponds to three close-packed layers staggered relative to each other. It is not until the fourth layer, when the sequence is repeated. This is known as *cubic close packed* (ccp), and is represented as (...*ABCABCABC*...). Geometric considerations show that, for equal-sized spheres in both the ccp and hcp arrangements, 74.05% of the total volume are occupied by the spheres. The packing densities of other non-closed-packed structures are given in Table 2.1.

Because the atoms of an element are all equal sized, the structures of many elements correspond to the ccp or hcp array. By contrast, many ionic compounds can be described as a close-packed array of anions (large spheres), with cations (smaller spheres) located in the hollows between the anions. The hollows, which are called *interstitial sites*, come in two different sizes. Tetrahedral sites are coordinated by four anions, and octahedral sites are coordinated by six anions, as shown

TABLE 2.1   Packing Densities for Various Structures

Structure	Density (%)
Cubic close packed	74.05
Hexagonal close packed	74.05
Body-centered cubic	69.81
Tetragonal	71.87
Cubic	52.36
Hexagonal	60.46

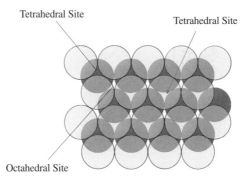

Figure 2.2   The octahedral and tetrahedral sites within two adjacent close-packed layers. The darkly shaded spheres are behind the plane of the page. The lightly shaded spheres are above the plane of the page.

in Figure 2.2. For every anion in two adjacent close-packed layers, there are two tetrahedral sites (an upper and a lower) and one octahedral site. The octahedral sites are larger and can thus accommodate larger cations. When the cations are large, they effectively push the anions further apart. The term *eutactic* is useful for describing such systems, where the arrangement of atoms is the same as in a close-packed structure, but in which the atoms are not touching. Some covalent compounds (e.g., diamond) can also be described in terms of close-packed arrays. Diamond can be considered two interlocking ccp lattices displaced a quarter of the body diagonal.

2.1.2 Space-Filling Polyhedra

A second approach to describing structures emphasizes the coordination around specific types of atoms in the structure. In this method, the atoms are omitted from the representation and replaced by coordination polyhedra. The vertices of the polyhedra represent the centers of the coordinating atoms, typically anions, and in the center of the polyhedra reside the atoms that are coordinated. For example, sodium chloride may be regarded as an array of edge-sharing $NaCl_6$ octahedra, as

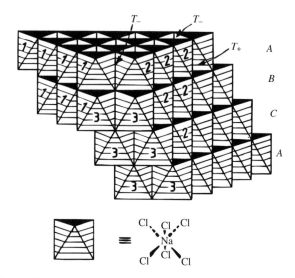

Figure 2.3   The rock salt structure depicted as an array of edge-sharing octahedra. (After West, 1984, *Solid State Chemistry and its Applications*. Copyright © John Wiley & Sons, Inc. Reproduced with permission.)

illustrated in Figure 2.3. Polyhedra can share corners, edges, or faces. The disadvantage in representing the atoms centered at the vertices as points is that one may perceive these atoms to be far from the central atom and from one another when, in fact, they often form a near-close-packed array.

2.1.3   The Unit Cell

The atoms in a crystalline substance occupy positions in space that can be referenced to *lattice points*, which crystallographers refer to as the *asymmetric unit* (physicists call it the *basis*). Lattice points represent the smallest repeating unit, or chemical point group. For example, in NaCl, each Na and Cl pair represent a lattice point. In structures that are more complex, a lattice point may represent several atoms (e.g., polyhedra) or entire molecules. The repetition of lattice points by translations in space form a *space lattice*, representing the full crystal structure.

Johan Hessel (1796–1872) showed that lattice points must belong to one of the 32 crystallographic point groups. These are given in Table 2.2 (by the international symbols) along with their symmetry elements. Note that only those point groups with 1-, 2-, 3-, 4-, or 6-fold rotations are found in 3D lattices. This can be understood by considering an analogous task of completely tiling a floor with regular polygon tiles. Rhombuses, rectangles, squares, triangles, and hexagons may be used, but not pentagons, heptagons, or higher polygons. However, *quasi-crystalline* solids have been discovered that contain 5-, 8-, 10-, and 12-fold rotational symmetry! These are discussed in Section 2.4.2.

If we connect lattice points together with translation vectors, $\boldsymbol{R} = u\boldsymbol{a} + v\boldsymbol{b} + w\boldsymbol{c}$ (in which u,v,w are integers) we can define a *primitive unit cell* that, when repeated

TABLE 2.2   The 32 Crystallographic Point Groups and Their Symmetry Elements

Point Group	Symmetry Operations and/or Elements
Triclinic	
1	E
1	E, i
Mononclinic	
2	E, C_2
m	E, σ_h
2/m	E, C_2, i, σ_h
Orthorhombic	
222	E, C_2, C_2', C_2'
mmm	$E, C_2, C_2', C_2', i, \sigma_h, \sigma_v, \sigma_v$
mm2	$E, C_2, \sigma_v, \sigma_v$
Tetragonal	
4	$E, 2C_4, C_2$
4	$E, 2S_4, C_2$
4/m	$E, 2C_4, C_2, i, 2S_4, \sigma_h$
4mm	$E, 2C_4, C_2, 2\sigma_v, 2\sigma_d,$
422	$E, 2C_4, C_2, 2C_2', 2C_2''$
4/mmm	$E, 2C_4, C_2, 2C_2', 2C_2'', i, 2S_4, \sigma_h, 2\sigma_v, 2\sigma_d$
42m	$E, C_2, 2C_2', 2\sigma_d, 2S_4$
Trigonal	
3	$E, 2C_3$
3	$E, 2C_3, i, 2S6$
32	$E, 2C_3, 3C_2'$
3m	$E, 2C_3, 3\sigma_v$
3m	$E, 2C_3, 3C_2', i, 2S6, 3\sigma_v$
Hexagonal	
6	$E, 2C_6, 2C_3, C_2$
6	$E, 2C_3, \sigma_h, 2S_3$
6/m	$E, 2C_6, 2C_3, C_2, i, 2S_3, 2S_6, \sigma_h$
622	$E, 2C_6, 2C_3, C_2, 3C_2', 3C_2''$
6mm	$E, 2C_6, 2C_3, C_2, 3\sigma_v, 3\sigma_d$
6m2	$E, 2C_6, 3C_2', \sigma_h, 2S_3, 3\sigma_v$
6/mmm	$E, 2C_6, 2C_3, C_2, 3C_2', 3C_2'', i, 2S_3, 2S_6, \sigma_h, 3\sigma_v, 3\sigma_d$
Cubic	
23	$E, 8C_3, 3C_2$
m3	$E, 8C_3, 3C_2. i, 8S_6, 3\sigma_h$
432	$E, 8C_3, 3C_2, 6C_2, 6C_4$
43m	$E, 8C_3, 3C_2, 6\sigma_d, 6S_4$
m3m	$E, 8C_3, 3C_2, 6C_2, 6C_4, i, 8S_6, 3\sigma_h, 6\sigma_d, 6S_4$

Notes: E = identity operation, C_n = n-fold proper rotation axis; S_n = n-fold improper rotation axis; σ_h = horizontal mirror plane; σ_v = vertical mirror plane; σ_d = dihedral mirror plane; i = inversion center.

by translations in space, generates the space lattice. There are other ways—but only 14 unique ones—of connecting lattice points in three dimensions, which define unit cells. The various types of 3D unit cells were worked out by Auguste Bravais (1811–1863). Each is described by six parameters: three translation vectors and three interaxial angles. These are shown in Figure 2.4, subdivided into the seven crystal systems. When multiple unit cells are possible, by convention, the unit cell with maximum symmetry is the one chosen.

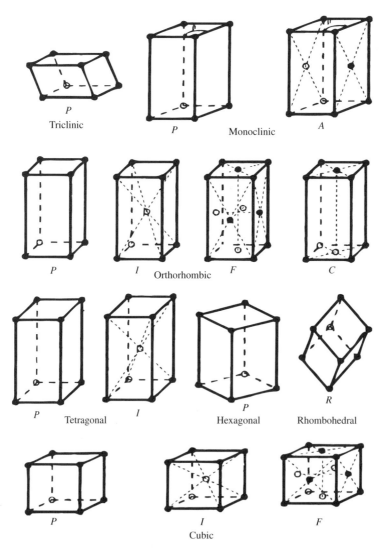

Figure 2.4 The 14 Bravais lattices in three dimensions arranged into the six crystal systems. (After Jenkins and Snyder, 1996, *X-Ray Powder Diffractometry.* Copyright © John Wiley & Sons, Inc. Reproduced with permission.)

It was shown independently in 1891 by William Barlow (1845–1934), Evgraf Stepanovich Federov (1853–1919) and Arthur Moritz Shönflies (1853–1928) that 230 *space groups* result from arranging the 32 crystallographic point groups in the patterns allowed by the 14 Bravais lattices. Although Schönflies' point group symmetry notation is still widely used by spectroscopists, crystallographers use the "International" notation for space group symmetry developed by Carl Hermann (1898–1961) and Charles Mauguin (1878–1958). Each space group is isogonal with one of the 32 crystallographic point groups. However, space group symbols reveal the presence of two types of combinational symmetry - the *glide plane* and *screw axis*. The first character of an international space group symbol is a capital letter designating the Bravais lattice centering type. This is followed by a modified point group symbol giving the symmetry elements (axes and planes) that occur for each of the lattice symmetry directions.

Often, it is necessary to refer to a specific crystallographic direction or lattice plane. For example, the spacing between adjacent planes in a family of planes determines the Bragg condition for diffraction of an electromagnetic wave. Lattice planes are sets of imaginary parallel planes that intersect the unit cell edges. Each family of planes is identified by integers called *Miller indices*, after the British mineralogist and crystallographer William Hallowes Miller (1801–1880). The Miller indices are the reciprocals of the fractional coordinates of the three points where the first plane away from the origin intercepts each of the three axes. The letter h refers to the intersection of the plane on a; k the intersection on b; and l the intersection on c. Some examples are illustrated in Figure 2.5. When referring to a plane, the numbers are grouped together in parentheses, (hkl). Any family of planes always has one member that passes through the origin of the unit cell. The plane used in determining the Miller indices is always the first one away from the origin, which may be obtained by moving in either direction.

Note that a Miller index of zero implies that the plane is parallel to that axis, since it is assumed that the plane will intersect the axis at $1/\infty$. A complete set of equivalent planes is denoted $\{hkl\}$. For example, in cubic systems (100), (010), and (001) are equivalent and represented as $\{100\}$. In hexagonal cells, four indices are sometimes used, $(hkil)$, where the relation $h + k + i = 0$ always holds. The value of the i index is the reciprocal of the fractional intercept of the plane on the a_3 axis, as illustrated in Figure 2.5. It is derived in exactly the same way as the others. Sometimes hexagonal indices are written with the i index as a dot, and in other cases it is omitted entirely.

Each plane in a family cuts through many unit cells, continuing out to the surface of the crystal. Likewise, other (parallel) planes in the family pass through adjacent unit cells. The interplanar spacing, or d-spacing, in a family of planes can be expressed in terms of the Miller indices and the unit cell distances and angles. Computation of the value of d becomes more difficult with decreasing symmetry of the crystal class. For example, for cubic systems, d is given by the simple expression:

$$d_{hkl} = a/(h^2 + k^2 + l^2)^{1/2} \qquad (2.1)$$

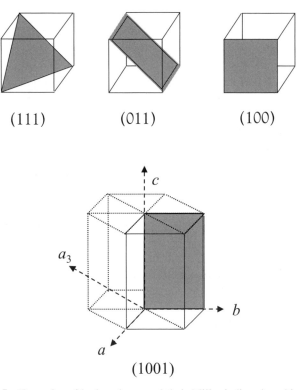

(111)   (011)   (100)

(1001)

Figure 2.5 Examples of lattice planes and their Miller indices in cubic crystals.

However, for triclinic cells, d is given by the rather formidable looking expression:

$$1/d^2 = (1/V^2)[h^2 b^2 c^2 \sin \alpha + k^2 a^2 c^2 \sin^2 \beta + l^2 a^2 b^2 \sin^2 \gamma + 2hkabc^2(\cos \alpha \cos \beta - \cos \gamma)$$
$$+ 2kla^2 bc(\cos \beta \cos \gamma - \cos \alpha) + 2hlab^2 c(\cos \alpha \cos \gamma - \cos \beta)] \tag{2.2}$$

where α, β, and γ are the angles between the unit cell axes and V is the unit cell volume. A complete listing of d formulas for each crystal class can be found in many textbooks (e.g., West, 1984). Once d_{hkl} is known for a set of planes, Bragg's law (Eq. 2.3) permits calculation of the diffraction angle, θ, for an electromagnetic wave with wavelength λ:

$$\lambda = 2d_{hkl} \sin \theta \tag{2.3}$$

In order to specify a crystal direction, a vector is drawn from the origin to some point P. This vector will have projections u' on the **a**-axis, v' on the **b**-axis, and w' on the **c**-axis. The three numbers are divided by the highest common denominator to give the set of smallest integers u, v, w. The direction is then denoted [uvw]. Sets of equivalent directions are labeled $\langle uvw \rangle$. Note that for cubic systems, the [hkl] direction is always orthogonal to the (hkl) plane of the same indices.

2.1.4 Pearson Symbols

William B. Pearson developed a shorthand system for identifying alloy and intermetallic structure types (Pearson, 1967). It is now widely used for ionic and covalent solids, as well. The Pearson symbol consists of a small letter that denotes the crystal system followed by a capital letter to identify the space lattice. To these is added a number equal to the number of atoms in the unit cell. Thus, the Pearson symbol for wurtzite (hexagonal, space group $P6_3mc$), which has four atoms in the unit cell, is $hP4$. Similarly, the symbol for sodium chloride (cubic, space group $Fm3m$), with eight atoms in the unit cell, is $cF8$.

2.2 COHESIVE FORCES IN SOLIDS

The cohesive forces, holding the atoms in a crystal together, can be ionic, covalent, or metallic. Although distinguishing compounds based on this classification scheme may seem like a straightforward matter, it is complicated by the fact that, typically, the bonding within a nonmolecular inorganic material is a mixture of bonding types. Not only does the attraction between two atoms in any solid, with the exception of pure elements, contain both ionic and covalent character, but many solids posses some predominantly ionic bonds as well as some predominantly covalent ones.

Coordination numbers of the constituent atoms are not always helpful for differentiating bonding types. The two most common coordination geometries observed in covalent compounds of the p-block and d-block elements are tetrahedral and octahedral coordination, respectively. These happen to be the same coordination numbers around the interstitial sites in the close-packed structures of many metallic elements and ionic compounds.

2.2.1 Ionic Bonding

The ionic bonding model was first suggested by the British physicist Joseph John (J.J.) Thomson (1856–1940), who also discovered the electron (Thomson, 1897; 1904). An ionic bond is the *nondirectional* electrostatic attraction between oppositely charged ions. For an isolated ion pair, the Coulomb potential energy, U, is simply

$$U_{\text{ion pair}} = -(1/4\pi\varepsilon_0)q_+q_-e^2/r \tag{2.4}$$

In this expression, the $q_{+/-}$ terms are the ion charges, r is the distance separating them, e is the electron charge (1.602×10^{-19} C), and $(1/4\pi\varepsilon_0)$ is the free-space permittivity (1.11265×10^{-10} $C^2J^{-1}m^{-1}$). The units of U are joules. There is also a strong repulsive force, due to the close proximity of the nuclei, which Max Born (1882–1970) suggested was

$$V = B/r^n \tag{2.5}$$

where B is a constant determined from the interatomic distance, r. The Born exponent, n, is normally between 5 and 12.

Arrays of ions tend to maximize the net electrostatic attraction between ions, while minimizing the repulsive interactions. The former ensures that cations are surrounded by anions, and anions by cations, with the highest possible coordination numbers. In order to reduce repulsive forces, ionic solids maximize the distance between like charges. At the same time, unlike charges cannot be allowed to get too close, or short-range repulsive forces will destabilize the structure. The balance between these competing requirements means that *ionic solids are highly symmetric structures with maximized coordination numbers and volumes*.

A purely ionic bond is an extreme case that is never actually attained in reality. Between any two bonded species, there is always some shared electron density, or *partial covalence*, however small. An anion has a larger radius than the neutral atom because of increased electron–electron repulsion. Its valence electron density extends out well beyond the nucleus, and it is thus more easily polarized by the positive charge on an adjacent cation. One can usually presume predominately ionic bonding when the (Pauling) electronegativity difference, $\Delta\chi$, between the atoms in their ground-state configurations is greater than about two. However, on polarizability grounds alone, we should anticipate circumstances that challenge the validity of such a simplistic viewpoint based on a single parameter. Despite the fact that atoms are frequently assigned oxidation numbers, as well as the quite common usage of terms like "cation" and "anion" in reference to solids, one should not take these to necessarily imply an ionic model.

There are some complementary concepts that allow us to predict when an ionic model is unlikely, even if the $\Delta\chi$ condition is met. First, rules by the Polish-born American chemist Kasimir Fajans (1887–1975) state that the polarizing power of a cation increases with increasing charge and the decreasing size of the cation. Likewise, the electronic polarizability of an anion increases with increasing charge and the increasing size of the anion (Fajans, 1915a, 1915b, 1924). The electronic polarizabilities of some ions in solids (Pauling, 1927) are listed in Table 2.3. It is

TABLE 2.3 Electronic Polarizability of Some Ions in Solids (10^{-24} cm^3)

Li^+	Be^{2+}	B^{3+}	C^{4+}	O^{2-}	F^-
0.029	0.008	.003	0.0013	3.88	1.04
Na^+	Mg^{2+}	Al^{3+}	Si^{4+}	S^{2-}	Cl^-
0.179	0.094	0.052	0.0165	10.2	3.66
K^+	Ca^{2+}	Sc^{3+}	Ti^{4+}	Se^{2-}	Br^-
0.83	0.47	0.286	0.185	10.5	4.77
Rb^+	Sr^{2+}	Y^{3+}	Zr^{4+}	Te^{2-}	I^-
1.40	0.86	0.55	0.37	14.0	7.1
Cs^+	Ba^{2+}	La^{3+}	Ce^{4+}		
2.42	1.55	1.04	0.73		

Source: After Frederikse, H. P. R. 2001, *CRC Handbook of Chemistry and Physics*, 82nd ed., Lide D. R., ed. Data from Pauling, L. 1927. *Proc. R. Soc. Lond., A114*. Copyright © CRC Press. Reproduced with permission.

expected that valence electron density on an anion will be polarized toward a highly charged cation, leading to some degree of covalence. Exactly how much charge is "highly charged"? It has been suggested that no bond in which the formal charge of the cation exceeds about 3^+ can possibly be considered ionic (Porterfield, 1993).

If we accept this, then we must concede that many so-called ionic solids, in fact, may not be so ionic after all. For example, in rutile (TiO_2) the nominal formal 4^+ charge on titanium would ensure a substantial amount of covalency in the Ti–O bonds, despite a $\Delta\chi$ of almost two. Indeed, band structure calculations using density functional theory indicate that titanium bears a $+1.2$ charge and oxygen a -0.6 charge (Thiên-Nga and Paxton, 1998), while Hartree-Fock calculations on the series $TiO/Ti_2O_3/TiO_2$ have shown that Ti–O bond ionicity decreases and bond covalence increases with increasing titanium valency (Evarestov et al., 1997).

In 1951, R. T. Sanderson introduced the *principle of electronegativity equalization*, which proposes that when two or more atoms combine, the atoms adjust to the same intermediate Mulliken electronegativity (Sanderson, 1951). Density functional theory tells us that the Mulliken electronegativity is the negative of the chemical potential (Parr et al., 1978). Sanderson's principle then becomes very appealing in that it can be considered analogous to a macroscopic phenomenon—the equalization of chemical potential. When atoms interact, the electronegativity, or chemical potential, must equalize.

Sanderson used his theory to calculate partial charges and ionic radii of atoms in solids and molecules; the partial charges in TiO_2 are evaluated as $+0.78$ and -0.39 for titanium and oxygen, respectively. Although these values are not in exceptional agreement with those obtained from band-structure calculations, both sets of results reinforce the idea that high formal oxidation states derived from simple valence rules (e.g., Ti^{4+}, O^{2-} in the case of TiO_2) do not reflect true charge—or bond ionicity!

2.2.2 Covalent Bonding

Covalent bonds form between atoms with similar electronegativities. There is electron-density concentrated in the region between covalently bonded atoms, both of which have "ownership" in the electron pair. The American chemist Gilbert Newton Lewis (1875–1946) is generally credited with first describing the covalent bond as a shared electron pair in 1916, before the advent of quantum theory. While there is no such thing as a completely ionic bond, the same cannot be said for covalent bonds. In elemental (homonuclear) silicon, for example, the bonding *must* be purely covalent. Such bonds are comparable in strength to predominately ionic bonds. Heteronuclear bonds, on the other hand, have a degree of ionicity, or charge transfer that is dependent on the electronegativity difference. The additional ionic interaction further strengthens the bond.

Of course, an adequate treatment of the covalent bond, in contrast to the ionic bond, requires that we invoke quantum mechanics, as the electron wave functions are solutions to Fock's equations. The first satisfactory explanation of the stability of the chemical bond was due to Walter Heinrich Heitler (1904–1981) and Fritz Wolfgang London (1900–1954) in 1927 (Heitler and London, 1927). It became

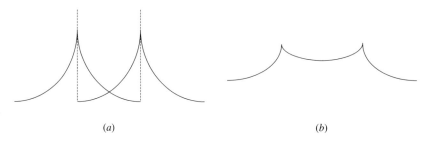

(a)   (b)

Figure 2.6 (a) The overlap of two one-electron atomic wave functions, each centered on a different atom, constitutes the Heitler–London (valence-bond) theory. (b) A one-electron molecular wave function, or *molecular orbital*, in the MO theory of Hund and Mulliken.

known as *valence-bond theory*. In valence-bond theory, a chemical bond is essentially regarded as being due to the overlap of neighboring valence-level hydrogen-like atomic orbitals, each of which is singly occupied by an electron with spin opposite to the other. Each electron belongs to its own individual atom, as shown in Figure 2.6a. However, because they are identical (except for spin), each electron could belong to either atom. Hence, the total wave function for the molecule is the linear combination of the wave function for both cases.

Linus Carl Pauling (1901–1994) and John Clarke Slater (1900–1976) later showed independently that n p atomic orbitals and one s atomic orbital could be combined mathematically to give a set of equivalent, or degenerate, singly occupied sp^n hybrid orbitals (Pauling, 1931; Slater, 1931). The directional effects of the hybridization on the set of orbitals were consistent with the known geometries of several polyatomic molecules of fluorine, oxygen, nitrogen, and carbon. For example, sp^3 hybridization of the carbon atom in methane, CH_4, explained that molecule's tetrahedral geometry. Pauling then extended this scheme to transition-metal compounds with the inclusion of d atomic orbitals (Pauling, 1932). One of the major shortcomings of the use of equivalent orbitals, however, is the inconsistency with spectroscopic findings. For example, photoelectron spectroscopy provides direct evidence showing that the equivalent electron charge densities in the four bonds of CH_4 actually arise from the presence of electrons in nonequivalent orbitals.

Although the concept of localized bonds proved useful for formalizing chemists' intuitive ideas about molecular geometry and Lewis's idea of shared electron pairs, it is not entirely correct. The approach regarded as being the most accurate explanation of covalent bonding is *molecular orbital (MO) theory*. In the MO approach, one-electron wave functions, or MOs, extend over all the nuclei in the entire molecule, having equal amplitude at equivalent atoms (Figure 2.6b). This theory originated from the theoretical work of German physicist Friederich Hund (1896–1997), and its application to the interpretation of the spectra of diatomic molecules by American physical chemist Robert S. Mulliken (1896–1986) (Hund, 1926, 1927a, 1927b; Mulliken, 1926, 1928a, 1928b, 1930). Inspired by the success of Heitler and London's approach, Finklestein and Horowitz introduced the linear combination of atomic orbitals (LCAO) method for approximating the MOs

(Finkelstein and Horowitz, 1928). The British physicist John Edward Lennard-Jones (1894–1954) later suggested that only valence electrons need be treated as delocalized; inner electrons could be considered as remaining in atomic orbitals (Lennard-Jones, 1929).

Simultaneous with, and independent of, the work to explain molecular electronic structure, Felix Bloch (1905–1983), who was a graduate student under Werner Heisenberg trying to explain the conductivity of metals, used Fourier analysis to obtain wave solutions to Schrödinger's equation for a 1D periodic potential. He showed that the wave functions have equal amplitude at equivalent positions. His thesis, "The Quantum Mechanics of Electrons in Crystal Lattices," was published in *Zeitschrift fur Physik* (Bloch, 1928). In a crystalline solid, therefore, we can think of "crystal orbitals" (COs) extending over the entire crystal, in an analogous fashion to MOs. This is termed the "Bloch scheme" or "band scheme."

The valence-bond and MO (band) models both have been used to explain the cohesive forces in solids. The two theories can be made equivalent in the limit of including additional ionic plus covalent terms in the former and excited-state configurations (*configuration interaction*) in the latter. Each has its advantages and disadvantages, and the choice of one over the other may be dictated by the type of experimental data (e.g., magnetic measurements) at hand. It seems reasonable to expect the valence-bond approach to most accurately reflect the bonding in ionic solids (Seitz, 1940). Many have advocated the argument that, in solids with internuclear distances greater than a critical value, the valence d electrons are also best described with localized atomic orbitals instead of (delocalized) COs (Mott, 1958; Goodenough, 1966, 1967). This is equivalent to saying that valence-bond, or Heitler–London, theory is more appropriate than MO theory in these cases. For many transition-metal compounds, this picture is supported by magnetic data consistent with localized magnetic moments. Localization is a particularly crucial concept for $3d$ electrons, since they do not range as far from the nucleus as $4s$ or $4p$ electrons. Valence s and p orbitals, by contrast, are always best described by Bloch functions, while $4f$ electrons are localized and $5f$ are intermediate. At any rate, whether we use the valence-bond method or construct delocalized orbitals from atomic functions, it is found that covalent bonding is directional since, due to symmetry constraints, only certain atomic orbitals on neighboring atoms are appropriate for combination.

2.2.3 Metallic Bonding

The simplest model for a metal is the Drude–Sommerfield picture of ions occupying fixed positions in a "sea" of mobile electrons. Thus, a close-packed array of monovalent metal atoms constitutes metallic bonding. The cohesive forces in this model are readily understood in terms of the Coulomb interactions between the ion cores and the sea of electrons. The ion cores are screened from mutually repulsive interactions by the mobile electrons. On the other hand, there are repulsive forces between electrons—electron motion is correlated. Two electrons with parallel spin tend to avoid one another (the Pauli exclusion principle). The $e–e$

repulsive interaction, however, is outweighed by the strong Coulomb *attraction* between the oppositely charged electrons and ions, which is what bonds the atoms together in the solid.

Although the cohesive forces in such an idealized metal as just described would be nondirectional (as in ionic solids), the orientation effects of d orbitals contribute a directional-covalent component to the bonding in transition metals that requires a more sophisticated definition for metallic bonding. The internuclear distances in the close-packed, or nearly close-packed, structures of most metallic elements are small enough that the valence orbitals on the metal atoms can overlap (in the valence-bond model) or combine to form COs (in the MO or Bloch model). The bonding COs are lower in energy and, hence, more stable than the atomic orbitals from which they originate. However, since metal atoms have fewer valence electrons than valence orbitals, there are not enough electrons for the number of two-electron bonds required.

For a metallic solid containing an enormous number of orbitals, all the bonding COs (that comprise the valence band) are filled and the antibonding orbitals (that make up the conduction band) are only partially filled. The Fermi level [analogous to the highest occupied molecular orbital (HOMO)] thus lies in this partially filled conduction band. States very close in energy (the separation between them is infinitesimal), into which the electrons can be excited and accelerated by an electric field, are available. The electrons are said to be *itinerant*, which gives rise to a high electrical conductivity. Thus, we can consider the bonding in metallic solids to be the special case of covalent bonding in which the highest-energy electrons (the Fermi level) are in a partially filled conduction band, or delocalized Bloch orbitals. This was the original application of the Bloch scheme. We are not restricted from treating substances other than close-packed elements in the same manner. The band theory of solids predicts that *any* substance, be it an element or compound, close-packed or non-close-packed, in which the Fermi level lies in a partially filled band of delocalized states will be metallic.

2.3 STRUCTURAL ENERGETICS

Inorganic solid-state chemistry is rich in structure diversity. In this section, we present some guidelines for predicting *local* coordination geometries. The challenging part is predicting the way these coordination polyhedra interconnect to form the crystalline solid. It is a rather simple matter to predict the polyhedral connectivity in simple solids if one knows the local coordination. For example, an array of sp^3-hybridized carbon atoms *must* adopt the diamond structure in which each vertex of every CC_4 tetrahedron is shared with three other tetrahedra. Our predictive power decreases, however, with an increase in the number of constituents, each with its own coordination preference. In some cases, a particular constituent may actually be found in more than one coordination environment. For example, it has been reported that the Ti^{4+} ion takes on both octahedral and tetrahedral coordination by O^{2-} in $Ba_6Nd_2Ti_4O_{17}$ (Kuang et al., 2002); that is, this oxide exhibits layers of face-sharing TiO_6 octahedra as well as TiO_4 tetrahedral layers.

It can be stated that the structure adopted by a solid at a particular temperature and pressure is the result of a competition between thermodynamics and kinetics to obtain the state with the lowest free energy via the route with the smallest energy expenditure, or formation barrier. Unfortunately, this is not particularly helpful. Empirical thermochemical-based approaches, such as Gibbs energy minimization, can be used to extrapolate the stability ranges of known phases as new components are added (Section 10.5.4), but they are not of great value for finding new structures or phases. Likewise, even state-of-the-art first principles techniques like molecular dynamics (MD), and MC simulations are still unable to predict definitely the structure, stability and properties of complex inorganic solids. In these methods, one investigates the potential energy surface (also called energy landscape) of a system. This is the potential energy as a function of all the atomic positions. In theory, the most stable structure can be obtained from a global energy minimization. Metastable structures that are kinetically stable can also be predicted by locating local minima of the potential energy surrounded by sufficiently high-energy barriers. Approximations have to be employed for the energy calculations on large systems with multiple minima, however. Thus, in general the systems readily amenable to computational techniques are limited to small numbers of atoms (a few hundred) by both the available resources (calculations are very much more computationally intensive for large systems) and the quality of the algorithms.

A detailed discussion of MD and MC techniques is beyond the scope of this book, but basically, they involve keeping track of the position and momenta of all the particles in an N-particle system. The potential energy component of the total energy calculation is often approximated as a sum over pair potentials involving interactions between two particles. Actually, we have already introduced one type of pair potential: Eq. 2.1, for the Coulomb interaction between an isolated pair of ions. Interatomic potential energy functions in solids must also include long-range interactions (Section 2.3.1) and may include many-body interactions. The latter are introduced in Section 9.2.3, where we relate cohesive forces to the elastic properties of solids.

The rational design of novel structures possessing predetermined physical properties thus remains an extremely daunting task. From the perspective of the synthetic chemist, it is often fruitful to take a compound with a known stable (thermodynamically or kinetically) crystal structure and alloy one of more of the sublattces to form a stable solid solution in which the product phase retains the same basic crystal structure of the parent phase (Section 2.6). In this way, the properties of the parent phase can be tuned to meet the requirements for a particular engineering function. It is reasonable to assume that atoms of similar size, valence, and coordination preference can be substituted on a sublattice. For now, we shall go ahead and examine some simple structural energetic factors that are useful to the synthetic chemist and that do not require sophisticated or time-consuming computational methods.

2.3.1 Lattice Energy

The lattice energy, U, of an ionic crystal is defined as the potential energy per mole of compound associated with the particular geometric arrangement of ions forming the structure. It is equivalent to the heat of formation from one mole of its ionic constituents in the gas phase. Equation 2.4 does *not* give the total energy of attraction for a 3D array of ions. Calculation of the total electrostatic energy must include summation of long-range attractions between oppositely charged ions and repulsions between like-charged ions, extending over the whole crystal, until a convergent mathematical series is obtained.

The sums may be carried out with respect to the atomic positions in direct (real) space or to lattice planes in reciprocal space, an approach introduced in 1913 by Paul Peter Ewald (1888–1985), a doctoral student under Arnold Sommerfeld (Ewald, 1913). In reciprocal space, the structures of crystals are described using vectors that are defined as the reciprocals of the interplanar perpendicular distances between sets of lattice planes with Miller indices (*hkl*). In 1918, Erwin Rudolf Madelung (1881–1972) invoked both types of summations for calculating the electrostatic energy of NaCl (Madelung, 1918).

Although direct space summations may be conceptually simpler, convergence can be time-consuming, if not problematic, even when one sums concentric, electrically neutral groups. Bertaut developed a method that achieves good convergence utilizing reciprocal space summations exclusively (Bertaut, 1952). The most generally accepted way, however, was presented by Ewald in 1921 (Ewald, 1921). In this method, an array of point charges neutralized by Gaussian charge distributions are summed in direct space, and an array of oppositely charged Gaussian distributions neutralized by a uniform charge density are summed in reciprocal space. A derivation can be found in the book by Ohno, Esfarjani, and Kawazoe (1999). Little would be gained in reproducing this arduous procedure here. Suffice it to say, when these mathematical summations are carried out over larger and larger crystal volumes, until convergence is achieved, a number known as the *Madelung constant* is generated, such that the final expression for the long-range force on an ion is

$$U_{ion} = [1/(4\pi\varepsilon_0)][-Mq_+q_-e^2/r] \tag{2.6}$$

in which $4\pi\varepsilon_0$ is the permittivity of free space (1.11265×10^{-10} C^2 $J^{-1}m^{-1}$), e is the electron charge (1.6022×10^{-19} C), q is the ion charge, and M is the Madelung constant. The Madelung constant is dependent only on the geometric arrangement of ions and the distance that r is defined in terms of (nearest neighbor, unit cell parameter, etc.; beware of different conventions when using Madelung constants from the literature!). The constant has the same value for all compounds within any given structure type. If r is in meters, the units of Eq. 2.6 will be in joules per cation.

Johnson and Templeton calculated values of M for several structure types using the Bertaut method (Johnson and Templeton, 1962). Their results are partially reproduced in Table 2.4, where the second column lists the Madelung constant based on the shortest interatomic distance in the structure. The third column gives

TABLE 2.4 Madelung Constants for Several Structure Types

Compound	$M(R_0)$	$M\langle R\rangle^a$
Al_2O_3 (corundum)	24.242	1.68
BeO	6.368	1.64
$CaCl_2$	4.730	1.601
CaF_2 (fluorite)	5.03879	1.68
$CaTiO_3$ (perovskite)	24.7550	—
$CdCl_2$	4.489	1.50
CsCl	1.76268	1.76
Cu_2O	4.44249	1.48
La_2O_3	24.179	1.63
LaOCl	10.923	—
$MgAl_2O_4$	31.475	—
MgF_2	4.762	1.60
NaCl (rock salt)	1.74756	1.75
SiO_2 (quartz)	17.609	1.47
TiO_2 (anatase)	19.0691	1.60
TiO_2 (brookite)18.066	1.60	—
TiO_2 (rutile)	19.0803	1.60
V_2O_5	44.32	1.49
ZnO	5.99413	1.65
ZnS (zinc blende)	6.55222	1.638
ZnS (wurtzite)	6.56292	1.641

[a]$M\langle R\rangle = M(R_0) * \langle R\rangle/R_0$, where R_0 is the shortest interatomic distance, and $\langle R\rangle$ is the average shortest distance.

Source: After Johnson, Q. C.; Templeton, D. H. 1962, *J. Chem. Phys.*, *34*, 2004. Copyright © American Institute of Physics. Reproduced with permission.

reduced Madelung constants based on the *average* shortest distance. For less symmetric structures, that is, when there are several nearest neighbors at slightly different distances, as in the ZnS polymorphs, the reduced Madelung constant is the more significant value.

In order to obtain the complete expression for the lattice energy of an ionic crystal we must: (1) add the term representing the short-range repulsive forces, (2) include Avogadro's number, N (6.022×10^{23} mol^{-1}), and (3) make provisions for ensuring that we do not overcount pairs of interactions. In so doing, the final expression for the lattice energy of an ionic crystal containing $2N$ ions was given by M. Born and A. Landé (Born and Landé, 1918) as:

$$U_{\text{lattice}} = [N/(4\pi\varepsilon_0)][-Mq_+q_-e^2/r + (B/r^n)] \qquad (2.7)$$

This equation gives the lattice energy in joules per mole. We can avoid having to determine a value for the parameter B by using the equilibrium interatomic distance as the value of r for which U is a minimum. This gives $dU/dr = 0$ and the following expression for U:

$$U_{\text{lattice}} = [N/(4\pi\varepsilon_0)][(-Mq_+q_-e^2/r)(1 - (1/n))] \qquad (2.8)$$

TABLE 2.5 Values of the Born Exponent^a

Cation–anion Electron Configurations	Example	n
$1s^2$–$1s^2$	LiH	5
$1s^2 2s^2 p^6$–$1s^2 2s^2 p^6$	NaF, MgO	7
$[Ne]3s^2 p^6$–$[Ne]3s^2 p^6$	KCl, CaS	9
$[Ar]3d^{10}4s^2 p^6$–$[Ar]3d^{10}4s^2 p^6$	RbBr, AgBr	10
$[Kr]4d^{10}5s^2 p^6$–$[Kr]4d^{10}5s^2 p^6$	CsI	12

^{a}For mixed-ion types, use the average (e.g., for NaCl, $n = 8$).

If the value of n is not known, an approximate value may be obtained from Table 2.5. The interatomic potential between a *pair* of ions in the lattice is given by Eq. 2.7 minus Avogadro's number. We see in Chapter 9 how the elastic modulus for an ionic solid can be estimated from such an expression by taking the second derivative with respect r. In cases where there are significant contributions from covalent bonding, Eqs. 2.7 and 2.8 will not reflect the true binding energy of the crystal. Nevertheless, they are still useful in comparing relative energies for different compounds with the same structure, as illustrated in the following worked example.

Example 2.1 Some transition-metal oxides contain ion-exchangeable layers. In many cases, the M^{n+} cations in these layers are amenable to aliovalent ion exchange with $M^{(n+1)+}$ ions of similar size. No structural change other than possibly a slight expansion or contraction of the unit cell occurs. In order to maintain charge neutrality, aliovalent exchange requires the introduction of a vacancy for every M^{n+} ion exchanged. Based on lattice energy considerations, would you expect the ion-exchanged product to be favored?

Solution To simplify the solution, imagine an analogous 1D array (Figure 2.7). We can carry out a direct-space summation of the long-range attractive and repulsive forces felt by any arbitrary ion, extending across this 1D array, before and after ion exchange. Before aliovalent ion exchange, we have:

$$U = -2q^2/r + 2q^2/2r - 2q^2/3r + 2q^2/4r + \cdots$$

which can be written as

$$U = (-2q^2/r)(1 - \tfrac{1}{2} + \tfrac{1}{3} - \tfrac{1}{4} + \cdots)$$

The second term in parenthesis may be recognized as an alternating series:

$$(1 - \tfrac{1}{2} + \tfrac{1}{3} - \tfrac{1}{4} + \cdots) = \sum_{n=1}^{8}(-1)^{n+1}[1/n]$$

This series converges to ln 2, or ~ 0.69.

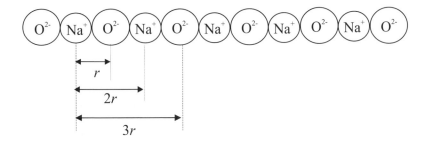

$$U = -2q^2/r[1 - 1/2 + 1/3 - 1/4 + 1/5 - 1/6\cdots]$$

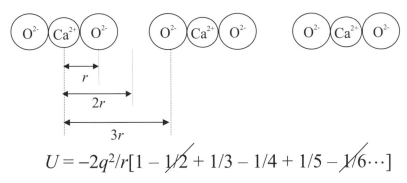

$$U = -2q^2/r[1 - \cancel{1/2} + 1/3 - 1/4 + 1/5 - \cancel{1/6}\cdots]$$

Figure 2.7 Calculation of the Madelung constant for a 1D array of cations and anions, before and after aliovalent ion exchange. See Example 2.1.

The Madelung constant for the array thus corresponds to:

$$M = (2 \times 0.69) = 1.38.$$

For the ion-exchanged array, each M^+ ion is exchanged with one M^{2+} cation and one vacancy. If the M^{2+} and M^+ ions are approximately the same size (e.g., Na^+ and Ca^{2+}), r has essentially the same value. However, the absence of some of the cation–cation repulsion terms (every fourth term, beginning with $\frac{1}{2}$) now forces the new alternating series to converge to ~2.2, resulting in a Madelung constant of $(2 \times 2.2) = 4.4$. The aliovalent ion-exchanged product is thus favored, because of both a larger Madelung constant and a larger product of ion charges in the lattice energy expression. A 1D analogy was chosen simply to illustrate the general summation procedure. To extend the summation to a real 3D crystal requires the inclusion of a considerable number of additional terms, in which case a hand calculation can become significantly less tractable.

We have just seen that it is possible to estimate the total binding energy, or lattice energy, of an ionic crystal by summing the long-range electrostatic interactions.

The interatomic potential energy between a pair of ions in the lattice may be obtained from this expression. In covalent solids, the binding energy of the crystal corresponds to the difference between the total energy of the electrons in the crystal and the electrons in the same, but isolated, atoms. One might reason then that the binding energy of a covalent crystal could somehow be estimated from tabulated bond energies. However, such an approach is inherently flawed because bond-strength values obtained from bond-energy tables generally represent an average evaluation for the bonds in gaseous diatomic molecules, in which case the bonding is substantially different from that in a solid. The calculation of interatomic potentials in covalent solids must account for many-body effects, since the interaction between a pair of atoms is modified by the surrounding atoms. Total energy estimates are possible using tight-binding (Hartree–Fock) calculations and density functional theory (Kohn and Sham, 1965), or on very small systems (the nanoscale range) and over ultrashort time intervals (nanoseconds) by an *ab initio* molecular dynamics technique (Car and Parrinello, 1985).

2.3.2 The Born–Haber Cycle

Lattice energies cannot be measured experimentally since they represent hypothetical processes:

$$M^{n+}(g) + X^{n-}(g) \rightarrow MX(s)$$

However, the following reaction sequence, relating the heat of formation, ΔH_f of a crystal $[M(s) + \frac{1}{2} X_2(g) \rightarrow MX(s)]$ to U $[M^+(g) \rightarrow MX(s)]$ is thermochemically equivalent (and ΔH_f can be measured):

$$M^0(g) + X^0(g) \xrightarrow[+EA]{IE} M^+(g + X^-(g)$$

$$\uparrow \Delta H_s^0 \qquad \uparrow D \qquad \downarrow U$$

$$M(s) \ + \tfrac{1}{2}X_2(g) \xrightarrow{\Delta H_F} MX(s)$$

In this diagram ΔH_s^0 gives the enthalpy of sublimation of the metal $[M(s) \rightarrow M^0(g)]$, D gives the dissociation energy, or bond energy of the diatomic gas $[\frac{1}{2} X_2(g) \rightarrow X^0(g)]$, *IE* gives the ionization energy of the gaseous metal $[M^0(g) \rightarrow M^+(g)]$, and *EA* gives the electron affinity for the formation of the gaseous anion $[X^0(g) \rightarrow X^-(g)]$. The lattice energy is obtained through the relation:

$$U = \Delta H_f - (\Delta H_s + \tfrac{1}{2}D + IE + EA) \tag{2.9}$$

One difficulty with using a Born–Haber cycle to find values for U is that heats of formation data are often unavailable. Perhaps the greatest limitation, however, is that we cannot experimentally obtain electron affinities for multiply charged anions (e.g., O^{2-}) or polyanions (e.g., SiO_4^{4-}). Such anions simply do not exist as gaseous species. No atom has a positive second electron affinity; energy must be added to a negatively charged gaseous species in order for it to accommodate additional electrons. In some cases, thermochemical *estimates* for second and third electron affinities are available from *ab initio* calculations. Even so, if there are large covalent forces in the crystal, one can expect poor agreement between the values of U obtained from a Born–Haber cycle and Madelung calculations.

2.3.3 Goldschmidt's Rules and Pauling's Rules

Some guiding principles, enunciated by the Swiss-born Norwegian geochemist Victor Moritz Goldschmidt (1888–1947) and Linus Pauling, make possible the rationalization and prediction of the structures of simple ionic solids. Three rules were devised by Goldschmidt to explain element distributions in minerals (Goldschmidt et al., 1925, 1926a, 1926b, 1926c). The basis of these rules is that ionic substitution of one cation by another is governed by the sizes and charges of the cations. The first rule is that extensive substitution of one cation for another can only occur with cations of the same size and charge. The second and third rules are that cations of smaller size and same charge, or same size and higher charge will preferentially incorporate into a growing crystal.

Pauling subsequently introduced three rules governing ionic structures (Pauling, 1928, 1929). The first is known as the *radius ratio rules*. The idea is that the relative sizes of the ions determine the structure adopted by an ionic compound. Pauling proposed specific values for the ratios of the cation radius to the anion radius as *lower* limits for different coordination types. These values are given in Table 2.6. Unfortunately, the radius ratio rules are incorrect in their prediction of coordination numbers about as often as they are correct. Usually, it overestimates the coordination number of the cation. This model essentially regards ions as hard, incompressible spheres, in which covalent bonding is not considered. The directionality, or overlap requirements, of the covalent bonding contribution probably plays as significant a role as ion size fitting.

Pauling's second rule is the *electrostatic valence rule*. It states that the charge on an ion must be balanced by an equal and opposite charge on the surrounding ions. A cation, M^{m+}, coordinated by n anions, X^{x-}, has an *electrostatic bond strength* (ebs) for each bond defined as:

$$\text{ebs} = m/n \tag{2.10}$$

Charge balance, then, is fulfilled if:

$$\sum m/n = x \tag{2.11}$$

TABLE 2.6 Radius Ratio Rules

r_c/r_a		
	< 0.16	3-fold
$0.16 <$	$r_c/r_a < 0.41$	4-fold
$0.41 <$	$r_c/r_a < 0.73$	6-fold
$0.73 <$	$r_c/r_a < 1.00$	8-fold
$1.00 >$	r_c/r_a	12-fold

The third rule by Pauling is that the presence of shared polyhedron edges and faces destabilize a structure. Polyhedra tend to join at the vertices (corners). Cations strongly repel each other as edges and faces are shared because the cation–cation distance decreases. The smallest decrease occurs for octahedral edge sharing, followed by tetrahedral edge sharing and octahedral face sharing, and the largest decrease is for tetrahedral face sharing, which makes this particular configuration quite unfavorable. Vertex-, edge-, and face-sharing octahedra are all commonly observed stable arrangements, as are vertex-sharing tetrahedra. Edge-sharing tetrahedra are not very common and the tetrahedra are usually distorted, particularly if the cations are highly charged. Face-sharing tetrahedra are not generally observed.

Linus Pauling (Courtesy of AIP Emilio Segrè Visual Archives, W. F. Meggers Gallery of Nobel laureates. © The Nobel Foundation. Reproduced with permission.)

LINUS CARL PAULING

(1901–1994) earned a Ph.D. in chemistry from the California Institute of Technology, Pasadena, in 1925 under Roscoe G. Dickinson. In 1926, Pauling accepted a position under physicist Arnold Sommerfield at the University of Munich, where he first applied quantum mechanics to chemical bonding. This was the beginning of Pauling's extraordinary career, which spanned nearly 70 years. He was a scientist of great versatility, having carried out research in numerous areas, including crystallography, inorganic and physical chemistry, the theory of ferromagnetism, and molecular biology. He was awarded the Nobel Prize in Chemistry in 1954 for his work on chemical bonding and molecular structure. Solid-state chemists are also indebted to Pauling for rules predicting the structures of ionic solids, and for his work on the structures of metals and intermetallic compounds. Pauling's work was not without disputes. W. L. Bragg accused him of stealing ideas on chemical bonding. Pauling debated his contemporaries on the merits of both MO theory and band theory as opposed to the more simplistic valence-bond theory, and he expressed his disbelief of the existence of the quasi-crystalline state. Pauling also endured great controversy in his personal life. His opposition to nuclear weapons and outspokenness on other war-related issues were regarded with suspicion by the government. He was forced to appear twice before a senate subcommittee to defend his views. He was even temporarily denied the right to travel abroad. However, in 1963, Pauling was awarded the Nobel Peace Prize for his efforts to ban nuclear testing. Pauling was elected to the U.S. National Academy of Sciences in 1933.

(*Source*: "My Memories and Impressions of Linus Pauling," by David Shoemaker, 1996. Coutesy Ava Helen and Linus Pauling Papers, Oregon State University Libraries, Corvallis.)

2.3.4 Electronic Origin of Coordination Polyhedra in Covalent Crystals

In valence-bond theory, the coordination number of an atom in a molecule or covalent solid is generally limited to the number of valence orbitals on the atom. Likewise, only certain combinations of atomic orbitals on the atoms involved are suitable for forming MOs possessing the point-group symmetry of the molecule, or COs with the proper space-group symmetry of the crystal. Furthermore, molecules are most stable when the bonding MOs or, at most, bonding plus nonbonding MOs, are filled with electrons and the antibonding MOs are empty. These principles form the quantum mechanical basis of G. N. Lewis's and Irving Langmuir's *octet rule* for compounds of the *p*-block elements and the *18-electron rule* for *d*-block elements. In covalent solids, a pair of electrons with opposing spins each occupies a two-center bonding site in the crystal orbital. The geometries of molecules and coordination polyhedra in covalent solids are thus determined by the types of valence orbitals contributed by the atoms involved.

Knowing the molecular or crystal geometry allows us to draw overlap sketches involving the atomic orbitals that will show the net overlap. This, in turn, enables construction of a qualitative energy-level diagram, from which we can verify the principles just discussed. There is, of course, a prescribed group theoretical treatment to be followed. This procedure is amply covered in many specialized textbooks on MO theory (e.g., Cotton, 1990). It will be beneficial for us to very briefly review the basic methodology applied to molecules for later comparison to solids. With molecules, one first finds the irreducible representations to which the central-atom atomic orbitals belong and then constructs ligand-group orbitals, which are symmetry-adapted linear combinations (SALCs) of the ligand atomic orbitals. For example, if the coordinate system for the ligand atomic orbitals in a molecule of T_d symmetry are as shown in Figure 2.8, group orbitals can be found that transform according to the same rows of the same irreducible representations as

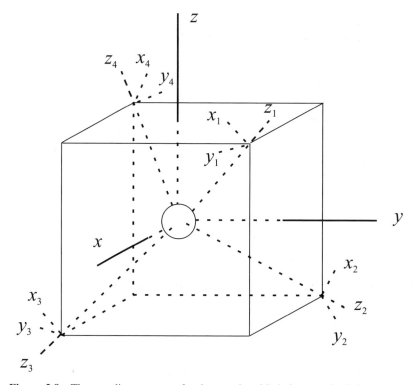

Figure 2.8   The coordinate system for the atomic orbitals in a tetrahedral molecule.

the central-atom orbitals. Only orbitals that have the same symmetry around the bond axes can form MOs.

Figure 2.9 shows a generalized energy-level diagram for a tetrahedral molecule with σ- and π-bonds involving only s and p atomic orbitals (e.g., CCl_4, $ClO_4{}^-$). The Mulliken symmetry labels are shown for the MOs, which are the same as the labels for the irreducible representations to which they belong. J. H. Van Vleck (1899–1980) showed in 1935 that the electron wave functions (the MOs) can be chosen to transform like, or "belong to," irreducible representations of the point group of the molecule. We then place electrons in the MOs beginning with the lowest energy orbital (the "buiding-up," or "aufbau" principle) as in atomic orbitals. Each MO can hold two electrons (of opposite spin). Hund showed that the state with maximum spin multiplicity is the lowest in energy (Hund, 1926). Thus, in degenerate sets, electrons are added singly, one to each orbital, before double occupancy occurs. In CCl_4, for example, there are 32 electrons available for filling the MOs. We see that all of the bonding and nonbonding MOs in Figure 2.9 will be filled with electrons, the antibonding MOs will be empty, and the octet rule is obeyed around the central carbon atom. Note also, an energy gap separates the HOMO and the lowest unoccupied molecular orbital (LUMO). It is important to realize, however, that our symmetry considerations alone provide no quantitative information on the actual energy levels.

M   MX$_4$ 4 X

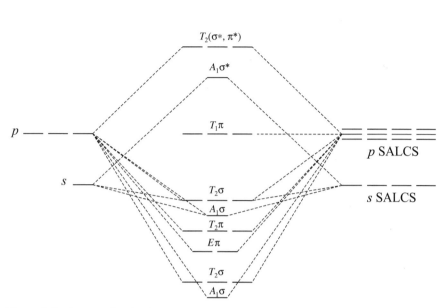

Figure 2.9   A generalized MO energy-level diagram for a tetrahedral molecule with σ- and π-bonding involving only *s*- and *p*-atomic orbitals.

Now consider the nonmolecular solid diamond, which may be considered built of two interlocking carbon fcc sublattices, displaced by a quarter of the body diagonal. The structure is shown in Figure 2.10a. There are two atoms associated with each diamond lattice point, or *basis*, corresponding to a carbon atom on one fcc sublattice and one of its nearest-neighbor carbon atoms on the other fcc sublattice. These are the atoms at the points labeled "0" and "3/4" in the bottom portion of Figure 2.10a. Every carbon has four such nearest neighbors that form a tetrahedron (T_d point group symmetry). Thus, diamond may also be described as a 3D network of vertex-sharing CC$_4$ tetrahedra. The tetrahedral coordination can be more easily seen in Figure 2.10b. The electronic structure of diamond is well described by considering both nearest-neighbor and second-nearest-neighbor interactions of the following type: $ss\sigma$, $pp\sigma$, $pp\pi$, and $sp\sigma$.

Compliance with the octet rule in diamond could be shown simply by using a valence-bond approach in which each carbon atom is assumed sp^3 hybridized. However, using the MO method will enable us to more clearly establish the connection with band theory. In solids, the extended electron wave functions analogous to molecular orbitals are called *crystal orbitals*. COs must belong to an irreducible representation, not of a *point group*, but of the *space group* reflecting the translational periodicity of the lattice.

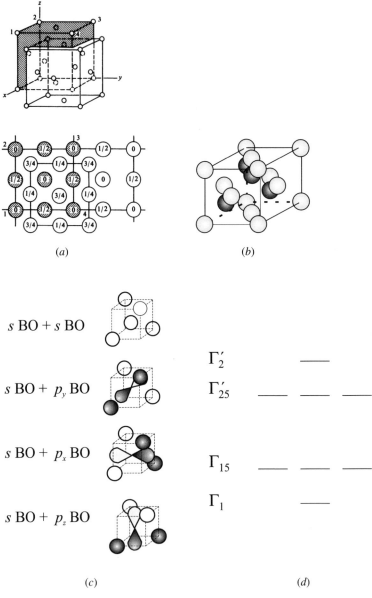

(a)

(b)

s BO $+ s$ BO

Γ_2' ———

s BO $+ p_y$ BO

Γ_{25}' ——— ——— ———

s BO $+ p_x$ BO

Γ_{15} ——— ——— ———

Γ_1 ———

s BO $+ p_z$ BO

(c)

(d)

Figure 2.10 (a) The diamond structure viewed as two interlocking fcc sublattices displaced by $\frac{1}{4}a$ along $\langle 111 \rangle$. (After Runyan and Bean, 1990, *Semiconductor Integrated Circuit Processing Technology.* Copyright © Addison-Wesley Publishing Company. Reproduced with permission.) (b) Another view, in which one sublattice is shaded gray. (c) Some possible sign combinations of the basis atomic orbitals used to construct LCAO COs from two Bloch sums. (d) A qualitative CO energy-level diagram for the center of the Brillouin zone, $\Gamma = k(0, 0, 0)$.

COs are built by combining different Bloch orbitals (BOs) (which we will henceforth refer to as Bloch sums), which themselves are linear combinations of the atomic orbitals. There is one Bloch sum for every type of valence atomic orbital contributed by each atom in the basis. Thus, the two-carbon atom basis in diamond will produce eight Bloch sums, one for each of the s- and p-atomic orbitals. From these eight Bloch sums, we obtain eight COs, four bonding and four antibonding. For example, a Bloch sum of s-atomic orbitals at every site on one of the interlocking fcc sublattices in the diamond structure can combine in a symmetric or antisymmetric fashion with the Bloch sum of s-atomic orbitals at every site of the other fcc sublattice. Alternatively, we could have a Bloch sum of s-atomic orbitals combine with a Bloch sum of p-atomic orbitals. The symmetric (bonding) combinations of the basis atomic orbitals for the latter case are illustrated for one CC_4 subunit in Figure 2.10c. The actual COs are delocalized over all the atoms with the space-group symmetry of the diamond lattice. LCAO–CO construction from Bloch sums is thus completely analogous to LCAO–MO construction from atomic orbitals. We should point out that some authors refer to COs as BOs, too, since a linear combination of BOs is also a BO.

It will be worthwhile to look at this in yet a little more detail. First, note that we do not rotate the coordinate system of the atomic orbitals at the vertices of the tetrahedra (representing one of the fcc sublattices) in constructing Bloch sums for diamond, as we did for the CCl_4 molecule. In fact, as mentioned earlier, our basis for the diamond structure consists of only the two atoms in the chemical point group corresponding to the diamond lattice point, *not* the five atoms in the CC_4 tetrahedron. The Bloch sums in diamond are SALCs adapted not to a molecule with T_d point-group symmetry, but to the cubic diamond lattice. Conversely, MO theory is equivalent to the band scheme *minus* consideration of the lattice periodicity. A qualitative MO-like treatment of the diamond lattice point, however, will suffice for obtaining the relative placement of the energy bands in diamond at one special k-point. A k-point corresponds to a specific value of the quantum number $k(x, y, z)$, which gives the wavelength, or number of nodes, in the BOs that combine to give a CO. In a band structure diagram, the CO energies are plotted as one moves from one k-point to another.

At the special point $k(0, 0, 0) = \Gamma$, known as the center of the Brillouin zone, there are no nodes in the BOs. At this k-point, the lowest-energy CO in diamond arises from totally symmetric $ss\sigma$ and $sp\sigma$ interactions (symmetric with respect to the product JC_4^2, where J is the inversion and C_4 is a proper rotation about a fourfold rotation axis of the cubic lattice). This CO, of course, transforms as the totally symmetric irreducible representation of the cubic lattice. That irreducible representation is designated Γ_1 (analogous to the Mulliken symbol A_1 for the cubic point groups). Next lowest in energy is a triply degenerate set of COs with both $pp\sigma$ and $pp\pi$ interactions. The set is of symmetry designation Γ_{15} (also symmetric to JC_4^2). The reader should easily be able to sketch these. Immediately above this is the triply degenerate antibonding set of symmetry Γ_{25}' (antisymmetric with respect to JC_4^2). The highest CO is the antibonding CO, which is labeled Γ_2' (again, antisymmetric with respect to JC_4^2). The relative energy levels of the various COs at Γ are shown in Figure 2.10d.

Now, every CO can hold two electrons (of opposite spin) per two-center bonding site and every carbon contributes four electrons. If we add electrons to the aforementioned COs of diamond, in accordance with the procedure used for MOs, we find the four bonding COs (collectively termed the *valence band*) are completely filled and the four antibonding COs (collectively termed the *conduction band*) are empty. The octet rule is therefore obeyed around each carbon atom. Because a sizable band gap separates the full valence band and empty conduction band, diamond is an insulator.

This same ordering of the energy bands is also found in other elements that adopt the diamond lattice (e.g., Si). However, in some of these substances (e.g., Ge), as well as in some compounds with the isostructural zinc blende lattice (e.g., InSb), the reverse ordering is observed (i.e., the Γ_2' band may be lower in energy than the Γ_{25}' band). Nonetheless, the general picture of a full valence band and an empty conduction band still holds. Furthermore, our MO treatment correctly predicts the formation of a pair of singly degenerate MOs and a pair of triply degenerate sets. However, the relative energy levels and degeneracies of COs will change in moving between k-points (giving rise to the band dispersion in a band structure diagram), which are not accounted for by the simple MO treatment.

2.4 COMMON STRUCTURE TYPES

Hundreds of inorganic structure types are known. Unfortunately, it is only possible to present a limited number of them here. We have chosen to describe the structures of several nonmolecular solids that are of historical or pedagogical significance, or which are currently of substantial technological interest. Unfortunately, omissions are inevitable. There are examples of ionic, covalent, and metallic compounds that exist for almost every structure type. Thus, the common practice of classifying the structure types themselves as ionic, covalent, or metallic is not followed in this text. It should also be noted that many structure types are common to both ionocovalent and intermetallic compounds.

We follow the convention here that ionocovalent compounds as those formed between a metal (or metalloid) and a nonmetal, that is, halide, chalcogenide, light pnictide (N, P, As), silicon, carbon, or boron. Any compound formed between an element from this group and a metallic element falls in this category. Compounds formed between two or more nonmetals are classified as ionocovalent. Similarly, included in the category of intermetallic compounds are those compounds formed between different metals. Any compound containing a nonmetallic element is excluded from this category. When referring to a generic structure type, we use the convention of noting the metallic element as *A*, *B*, or possibly *C*, and the nonmetal or metalloid as *X*. For example, NaCl has *AX* stoichiometry, TiO_2 has AX_2 stoichiometry, $SrTiO_3$ is of ABX_3 stoichiometry, and so on.

2.4.1 Ionocovalent Solids

AX Compounds Many solids of *AX* stoichiometry possess the rock-salt structure including alkali halides (with the exception of cesium) and alkaline–earth chalcogenides (e.g., BaO, CaO, MgO). The arrangement of atoms in the rock-salt structure is very favorable for ionic compounds. However, there are examples of more covalent and even metallic compounds that also adopt this structure. These include SnAs, TiC, and TiN. Titanium oxide, TiO, is metallic and nickel oxide, NiO, is a *p*-type (hole) semiconductor, although the carrier mobility is extremely low. The rock-salt unit cell is shown in Figure 2.11. It consists of two interlocking fcc sublattices (one of *A* cations, the other of *X* anions) displaced relative to one another by $1/2a$ along $\langle 111 \rangle$, where *a* is the cubic cell dimension. Both the cations and anions are situated at sites with full O_h point symmetry. That is, every ion is octahedrally coordinated. The rock-salt lattice is a Bravais lattice since every lattice point, which consists of a cation–anion pair, is identical.

Other *AX* structure types include cesium chloride, CsCl (Figure 2.12); two polymorphs of zinc sulfide–wurtzite, and zinc blende; and NiAs. Although these structure types are often classified as ionic, many substantially covalent compounds adopt them as well. For example, in γ-CuI, which has the zinc-blende structure, the radius rules for ionic solids correctly predict that copper should be tetrahedrally coordinated $(r_+ : r_- = 0.60/2.20 = 0.273)$. However, the electronegativity difference between copper and iodine is less than one unit and CuI is insoluble in water and dilute acids, which would be quite unexpected for an ionic $1 + /1-$ salt.

Zinc blende can be considered isostructural with diamond, but with zinc cations residing at the centers of the same tetrahedra and the sulfide anions at the vertices. For zinc blende, an alternative description is in terms of a ccp-like array of S^{2-} anions with one-half of the tetrahedral sites occupied by Zn^{2+} cations. The polyhedral representation is depicted in Figure 2.13. Both diamond and zinc blende are best considered as two interlocking fcc sublattices displaced by a quarter of

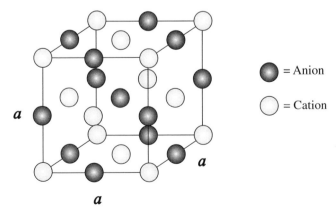

Figure 2.11 The cubic rock-salt or sodium chloride unit cell consists of two interlocking fcc sublattices displaced by $\frac{1}{2}a$ along $\langle 111 \rangle$.

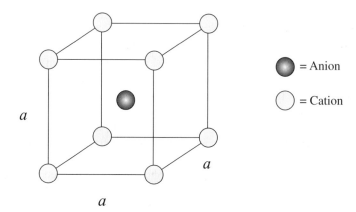

Figure 2.12 The cubic cesium chloride unit cell is not a body-centered cubic Bravais lattice since there are two nonequivalent lattice points.

the body diagonal. Many compounds comprised of main group *p*-block elements, with $\Delta\chi < 1$, adopt the zinc-blende structure. Some of these include β-SiC (α-SiC has the wurtzite structure), BeSe, and the majority of binary compounds between group 13 and group 15 elements (e.g., GaAs), and binary compounds of group 12 with group 16 (e.g., CdTe, ZnSe, HgSe).

The other polymorph of ZnS is wurtzite (Figure 2.14). The zinc atoms are tetrahedrally coordinated as in zinc blende, but the anions in wurtzite form an hcp-like array instead of a ccp-like array. Indeed, the wurtzite structure is often thought of as an hcp-like array of S^{2-} anions with one-half the tetrahedral sites occupied by Zn^{2+} cations. Hence, the next nearest and third nearest coordination in the two polymorphs are quite different. Nevertheless, wurtzite and zinc blende are

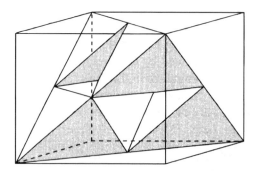

Figure 2.13 The diamond and zinc-blende structure depicted as a cubic array of vertex-sharing tetrahedra. In ZnS, the zinc cations reside at the center of the tetrahedra and the sulfide anions at the vertices. (After Elliot, 1998, *The Physics and Chemistry of Solids*. Copyright © John Wiley & Sons, Inc. Reproduced with permission.)

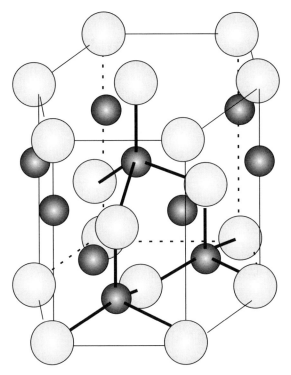

Figure 2.14   The hexagonal ZnS or wurtzite unit cell. Cations are the dark shaded circles.

almost energetically degenerate. Given their nearly identical Madelung constants (see Table 2.4), the overall Coulombic forces must be roughly comparable, wurtzite being only very slightly favored. Electronic structure calculations have also shown that the two ZnS polymorphs are essentially energetically equivalent (Saitta, 1997).

The wurtzite structure is adopted by most of the remaining compounds comprised of elements from the same groups as zinc blende, but that do not take the zinc-blende structure, that is, AlN, InN, CdSe. The structure seems to be able to accommodate larger electronegativity differences between the constituent atoms, as in BeO, GaN, and ZnO. For these more ionic compounds, the wurtzite unit cell must be more stable than that of zinc blende, to an extent governed by the specific bonding forces in each case.

In contrast to wurtzite, the structure of nickel arsenide, NiAs (Figure 2.15a), contains vacant tetrahedral sites but a completely occupied set of octahedral sites. In NiAs, the $NiAs_6$ octahedra share edges in one direction (the *ab* plane) and faces in another (along the *c* direction). Many transition metal chalcogenides with a 1:1 cation to anion ratio have this structure, for example, NiS, FeS, FeTe, CoTe, and CrSe. Some of these cannot possibly be considered ionic. For example, below 260 K, NiS is a semimetal (the resistivity is 10^{-3} Ω cm and temperature independent) and metallic above 260 K (with a resistivity as low as 10^{-5} Ω cm that increases with

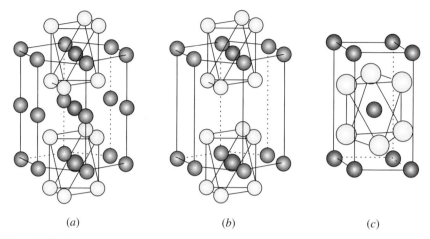

(a)   (b)   (c)

Figure 2.15 Some simple structure types containing octahedrally coordinated metal atoms. (a) NiAs. (b) CdI_2. (c) rutile (TiO_2). The anions are the light circles. The cations at the center of the octahedra are the darkest shaded circles.

temperature), the transition not being accompanied by a change in the symmetry of the crystal structure (Imada et al., 1998).

AX_2 Compounds One structure with AX_2 stoichiometry is that of CdI_2 (Figure 2.15b), which can be considered an hcp-like array of anions with cations occupying one-half of the octahedral sites. It is very similar to the NiAs structure with alternating layers of nickel atoms missing. The CdI_2 structure is very commonly observed with transition metal halides. Another structure type is that of rutile, TiO_2 (Figure 2.15c), in which chains of edge-sharing octahedra run parallel to the c-axis. The chains are linked at their vertices to form a 3D network.

Many ionic compounds of AX_2 stoichiometry possess the CaF_2 (fluorite), or Na_2O (antifluorite) structures shown in Figure 2.16. Fluorite is similar to CsCl, but with every other eight coordinate cation removed. Each fluoride anion is tetrahedrally coordinated by calcium ions. This structure is adopted by several fluorides and oxides. In the antifluorite structure, the coordination numbers are the inverse. Most oxides and other chalcogenides of the alkali metals (e.g., Na_2Se, K_2Se) possess the antifluorite structure, but so do some more covalent compounds, such as the silicides of Mg, Ge, Sn, and Pb.

An important oxide with the fluorite structure is ZrO_2. At room temperature, zirconia has a monoclinic structure in which zirconium is seven-coordinate. This transforms to a tetragonal structure at 1100°C and, at 2300°C, to the cubic fluorite structure. Aliovalent substitution of Zr^{4+} by the trivalent ion Y^{3+} stabilizes the fluorite structure at low temperatures by the creation of oxygen vacancies. One vacancy is required for every two yttrium atoms introduced. Yttria (Y_2O_3)-stabilized zirconia has the general formula $Zr_{1-x}Y_xO_{2-(x/2)}$. The Y^{3+} cations are randomly distributed and there is some experimental evidence that suggests they

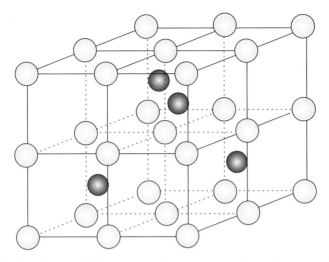

Figure 2.16 The fluorite (CaF_2) structure. The cations are the dark shaded circles in the octant centers.

are *next*-nearest neighbors to the vacancies (Fabris et al., 2002). The Y^{3+} cations thus have 8-fold coordination, as in the ideal fluorite structure. The presence of oxygen vacancies around the Zr^{4+} ions reduces the average coordination number of zirconium to values closer to seven, as in the stable monoclinic structure. The oxygen vacancies not only stabilize the fluorite structure in $Zr_{1-x}Y_xO_{2-(x/2)}$, they give rise to a mechanism for oxide ion conduction. Thus, there is interest in this material for use as an oxide ion-conducting electrolyte in solid oxide fuel cells.

Another important group of compounds with AX_2 stoichiometry are the metal–metalloid diborides, AB_2, where A = Mg, Al, Sc, Y, Ti, Zr, Hf, V, Nb, Ta, Cr, Mo, W, Mn, Tc, Re, Ru, Os, U, P. or Pu. These compounds form graphite-like hexagonal layers of boron atoms that alternate with layers of A atoms. Sometimes the boron layers are puckered. Many of these materials have the distinction of being among the hardest, most chemically inert, highest melting, and heat-resistant substances. Many are also better electrical conductors than the constituent elements. For example, TiB_2 has five times the electrical conductivity of titanium metal (Holleman and Wiberg, 2001).

AX_6 Compounds There also exist several binary borides with the AB_6 formula. The structure can be visualized as a body-centered CsCl lattice with the Cl^- ions replaced by B_6 octahedra, while the body-center cation may be Na, K, Rb, Cs, Ca, Sr, Ba, Sc, Y, Zr, La, lanthanide, or actinide. The $A^{III}B_6$ and $A^{IV}B_6$ borides have a high metallic conductivity (10^4–10^5 ohm^{-1} cm^{-1}) at room temperature, but the other borides are semiconductors. Other boron-rich binary borides include: AB_3, AB_4, AB_{10}, AB_{12}, and AB_{66}. The AB_{12} structure, like AB_6, is comparatively simple, possessing the NaCl structure, in which the A atoms alternate in the lattice with B_{12} cubo-octahedra. The other boron-rich compounds are often very complex, containing interconnected B_{12} icosahedra. The icosahedron is a special kind of polyhedron, sometimes called a *deltahedron*, whose faces are equilateral triangles. Other deltahedra include the octahedron, trigonal bipyramid, and tetrahedron. The particular deltahedron observed for a given compound can be predicted with Wade's rules (Wade, 1976). It can be stated as follows: *The number of vertices the deltahedron must have is equal to the number of cluster bonding (skeletal) electron pairs minus one.* Examples of the use of Wade's rules are given in Section 2.4.2.

ABX_2 Compounds   Many oxides with the ABX_2 delafossite structure, in which $A =$ Cu, Pd, Ag, or Pt; $B =$ Al, Sc, Cr, Fe, Co, Ni, Rh, or Ln; and $X =$ O ($CuFeO_2$ is the mineral delafossite), have been studied by Shannon, Prewitt and co-workers (Shannon et al., 1971, Prewitt et al., 1971; Rogers et al., 1971) and others (Seshadri et al., 1998). These oxides contain BO_2^- layers of edge-connected BO_6 octahedra tethered to one another through two-coordinate A^+ cations, that is, interlayer cohesion is due to electrostatic forces. Every A^+ cation has six neighboring A^+ cations in the same plane, which can be considered a close packed layer. The BO_2^- and A^+ layers can show stacking variants, one of which is illustrated in Figure 2.17, where the similarity with the octahedra of CdI_2 (Figure 2.15b) can be

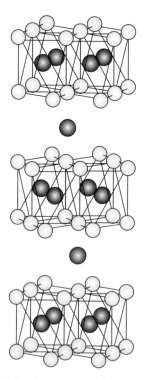

Figure 2.17   The hexagonal delafossite structure. Lightly shaded circles are oxygen atoms. The dark circles in the centers of the octahedra are the B atoms. The A atoms are the gray circles located between the slabs of edge-sharing BO_6 octahedra.

seen. Many delafossites possess interesting properties. For example, some of the oxides with palladium or platinum as A have very high in-plane electronic conductivities (only slightly smaller than copper metal). However, the same oxides with copper or silver as A, are insulating. Delafossites are also of interest magnetically, since the 2D triangular lattice enhances geometrical spin (magnetic) frustration.

Hagenmuller and co-workers have investigated the synthesis, structure, and properties of many other layered oxides with ABO_2 stoichiometry (Fouassier et al., 1975; Mendiboure, 1985; Delmas et al., 1975; Olazcuaga et al., 1975). Just as with the delafossites, the octahedral slabs in these oxides are separated by ionic A^+ layers. However, they differ in that the A^+ cation is octahedrally (or prismatically) coordinated by oxygen, whereas in delafossite the A^+ cations are linearly coordinated by only two oxide anions. For example, in α-NaMnO$_2$, each Na$^+$ cation has six equidistant oxygen anions because every edge-sharing MnO$_6$ octahedron in each single-layer MnO$_2$ slab coordinates one face to a Na$^+$ cation above it and one face to a Na$^+$ cation below it. The conventional notation used for this structure of α-NaMnO$_2$ is O$'$3, where the O signifies octahedral coordination around the alkali metal, the prime designates a monoclinic distortion, and the number three refers to the number of sheets in the unit cell.

The similar O2-type layered ABO_2 oxides, including α-NaFeO$_2$, NaNiO$_2$, LiFeO$_2$, LiCoO$_2$, and LiNiO$_2$, can be considered ordered derivatives of rock salt, the ordering occurring along alternate 111 layers. LiNiO$_2$ and LiCoO$_2$ are *mixed conductors*, exhibiting fast ionic (Li$^+$) conductivity as well as electronic conduction. Thus, they find use as cathode materials in rechargeable lithium batteries. During cell charging, lithium ions are extracted from the cathode and inserted into the anode. In the discharge cycle, the reverse reactions occur. LiNiO$_2$, however, suffers from severe capacity loss during recharging. Another oxide, LiMnO$_2$, also has been under consideration as a cathode material, but it is metastable and transforms to the spinel LiMn$_2$O$_4$ on cycling.

AB_2X_4 Compounds (Spinels) The AB_2X_4 spinels, based on the mineral MgAl$_2$O$_4$, and the inverse spinels, $B[AB]O_4$, are predominately ionic mixed oxides, containing a ccp-like array of X^{2-} anions (Figure 2.18). Most of the chalcogenides (O, S, Se, Te) can serve as the anion. More than 30 different cations can be incorporated into the spinel structure, with various combinations of charges: $A^{II}B^{III}_2X_4$, $A^{IV}B^{II}_2X_4$, and $A^{VI}B^{I}_2X_4$. In spinel, the A cations reside in one-eighth of the 64 tetrahedral sites and the B cations in one-half of the 32 octahedral sites. In the other extreme, the inverse spinels, the A cations and half the B cations swap positions. Many intermediate cation distributions have been observed between these two extreme cases.

Spinels have been studied intensively because of the sensitive dependence of their electronic and magnetic properties on the cation arrangement. For example, Fe$_3$O$_4$ is a mixed-valent (Fe^{II}/Fe^{III}) oxide with the inverse spinel structure. It is highly ferrimagnetic and has a high electronic conductivity, which can be attributed to electron transfer between Fe^{II} and Fe^{III}. It was recognized two decades ago that

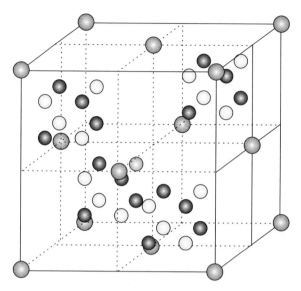

Figure 2.18 The spinel structure. The *A* cations are the light-gray circles located at the corner and face-center positions of the unit cell. The *B* cations (dark-gray circles) and the anions (lightly shaded circles) are located at the corners of the four cubes contained in the octants.

the spinel structure might also exhibit ion conduction (Thackery et al., 1982). In fact, due to the high lithium-ion conductivity in the oxide spinel $LiMn_2O_4$, this material has been used as a cathode replacement for $LiCoO_2$ in rechargeable lithium batteries. Unfortunately, $LiMn_2O_4$ exhibits less capacity and poorer cycling stability than $LiCoO_2$, especially at elevated temperatures. This is believed to be due, in part, to the cooperative Jahn–Teller distortion of Mn^{3+}, which causes the material to undergo a cubic-to-tetragonal phase transition.

ABX₃ Compounds (Perovskite and Related Phases) The perovskites, ABX_3, are 3D cubic networks of vertex-sharing BX_6 octahedra. Perovskite itself is $CaTiO_3$. The more ionic twelve-coordinate *A* cation sits in the center of the cube defined by eight vertex-sharing octahedra (Figure 2.19). It has been pointed out that the cubic perovskite structure can also be viewed as a four-sided octahedral channel structure in which the *A* cation resides in the channels (Rao and Raveau, 1998). Various combinations of *A* and *B* cation valences can be accommodated in the perovskites, including $A^{I}B^{V}X_3$, $A^{II}B^{IV}X_3$ and $A^{III}B^{III}X_3$. By far, most perovskites contain the oxide or fluoride anions, but the structure is also found for some compounds containing other anions as well, for example, S^{2-}, Cl^-, H^-, and Br^-.

Many closely related phases exist that can be considered distorted variants of perovskite. For example, when the *A*-site cation is too small for its cavity, octahedral tilting lowers the symmetry from cubic to orthorhombic (e.g., $GdFeO_3$). In Ba_2MgWO_6, a rock-salt-like ordering of the octahedral cations is found to accompany octahedral tilting. Perovskites can also tolerate vacancies, mixed

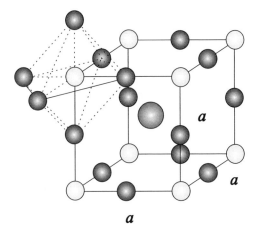

Figure 2.19 The cubic ABX_3 perovskite structure. The B cations (dark-gray circles) are at the vertices of the octahedra (the midpositions of the cell edges) and the A cation (light-gray circle) is located in the cener of the cube. Anions are at the corners of the cube.

valency, and/or oxygen deficiency. There are several important perovskite-related phases in which perovskite slabs are interleaved with other crystal structures. One is that of the Aurivillius phases, $(Bi_2O_2)(A_{n-1}B_nO_{3n+1})$, which result from the intergrowth of perovskite with Bi_2O_2 layers. Another intergrowth structure is the Ruddlesden–Popper (RP) series of oxides, $Sr_{n+1}Ti_nO_{3n+1}$. These can be considered $SrTiO_3$ perovskite layers interleaved with SrO rock-salt-like layers. The RP phases have been of renewed interest since the discovery that ion-exchangeable cations can replace strontium in the rock-salt-like layers. The ion-exchangeable oxides can be represented as $A'_2[A_{n-1}B_nO_{3n+1}]$, where $A' =$ alkali metal, $A =$ alkali, alkaline earth, rare earth, or main-group element. One such $n = 3$ phase is $Na_2La_2Ti_3O_{10}$, shown in Figure 2.20. A related series of oxides are the Dion–Jacobson phases, $A'[A_{n-1}B_nO_{3n+1}]$.

Because of their large structural and compositional flexibility, perovskites and perovskite-related compounds as a structure class exhibit perhaps the richest variety of magnetic and electrical transport (electronic and ionic conductivity) properties in solid-state chemistry. Fortunately, because of their relatively simple structures, perovskites are rather easily amenable to theoretical treatment.

$A_2B_2O_5$ ($ABO_{2.5}$) Compounds—Oxygen Deficient Perovskites

Brownmillerite is the name given to the mineral Ca_2FeAlO_5 (Hansen et al., 1928). Several $A_2B_2O_5$ compounds ($A =$ Ca, Ba, Sr; $B =$ Fe, Al, Ga, Mn, In) isostructural with brownmillerite are known to exist. The brownmillerite structure (Figure 2.21) can be thought of as an oxygen-deficient perovskite with the oxygen vacancies ordered into defect chains along the $\langle 110 \rangle$ direction. The structure can be described as an array of alternating layers of BO_3 octahedra and BO_2 tetrahedra, with the A^{2+} cations occupying the spaces between. In order to optimize the coordination around A^+, the tetrahedra are distorted and the octahedra are tilted (Section 2.5).

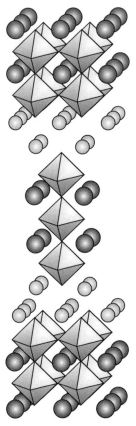

Figure 2.20 $Na_2La_2Ti_3O_{10}$, an $n = 3$ member of the ion-exchangeable Ruddlesden–Popper-like phases. The TiO_6 octahedra are shown. The O^{2-} anions are at the vertices of the octahedra and the Ti^{4+} cations in the centers of the octahedra. The La^{3+} cations are the large circles. The Na^+ cations are the small circles in between the triple-layer slabs.

In principle, any trivalent cation that can accept either tetrahedral or octahedral coordination can be incorporated into the brownmillerite structure. It is also possible to lightly dope brownmillerite with aliovalent cation pairs, for example, Mg^{2+}/Si^{4+} in $Ca_2Fe_{0.95}Al_{0.95}Mg_{0.05}Si_{0.05}O_5$. The usual strategy, however, is to incorporate pairs of cations in which one of the members has a distinct preference for a particular type of site. For example, in Ca_2FeAlO_5 the Al^{3+} cations are found in the tetrahedral sites and the Fe^{3+} in the octahedral sites. Similarly, in Ca_2MnGaO_5, Ga^{3+} cations exclusively occupy the tetrahedral sites, while Mn^{3+} occupies the octahedral sites. The ability to achieve this type of ordered cation arrangement offers the potential to realize 2D layers (e.g., MnO_2 and GaO in Ca_2MnGaO_5) that may have concomitant magnetic or electronic transport properties of interest.

ABO_3 $ABO_{2.5}$

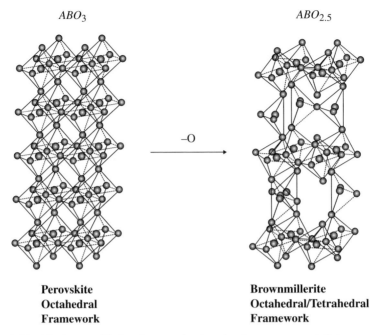

$\xrightarrow{-O}$

Perovskite
Octahedral
Framework

Brownmillerite
Octahedral/Tetrahedral
Framework

Figure 2.21   The brownmillerite phase can be thought of as an oxygen-deficient perovskite, in which the oxygen vacancies are ordered along the $\langle 110 \rangle$ direction of the perovskite cell. The tetrahedra are distorted and the octahedra are tilted in order to optimize coordination around the A cations. For clarity, the A cations are not shown in this figure.

Interest in the brownmillerites as fast oxide-ion conductors was first stimulated by Goodenough, who showed that $Ba_2In_2O_5$ displayed an abrupt increase in electrical conductivity above a certain temperature (Goodenough et al., 1990). Oxide-ion conductors must have a high concentration of oxygen vacancies (either intrinsically or from aliovalent doping as in Y-stabilized ZrO_2) for O^{2-} hopping to occur. However, high oxide-ion diffusivity is generally associated with a disordered oxygen sublattice (Norby, 2001). Accordingly, fast oxide-ion conduction has been observed only at high temperatures ($800°$–$1000°C$) in $Sr_2Fe_2O_5$ (Holt et al., 1999) and $Ba_2In_2O_5$ (Goodenough, et al., 1990). Atomistic modeling has supported the theory that the abrupt conductivity change in $Ba_2In_2O_5$ is due to an order–disorder phase transition. The computer simulations suggest that O^{2-} anions become displaced from their lattice sites into the open interstitial sites in the tetrahedral layer, forming Frenkel defect pairs (Section 2.5.1). As the number of Frenkel defects rises with increasing temperature, the anions at the equatorial positions in the octahedral and tetrahedral layers become indistinguishable and the displaced O^{2-} anions diffuse rapidly through the material (Fisher and Islam, 1999).

Some phases with the $ABO_{2.5}$ composition do not have the brownmillerite structure. This is possible when the B cation is a transition metal with a tendency to adopt geometries other than tetrahedral and/or octahedral. For example, the oxygen vacancies in $LaNiO_{2.5}$ order in such a way as to form alternating NiO_6 octahedra and NiO_4 square planes within the ab plane, which results in chains of octahedra along c (Vidyasagar et al., 1985). In still other (primarily mixed-valent) ABO_{3-x} phases, with $x < 0.5$, other polyhedra are observed. The oxide $CaMnO_{2.8}$, for instance, is built up of an ordered framework of $Mn^{(III)}O_5$ pyramids and $Mn^{(IV)}O_6$ octahedra.

$A_xB_yO_z$ Compounds (Bronzes)   The *bronzes* are channel structures with an openness that allows for the transport of atoms or ions into the crystal. Bronzes have the general formula $A_xB_yO_z$, in which A is an alkali, alkaline-earth, or rare-earth metal, and B can be Ti, V, Mn, Nb, Mo, Ta, W, or Re. The German chemist Friedrich Wöhler (1800–1882), who discovered Na_xWO_3 in 1824, called these materials bronzes due to their intense color and metallic luster. The introduction of sodium into the WO_3 perovskite structure chemically reduces a portion of the W^{6+} ions to W^{5+}. When $x \geq 0.28$, this results in metallic conductivity (Greenblatt, 1996; Goodenough, 1965). Nonstoichiometric sodium tungsten oxide ($x < 1$) has a distorted perovskite structure with unequal $W{-}O$ bond lengths and tilted WO_6 octahedra.

Some bronzes have lamellar structures. For example, Na_xMnO_2, which was already discussed, is a bronze. This phase consists of slabs of edge-sharing MnO_6 octahedra separated by layers of Na^+ cations. There are many structural variations among the bronzes. In fact, some bronzes with rather complex structures bare little or no resemblance to perovskite, although they are all generally built from vertex-sharing and/or edge-sharing octahedra. The open channels may have triangular, square, rectangular, diamond-shaped, pentagonal, hexagonal, or other polygonal cross-sections.

$A_2B_2X_7$ Compounds (Pyrochlores)   Another channel structure is that of the pyrochlores (Figure 2.22), with general formula $A_2B_2X_7$. The mineral pyrochlore is $(Ca, Na)_2Nb_2O_6(O, OH, F)$. The A and B cations form a face-centered cubic array with the anions occupying tetrahedral interstitial sites. The A ion has eightfold anion coordination, and B has sixfold anion coordination. Thus, the pyrochlore lattice consists of two sublattices: $(A_2X)B_2X_6$. It may also be thought of as an anion deficient derivative of the fluorite structure, but with an ordered arrangement of anion vacancies and an ordered cation arrangement. The size of the A cation has a large effect on the stability of the pyrochlore structure. As the size difference between the A and B cations decreases, the fluorite structure becomes favored over pyrochlore.

The B_2X_6 sublattice is a tetrahedral network of vertex-sharing BX_6 octahedra-containing channels with hexagonal cross sections. However, the channels are obstructed by the anions of the hexagonal A_2X sublattice of vertex-sharing AX_4 tetrahedra, which prohibit cationic mobility. In nonstoichiometric (anion-deficient) pyrochlores, $Gd_{1.8}Ca_{0.2}Ti_2O_{6.95}$, for example, cationic mobility and ion exchange of the A-site cations are possible at high temperatures.

The stoichiometric pyrochlore transition metal oxides exhibit a wide range of magnetic and electronic transport properties. These properties are, of course,

Figure 2.22 The $A_2B_2X_7$ pyrochlore structure. The B_2X_6 sublattice is a tetrahedral channel-forming network of vertex-sharing BX_6 octahedra. The channels are occupied by the anions of the A_2X sublattice of vertex-sharing AX_4 tetrahedra (not shown).

dependent on the d electron count of the B cation. Electrical conductivity may be insulating (e.g., $Gd_2Ti_2O_7$, $Tb_2Ti_2O_7$), semiconducting (e.g., $Tl_2Ru_2O_7$), or metallic (e.g., $Nd_2Mo_2O_7$, $Cd_2Re_2O_7$). The oxide $Cd_2Os_2O_7$ undergoes a transition from semiconducting to metallic conductivity at 226 K accompanied by a magnetic transition (paramagnetic to antiferromagnetic), and $Tl_2Mn_2O_7$ exhibits colossal magnetoresistance, a dramatic drop in its electrical resistivity in the presence of a magnetic field. The antiferromagnetic ground state of the pyrochlore oxides can have a very large spin degeneracy. This is because the A_2O sublattice that the magnetic rare-earth cations reside on is a geometrically frustrated system with competing magnetic interactions. Hence, spin glass behavior, as characterized by hysteresis and nonlinearity in the magnetic susceptibility, is observed in many pyrochlore oxides at low temperatures, even in the absence of chemical or bond disorder, including $Y_2Mo_2O_7$ and $Tb_2Mo_2O_7$ (Gardner et al., 2001).

Silicon Compounds Silicon, like carbon, has a propensity for tetrahedral coordination. The SiX_4 tetrahedron can be considered the building block of most silicon compounds. The two polymorphs of silicon carbide, for example, adopt the zinc-blende and wurtzite structures. In a similar fashion, silicon dioxide is found in nature in both crystalline and amorphous forms, in which the SiO_4 tetrahedra share all their vertices. There are eight different modifications of the crystalline form of SiO_2. The most stable is α-quartz, consisting of interlinked helical chains, with three tetrahedra per turn. Quartz crystals can be either right-handed or left-handed, so that they are nonsuperimposable on their mirror image, that is, they exhibit enantiomorphism (Greenwood and Earnshaw, 1997). The other crystalline forms contain interlinked sheets of six-membered rings of tetrahedra. It has been said that more is known about the chemical, structural, physical, and electrical properties of SiO_2 than any other oxide. This is, no doubt, in part due to the importance of this material as a dielectric for silicon-based semiconductor devices.

There is a wide variety of silicate mineral structure types. However, the connectivity of the tetrahedra in most silicates can be determined from their formulas (West, 1985). The smaller the Si:O ratio, the fewer the number of SiO_4 vertices shared with neighboring units. For example, Mg_2SiO_4 (Si:O = 1:4) contains no bridging oxygens, and thus has discrete SiO_4 tetrahedra. By contrast, SiO_2 (Si:O = 1:2) is a 3D network containing no nonbridging oxygens. In fact, one classification system for silicate structures is based on the number of oxygen atoms per tetrahedron that are shared. The notation scheme follows: *Neso-* (Si:O = 1:4); *soro-* (Si:O = 1:3.5); *cyclo-* (*ino-*) (Si:O = 1:3); *phyllo-* (Si:O = 1:2.5); and *tecto-* (Si:O = 1:2), corresponding to, respectively, 0, 1, 2 (closed-ring or continuous-chain), 3, and 4 shared oxygen atoms. Unfortunately, due to space constraints, we are not able to provide a more detailed study of the structures of silicates. However, we will discuss another binary silicon compound that is also important to the semiconductor industry, Si_3N_4, as well as touch on a few aspects of zeolites and other porous solids.

There are many other binary and ternary silicon compounds of commercial importance. For example, silicon nitride (Si_3N_4) occurs in two hexagonal forms: the α-form and a denser β-form. The crystal structures are very complex, but may be thought of as close-packed nitrogen atoms, with 3/8 of the tetrahedral vacancies occupied by silicon atoms. In this respect, it is like SiO_2, being a 3D network of tetrahedral units. However, although the nitrogen atoms do arrange roughly into a tetrahedron around the silicon atoms, the silicon atoms are arranged into planar triangles (not pyramids!) around the nitrogen atoms. The β-form contains small-diameter ($\sim$0.15 nm) one-dimensional channels. Silicon nitride is very hard, strong, and chemically inert toward most agents up to 1300°C (Holleman and Wiberg, 2001). It is relatively impermeable to sodium, oxygen, and other species (even hydrogen diffuses slowly through silicon nitride). Hence, (amorphous) Si_3N_4 films find wide use in silicon-based integrated circuits as diffusion barriers and passivation layers.

Porous Structures Many nonmetal oxides (e.g., silicates and phosphates) possess 3D framework structures, built from vertex-sharing polyhedra, containing large tunnels or cavities. The *microporosity* (pore diameters ≤ 20 Å) or *mesoporosity* (pore diameters = 20–500 Å), plus the large internal surface area in these materials, enable their use as small molecule sieves, adsorbents, ion-exchange media, and catalysts. As with the silicates, the structures and chemical formulas depend on the numbers of free and shared polyhedral vertices. Frameworks containing aliovalent substitutional ions, such as Al^{3+} substituting for Si^{4+} ion in SiO_4 tetrahedra, bear net negative charges. Cations must be accommodated within the framework to balance this charge. Examples include the felspars $NaAlSi_3O_8$ and

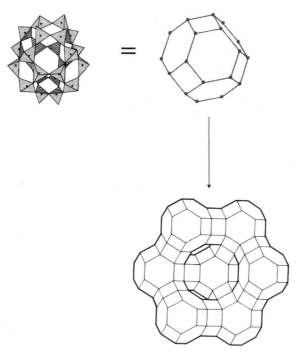

Figure 2.23 A zeolite is constructed from linked cuboctahedra, which in turn, are constructed from 4-membered squares of tetrahedra.

$CaAl_2Si_2O_8$. The structures of inorganic frameworks can be classified into two basic categories: zeolite types containing exclusively tetrahedra, and those types with mixed tetrahedral–octahedral (or bipyramidal) frameworks.

Figure 2.23 shows how six groups of four corner-sharing SiO_4 tetrahedra—six squares of tetrahedra giving 24 tetrahedra total—connect together in the zeolites to form a cuboctahedron (intermediate between a cube and octahedron), which unites with other cuboctahedra to form the large open network structure characteristic of these compounds. Zeolites are commercially prepared by a hydrothermal technique (high temperature and pressure) involving crystallization from strongly alkaline solutions of sodium silicate and aluminum oxide. The aluminophosphates possess an $(Al,P)O_4$ tetrahedra framework similar to that of the zeolites.

Many transition-metal silicates, germanates, and phosphates possess a mixed framework of tetrahedra and octahedra. These phases can be prepared by hydro-thermal techniques, but it is still not entirely clear how the structures assemble at the molecular level. Nonetheless, inorganic and organic moieties have found use as templates to control the pore size and shape in many cases, but the large phosphate structures often collapses if the template is removed on heating. Again, the channels or cavities possessed by these frameworks make them interesting for certain applications. For example, the silicates $Na_5BSi_4O_{12}$ (B = Fe, In, Sc, Y, La, Sm) and the nasicons, which are silicophosphates with formula $Na_{1+x}Zr_2P_{3-x}Si_xO_{12}$, are famous for superionic conduction. In each of these types of oxides the transition-metal-oxygen octahedra shares their six vertices with SiO_4 (or PO_4) tetrahedra, but in $Na_5BSi_4O_{12}$ channels are formed, whereas in the nasicons, cavities are formed (Rao and Raveau, 1998).

Periodic mesoporous silicas were reported for the first time in the literature by researchers at the Mobil Oil Corporation in the early 1990s, although a synthetic process that yields very similar reaction products was patented 20 years prior (Moller and Bein, 1998). The reactive internal surfaces of these solids have been used to attach functional groups that can act as complexing agents for metal cations. However, as well as complexation and similar uses, such as ion exchange and sorption, the channels have been used to grow metal clusters and wires.

Another class of porous materials is the pillared layered structures. In these phases, the pillars separate inorganic layers of interconnected polyhedra, to which they are covalently bonded on both ends. The pillars themselves may be organic or inorganic; when the former, the material is referred to as a *hybrid*. A schematic illustration of a hybrid structure is shown in Figure 2.24a. Porosity is introduced by spacing the pillars apart, which is most easily accomplished by interposing smaller R groups between the pillaring groups (Figure 2.24b). The smaller R groups are bonded on only one end. The study of organically pillared compounds began about 25 years ago with the synthesis of one of the first zirconium phenylphosphonates, $Zr(O_3PC_6H_5)_2$ (Alberti et al., 1978). When the inserted organic moieties are *not* covalently bonded to the inorganic layers, the material is considered a *nanocomposite* (Figure 2.24c). The hybrid organic–inorganic materials field, and the field of porous solids, in general, is vast and rapidly growing, but is, unfortunately, outside the main scope of this book.

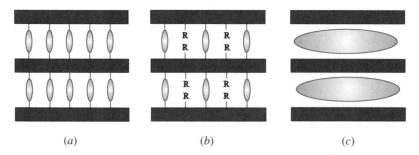

(a) $\qquad\qquad\qquad (b)$ $\qquad\qquad\qquad (c)$

Figure 2.24 Cross sections of hybrid organic–inorganic materials. (a) A pillared layered structure, in which organic moieties (ovals) are covalently bonded to the inorganic layers (rectangles). The separation between the layers can be controlled by changing the size of the organic molecules. (b) Porosity is introduced by interposing smaller R groups (e.g., OH, CH_3) between the pillars. (c) A nanocomposite formed by incorporating an organic molecule or polymer between two already separated inorganic layers. The organic moieties are not covalently bonded to the inorganic layers.

Alexander Wells (Courtesy of Terrell Vanderah. Reproduced with permission.)

ALEXANDER FRANK WELLS

(1912–1994) earned his B.S. and M.A. degrees in chemistry from Oxford University and his Ph.D. from Cambridge University under H. M. Powell in 1937. He received his D.Sc. in 1956, also from Cambridge University. From 1944 to 1968, he was Director of the Crystallographic Laboratory at Imperial Chemical Industries. The first edition of Wells' now classic book *Structural Inorganic Chemistry* was published in 1945. Four subsequent editions were eventually published, the last in 1985. This work constitutes a substantial portion of the body of knowledge on structural principles and space-filling patterns of inorganic solids. For many practitioners of solid-state chemistry, this book remains the standard reference for inorganic crystal structures. Wells also authored four other well-known books: *The Third Dimension in Chemistry*; *Models in Structural Inorganic Chemistry*; *Three-Dimensional Nets and Polyhedra*; and *Further Studies of Three-Dimensional Nets*. He was also among the first editorial advisors for the *Journal of Solid State Chemistry*. In 1968, Wells accepted a position as Professor of Chemistry at the University of Connecticut, becoming emeritus in 1982.

(*Source*: B. Chamberland, personal communication, March 5, 2004.)

2.4.2 Intermetallic Compounds

In many cases, there is a substantial degree of solid solubility of one metal or metalloid in another. For example, the binary systems Ag–Au, Ag–Pd, Bi–Sb, Ge–Si, Se–Te, and V–W, exhibit complete liquid and solid miscibility, with no atomic ordering at any temperature. This type of mixture is a particular class of alloy called a *solid solution*. At the other extreme, some systems exhibit negligible solid solubility, forming instead multiphase alloys that consist of grains of each pure component. Examples are As–Bi, Bi–Cu, Bi–Ge, Na–Rb, Sb–Si, and Ge–Zn.

Metallurgists consider intermetallics another type of alloy. The crystal structures of these phases consist of ordered arrangements of atoms, as opposed to the disordered arrangement in solid solutions. Intermetallics are actually compounds formed between two or more metallic (or metalloid) elements. They can have a fixed stoichiometry (line compounds) or exist over a narrow compositional range (intermediate phases). Despite the fact that intermetallics are made up of metallic elements, they usually have a mix of metallic and nonmetallic properties. Like ceramics, they are brittle at low temperatures. However, intermetallics generally exhibit higher thermal and electrical conductivities than ceramics, although less than those of their constituent elements. Intermetallics have been prepared by the mechanical alloying process known as ball milling, combustion synthesis (self-propagating combustion of a mixture of metal powders), rapid solidification techniques, and grown as single crystals. In some cases, it is also possible to deposit intermetallic thin films by sputtering techniques.

The circumstances under which intermetallics form were elucidated by William Hume-Rothery for compounds of the noble metals and the elements to their right in the periodic table (Hume-Rothery, 1934; Reynolds and Hume-Rothery, 1937). These are now applied to all intermetallic compounds, in general. The converse to an intermetallic, a solid solution, is only stable for certain valence electron count per atom ratios, and with minimal differences in the atomic radii, electronegativities, and crystal structures (bonding preferences) of the pure components. For example, it is a rule-of-thumb that elements with atomic radii differing by more than 15% generally have very little solid-phase miscibility.

Considering the fact that the majority of elements are metals, it would seem that there could be a vast number of intermetallic compounds and structures. Indeed there are, and the rules of bonding and valence in these materials are still largely unknown. Many intermetallic compounds are characterized by unusual and complex stereochemistries. In fact, it is often impossible to rationalize the stoichiometry using simple chemical valence rules, a situation reminiscent for the inorganic chemist of the hydrides, borides, and silicides.

Zintl Phases   Invoking Lewis' octet rule, Hume-Rothery published his "$8 - N$" rule in 1930 to explain the crystal structures of the p-block elements (Hume-Rothery, 1930, 1931). In this expression, N stands for the number of valence electrons on the p-block atom. An atom with 4 or more valence electrons forms $8 - N$ bonds with its nearest neighbors, completing its octet. The Bavarian chemist Eduard Zintl (1898–1941) later extended Hume-Rothery's $8 - N$ rule to ionic compounds (Zintl, 1939). In studying the structure of NaTl, Zintl noted that the Tl^{1-} anion has four valence electrons, so should bond to four neighboring anions.

The term *Zintl phase* is applied to solids formed between an alkali (or alkaline–earth) metal and a main-group p-block element from group 14, 15, or 16 in the periodic table. These phases are characterized by a network of homonuclear or heteronuclear polyatomic clusters (the Zintl ions), which carry a net negative charge, and that are neutralized by cations. Broader definitions of the Zintl phase are sometimes used. Group 13 elements have been included with the Zintl anions, and an electropositive rare-earth element or transition element with a filled d shell (e.g., Cu) or empty d shell (e.g., Ti) has replaced the alkali or alkaline–earth element in some reports. Although the bonding between the Zintl ions and the cations in the Zintl phases is markedly polar, by our earlier definition those compounds formed between the alkali or alkaline–earth metals with the heavier anions (i.e., Sn, Pb, Bi) can be considered intermetallic phases.

A diverse number of Zintl ion structures are formed, even among any same two elements. For example, the metal phosphides M_xP_y may contain discrete P^{3-} anions (K_3P), or negatively charged chains (K_4P_6), rings (KP), or cages (K_4P_{26}) of phosphorus atoms. Very commonly, the Zintl ions adopt structures consisting of polyhedra all of whose faces are equilateral triangles. These are sometimes called deltahedra. Examples include: the tetrahedron, trigonal bipyramid, octahedron, pentagonal bipyramid, dodecahedron, tricapped trigonal prism, bicapped square antiprism, octadecahedron, and icosahedron. Triangular networks make the most efficient use of a limited number of cluster bonding (skeletal) electrons, and so are electronically advantageous (Porterfield, 1993). The Zintl phases are often prepared by dissolving the constituent elements or alloys in liquid ammonia or $AlCl_3$.

The anion connectivity of many Zintl phases can be rationalized in terms of Hume-Rothery's $8 - N$ rule. For example, in $BaSi_2$ (with Si_4^{4-} clusters), the Si^{1-} anion is isoelectronic with the nitrogen group elements, that is, it has five valence electrons. The $8 - N$ rule correctly predicts that each silicon atom will be bonded to three other silicon atoms. Similarly, in Ca_2Si, Si^{4-} is isoelectronic with the noble-gas elements. Again, the $8 - N$ rule correctly predicts that silicon will occur as an isolated ion. Indeed, this compound has the anti-$PbCl_2$ structure, in which the silicon is surrounded by nine calcium ions at the corners of a tricapped trigonal prism.

A generalized $8 - N$ rule was derived by Pearson in order to address those cases in which fractional charges appear on the anion (Pearson, 1964). Fractional charges are usually indicative of multiple structures or anion connectivities existing in the compound. The generalized $8 - N$ rule is formulated as:

$$8 - VEC_A = AA - CC \qquad (2.12)$$

where VEC_A is the total number of valence electrons per anion, AA is the *average* number of anion–anion bonds per anion, and CC is the *average* number of cation–cation bonds per cation.

Unfortunately, neither Hume-Rothery's original rule nor the generalized $8 - N$ rule is valid for nonpolar intermetallics, or when an octet configuration is unnecessary for stability of the compound. In order to increase the domain of structures for which one can make predictions, extensions have been made to the generalized $8 - N$ rule (Parthé, 2000), but are not discussed here.

Traditionally, the Zintl phases are semiconductors that either obey the $8 - N$ or generalized $8 - N$ rule. Unfortunately, there appears to be controversy over the range of applicability for the empirical rules guiding what constitutes a Zintl compound. The distinguished professor John D. Corbett, a leading researcher in the field of intermetallics, has argued that some weakly metallic compounds containing anion arrays with slightly delocalized electrons can also be classified structurally as Zintl phases. The R_3In_5 ($R = Y$, La) phases are one such example.

In R_3In_5 ($R = Y$, La), the electropositive R is assumed to have a formal charge of $+3$. Each indium anion thus has a charge of $-(3 \times 3)/5 = -9/5 = -1.8$. Since a neutral indium atom has a valence electron count of 3, the VEC_A for $In^{-1.8}$ is equal to $3 + 1.8$, or 4.8. The R_3In_5 phases contain distantly interconnected square pyramidal In_5 clusters (Corbett and Zhao, 1995). Hence, four indium anions have threefold coordination and one indium has fourfold coordination, giving $AA = 16/5 = 3.2$. The generalized $8 - N$ rule correctly predicts that CC should be equal to zero, which is the case, as In anions are the nearest neighbors to each R cation.

The approach developed by K. Wade (Wade, 1976) for predicting the geometric structures of boranes and boron halides is also useful for many Zintl anions. For convenience, it is restated here: The number of vertices the deltahedron must have is equal to the number of cluster-bonding (skeletal) electron pairs minus one. In counting the cluster bonding electrons in Zintl ions, one uses the formula $v-2$ where v is the number of valence electrons on the element. A total of n cluster bonding electrons ($n/2$ pairs) requires a deltahedron with $n/2 - 1$ vertices, even with a different number of atoms in the anion cluster. For example, in R_3In_5 ($R = Y$, La) each indium atom contributes $3 - 2 = 1$ electron to the cluster, which has a charge of nine, In_5^{9-}. There is $9 + 5 = 14$ cluster bonding electrons, or seven pairs. Wade's rules correctly predict that the In_5^{9-} cluster will adopt an octahedral geometry $(7 - 1)$ with one vertex missing—the square pyramid.

Included among the interesting properties currently being pursued in some Zintl phases are the metal–nonmetal transition and colossal magnetoresistance. Zintl phases have also been used as precursors in the synthesis of novel solid-state materials. For example, a fullerene-type silicon clathrate compound, $Na_2Ba_6Si_{46}$ (clathrates are covalent crystals, whereas fullerides are molecular crystals), with a superconducting transition at 4 K was prepared from the Zintl phases NaSi and $BaSi_2$ (Yamank et al., 1995). In the final product, Ba atoms are located in the center of tetrakaidecahedral (Si_{24}) cages, while the Na atoms are in the centers of pentagonal dodecahedral (Si_{20}) cages. The $Na_2Ba_6Si_{46}$ clathrate compound was the first superconductor found for a covalent sp^3-hybridized silicon network.

Nonpolar Binary Intermetallic Phases Zintl phases are characterized by the presence of markedly heteropolar bonding between the Zintl ions (electronegative polyatomic clusters) and the more electropositive metal atoms. By contrast, the bonding between heteronuclear atoms within other intermetallic compounds is primarily covalent or metallic. A number of different structure types exist for any given type of stoichiometry, as indicated in Table 2.7. The largest single intermetallic structural class is commonly referred to as the *Laves phases*. They are named after the German mineralogist and crystallographer Fritz Laves (1906–1978). These AB_2 intermetallic compounds form dense tetrahedrally close-packed structures. In $MgCu_2$ (Pearson symbol $cF24$), the Mg atoms are ordered as in cubic diamond. However, in $MgZn_2$ (Pearson symbol $hP12$), the Mg atoms are located on sites corresponding to those in the hexagonal diamond (Londsdaleite) structure. Both are illustrated in Figure 2.25.

TABLE 2.7   Several Binary Intermetallic Structure Types

Stoichiometry	Prototype	Pearson Symbol
AB	NiTi	$mP4$
	ηAgZn	$hP9$
	CoSn	$hP6$
	AuCd	$oP4$
	CoU	$cI16$
	σCrFe	$tP30$
	ωCrTi	$hP3$
AB_2	Cu_2Sb	$tP6$
	$PdSn_2$	$oC24$
	Cu_2Mg (Laves)	$cF24$
	$MgNi_2$ (Laves)	$hP24$
	$MgZn_2$ (Laves)	$hP12$
	Al_2Cu	$tI12$
	$NiTi_2$	$cF96$
	$MoPt_2$	$oI6$
A_xB_y	$βCu_3Ti$	$oP8$
	Ni_3Sn	$hP8$
	Al_3Ni_2	$hP5$
	Al_3Zr	$tI16$
	$AlFe_3$	$cF16$
	Al_4Ba	$tI10$
	$PtSn_4$	$oC20$

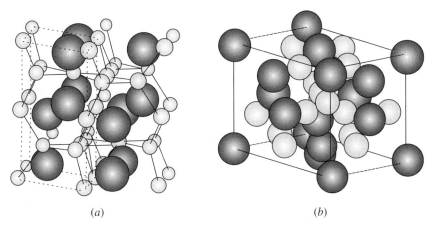

(a)   (b)

Figure 2.25   The AB_2 Laves phases. (a) The hexagonal structure (the unit cell is shown by the dashed line). (b) The cubic structure. The A atoms are the dark circles.

The crystal structure of $MgZn_2$ was first determined by James B. Friauf (Friauf, 1927). The phase is composed of intrapenetrating icosahedra (each one with six Mg atoms and six Zn atoms at the vertices) that coordinate Zn atoms and 16-vertex polyhedra that coordinate Mg atoms. The latter are actually interpenetrating tetrahedra (with Mg atoms at the four vertices) and 12-vertex truncated tetrahedra (with Zn atoms at the vertices). The 16-vertex polyhedron so formed from the two smaller polyhedra is called, appropriately, a Friauf polyhedron. In fact, the Laves phases are sometimes referred to as the Friauf phases or Laves-Friauf phases. A third Laves phase, with prototype $MgNi_2$ ($hP24$), has a dihexagonal structure. The Laves phases are the most common structure type of binary intermetallic compound between a $3d$ transition element and a $4d$, $5d$, or $6f$ element. Compounds crystallizing in the hexagonal structure include: $TaFe_2$, $ZrRe_2$, $NbMn_2$, and UNi_2. Those with the cubic structure include: $CsBi_2$ and $RbBi_2$, while compounds with the dihexagonal structure include: $NbZn_2$, $ScFe_2$, $HfCr_2$, UPt_2, and $ThMg_2$.

Because of their high-temperature deformability and good oxidation resistance, some Laves phases (e.g., $NbCr_2$) are being considered as structural materials in gas turbine engines. Others have been considered for functional applications. For example, the cubic ZrV_2 exhibits a superconducting transition below 8 K. It also has been considered as a hydrogen storage material, in which hydrogen can be absorbed into interstitial sites and reversibly desorbed at high temperatures. The Laves phases are a subset of the family of tetrahedrally close-packed structures known as the *Frank–Kasper phases*, in which atoms are located at the vertices and centers of various space-filling arrangements of polytetrahedra (Frank and Kasper, 1958a, 1958b). In the following section, we see that the Frank–Kasper phases have aided the understanding of quasi-crystal and liquid structures.

Ternary Intermetallic Phases   Many ternary intermetallics have interesting structures and properties. For example, of great theoretical interest to crystallographers in recent years are the ternary intermetallic systems that are among the few known *stable* solids with perfect long-range order but with no 3D translational periodicity. The former is manifested by the occurrence of sharp electron diffraction spots and the latter by the presence of a noncrystallographic rotational symmetry. It will be recalled that 3D crystals may only have one-, two-, three-, four-, or sixfold rotation axes; all other rotational symmetries are forbidden. However, the discovery of metastable icosahedral *quasi-crystals* of an Al–Mn alloy exhibiting fivefold rotational symmetry was reported 20 years ago (Shectman et al., 1984). Since that time, many thermodynamically stable quasi-crystals of ternary intermetallic compounds have been found. These have mostly been obtained by rapidly solidifying phases with equilibrium crystal structures containing icosahedrally packed groups of atoms (i.e., phases containing icosahedral point-group symmetry). The quasi-crystalline phases form at compositions close to the related crystalline phases.

The icosahedron is one of the five Platonic solids, or regular polyhedra, and is shown in Figure 2.26. A regular polygon is one with equivalent vertices, equivalent edges, and equivalent faces. The icosahedron has 20 faces, 12 vertices, 30 edges, and 6 fivefold proper rotation axes (collinear with six tenfold improper rotation

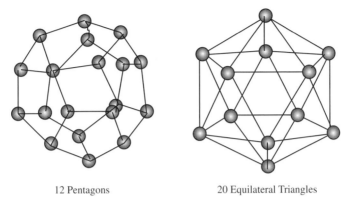

| 12 Pentagons | 20 Equilateral Triangles |

Figure 2.26 The dodecahedron (top) and icosahedron (bottom) are two of the five Platonic solids. The others (not shown) are the tetrahedron, the cube, and the octahedron.

axes). It is possible for crystal twinning to produce disallowed diffraction patterns, but, in order to produce the fivefold symmetry, twinning would have to occur five times in succession. The possibility of twinning was, in fact, the main point of contention after the first report on the discovery of icosahedral quasi-crystals However, numerous attempts to disprove the true fivefold symmetry failed and the icosahedral symmetry was confirmed as real.

It turns out that icosahedral coordination ($Z = 12$) and other coordination polytetrahedra with coordination numbers $Z = 14$, 15, and 16 are a major component of some liquid structures, more stable than a close-packed one, as was demonstrated by Frank and Kasper. When these liquid structures are rapidly solidified, the resultant structure has icoshedra threaded by a network of wedge disclinations, having resisted reconstruction into crystalline units with 3D translational periodicity (Turnbull and Aziz, 2000; Mackay and Kramer, 1985). Stable ternary intermetallic icosahedral quasi-crystals are known from the systems Al–Li–Cu, Al–Pd–Mn, and Zn–Mg–*Ln*. Several other ternary systems yield metastable icosahedral quasi-crystals.

Some stable ternary intermetallic phases have been found that are quasi-periodic in two dimensions and periodic in the third. These are from the systems Al–Ni–Co, Al–Cu–Co, Al–Mn–Pd). They contain decagonally packed groups of atoms (local tenfold rotational symmetry). It should be noted that there are also known metastable quasi-crystals with local eightfold rotational symmetry (octagonal) and twelvefold rotational symmetry (dodecagonal), as well. The dodecahedron is also one of the five Platonic solids.

Several crystalline ternary intermetallic compounds are currently used in engineering applications. The ternary phase Ni_2MnIn (Pearson symbol $cF16$) is illustrated in Figure 2.27a. Each atom is located on the site of a cesium chloride cubic lattice. The unit cell consists of a face-centered cubic arrangement of nickel atoms with one additional nickel atom located in the center of the unit cell, while a manganese or indium atom is in the center of each octant of the cell with eightfold

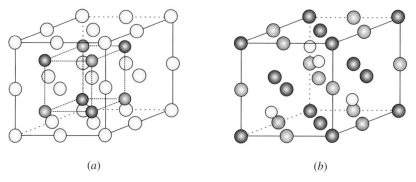

(a)   (b)

Figure 2.27 (a) The crystal structure of the Heusler alloys. (b) The crystal structure of the half-Heulser alloys. The *A* atoms are the very lightly shaded circles.

nickel coordination. Other compounds crystallizing with this structure include Ni_2MnGa, Ni_2MnAl, Cu_2MnAl, Co_2MnSi, Co_2MnGe, Co_2MnSn, and Fe_2VAl. These compounds are known as *Heusler alloy*, named after the German mining engineer and chemist Friedrich "Fritz" Heusler (1866–1947) at the University of Bonn, who studied the magnetic properties of Mn_2CuAl and Mn_2CuSn in the early 1900s (Heusler, 1903).

Heusler alloys have a rich variety of applications, owing to some of their unique properties. Some of these phases are "half-metallic" ferromagnets, exhibiting semiconductor properties for the majority-spin electrons and normal metallic behavior for the minority-spin electrons. Therefore, the conduction electrons are completely polarized. The Ni_2MnGa phase is used as a magnetic shape memory alloy, and single crystals of Cu_2MnAl are used to produce monochromatic beams of polarized neutrons.

If half of the nickel atoms in Figure 2.27a are removed, the "half-Heusler" *ABC* structure of Figure 2.27b is obtained. In the half-Heusler structure, each atom still resides on a cesium chloride lattice site. The rock-salt component (*B* and *C*) remains intact, but the *A* atoms form a zinc-blende lattice with *B* and *C*. Examples of compounds with this structure include MnNiSb, AuMgSn, BiMgNi, and RhSnTi.

Some intermetallics containing rare-earth elements are under consideration as magnetic refrigerants. The advantages of magnetic cooling over gas-compression technology include the removal of environmentally hazardous coolants (i.e., chlorofluorocarbons) and energy-consuming compressors. In the magnetic cooling process, a strong magnetic field is applied to the refrigerant, aligning the spins of its unpaired electrons. That provides for a paramagnetic–ferromagnetic phase transition (a magnetic entropy reduction) upon cooling, which causes the refrigerant to warm up. Upon removal of the field, the spins randomize and the refrigerant cools back down. This is termed the *magnetocaloric effect* (MCE), and is normally from 0.5°C to 2°C per Tesla change in the magnetic field. Other types of magnetic ordering (ferromagnetic, antiferromagnetic, and spin glass) absorb energy internally when the spins are aligned parallel by the applied field, thus reducing the MCE (Gschneider et al., 2000).

As of the present, gadolinium and its alloys have been the most studied materials for this application, since the $4f$ orbitals provide for a comparatively large magnetic entropy change. Recently, a *giant magnetocaloric* effect (a $3°-4°C$ per Tesla change) was reported in the intermetallic compounds $Gd_5(Si_xGe_{1-x})_4$, where $x \leq 0.5$ (Pecharsky and Gschneider, 1997a, 1997b). In $Gd_5(Si_{1.8}Ge_{2.2})$, the magnetocaloric effect has been associated with a field-induced first-order structural transition (Morellon et al., 1998). The structure of $Gd_5(Si_xGe_{1-x})_4$ can be described as a monoclinic distortion of the orthorhombic Gd_5Si_4 phase (Sm_5Ge_4 structure type), in which the Gd atoms occupy five independent fourfold sites, and Si and Ge occupy four independent fourfold sites, in a random manner.

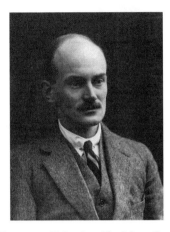

William Hume-Rothery (Courtesy of Mrs. Jennifer Moss. Reproduced with permission.)

WILLIAM HUME-ROTHERY

(1899–1968) was left deaf from a bout of cerebrospinal meningitis in 1917, which ended his pursuit of a military career at the Royal Military Academy. He went on to obtain a B.S. degree in chemistry from Oxford in 1922 and subsequently earned a Ph.D. in metallurgy under Sir Harold Carpenter from the Royal School of Mines in London in 1925. Hume-Rothery returned to the inorganic chemistry department at Oxford to carry out research on intermetallic compounds, bordering between metallurgy and chemistry. In 1956, he became the first chair of the newly founded Oxford metallurgy department, which evolved into the materials science department. Hume-Rothery was keen to use his chemistry training to bring a theoretical basis to the field of metallurgy. He wrote the now famous textbooks *The Structure of Metals and Alloys* (1936) and *Atomic Theory for Students of Metallurgy* (1946). The well-known "Hume-Rothery rules" established that intermetallic compounds result from differences in the atomic radii and electronegativities of alloy constituents, as well as certain valence electron count per atom ratios. Hume-Rothery was elected a Fellow of the Royal Society in 1937.

(*Source*: "William Hume-Rothery: His Life and Science," by D. G. Pettifor in P. E. A. Turchi, R. D. Shull, editors, *The Science of Alloys for the 21st Century: A Hume-Rothery Symposium Celebration*, The Minerals, Metals and Materials Society, Warrendale, PA, 2000.)

2.5   STRUCTURAL DISTURBANCES

Inhomogeneous structural disturbances, of course, can only be understood by comparison with a reference standard, or *ideal* structure. The types of disturbances we discuss here include defects and bond length/angle distortions. Defects may be intrinsic or extrinsic. Intrinsic defects are the result of thermal activation in an otherwise perfect crystal, where there is no reaction between the substance and the environment or other substance. By contrast, extrinsic structural defects may be defined as those introduced in a substance through reaction with an external agent, which may be another substance, or the environment (e.g., irradiation or a mechanical force). If another substance, the agent may or may not be native to the lattice.

Defects can be further classified into *point defects* and *extended defects*. Unassociated point defects are associated with a single atomic site and are thus zero-dimensional. These include vacancies, interstitials, and impurities, which can be intrinsic or extrinsic in nature. Extended defects are multidimensional in space and include dislocations and stacking faults. These tend to be metastable, resulting from materials processing. The mechanical properties of solids are intimately related to the presence and dynamics of extended defects. A discussion of extended defects is deferred until Chapter 9. For now, only point defects are covered. Their importance in influencing the physical and chemical properties of materials cannot be overemphasized.

2.5.1   Intrinsic Point Defects

The concept of a zero-dimensional *intrinsic point defect* was first introduced in 1926 by the Russian physicist Jacov Il'ich Frenkel (1894–1952), who postulated the existence of *vacancies*, or unoccupied lattice sites, in alkali halide crystals (Frenkel, 1926). Vacancies are predominant in ionic solids when the anions and cations are similar in size, and in metals when there is very little room to accommodate interstitial atoms, as in closed-packed structures. The *interstitial* is the second type of point defect. Interstitial sites are the small voids between lattice sites. These are more likely to be occupied by small atoms, or if there is a pronounced polarization to the lattice. In this way, there is little disruption to the structure. Another type of intrinsic point defect is the *antisite* atom (an atom residing on the wrong sublattice).

In ionic crystals, there is the requirement that charge neutrality be maintained *within* the crystal. Hence, if a cation vacates its lattice site and ends up on the crystal surface, an anion must also vacate its sublattice and end up on the surface. The individual vacancies need not be located near one another in the crystal. Interestingly, as pointed out by Seitz (Seitz, 1940), the occurrence of cation–anion vacancy pairs was first postulated by Frenkel on the grounds of excessively high activation energy for diffusion by interchange on the ideal alkali halide lattice. Nevertheless, the cation–anion vacancy pair is called the *Schottky defect*, after Walter Hans Schottky (1886–1976), who studied the statistical thermodynamics of point-defect formation (Schottky and Wagner, 1930). The *Frenkel defect*, on the other hand, is the term applied when a displaced cation in an ionic solid ends up in an interstitial site, rather than at the surface. Note that, in this case, charge neutrality in the crystal is still maintained, since the extra positive charge (the cation interstitial) is balanced by the negatively charged vacancy. The interstitial and vacancy-defect pair constitute the Frenkel defect.

A finite equilibrium concentration of intrinsic point defects can be found in any crystalline material because a small number of defects is thermodynamically favored. This can be seen by considering the *configurational entropy*, or the number of possible ways in which n defects can be distributed among N lattice sites. The number of ways, Ω, is given by combinatorics as:

$$\Omega = N!/[n!(N-n)!] \tag{2.13}$$

The configurational entropy is thus:

$$\Delta S = k \ln N!/[n!(N-n)!] \tag{2.14}$$

As discussed earlier, in ionic solids intrinsic point defects are really defect *pairs* due to the charge neutrality requirement. We need to account for this in the configurational entropy term. For example, for the interstitial-vacancy (Frenkel) defect pair, Eq. 2.14 would be

$$\Delta S = k \ln \left\{ N!/[n_i!(N-n_i)!] \cdot N!/[n_v!(N-n_v)!] \right\} \tag{2.15}$$

We can use the equlity $n_v = n_i$ to obtain:

$$\Delta S = k\{2 \ln N! - 2 \ln [n!(N-n)!]\} \tag{2.16}$$

Using Stirling's approximation:

$$\Delta S = 2k\{N \ln N - (N - n)] \cdot \ln(N - n) - n \ln n\} \qquad (2.17)$$

Finally, we can insert this term into an expression for the Gibbs energy to obtain:

$$G = G^0 + n\Delta g - 2k\{N \ln N - (N - n)] \cdot \ln(N - n) - n \ln n\} \qquad (2.18)$$

In Eq. 2.18, $n\Delta g$ is the free-energy change necessary to create n defects, and G^0 is the Gibbs energy of a perfect crystal. Equation 2.18 is a minimum with respect to n when $(\partial G/\partial n)_{T,P} = 0$. If we approximate $N - n \approx N$, the equilibrium concentration is found to be

$$n/N = \exp(-\Delta g/2kT) \qquad (2.19)$$

It is important to study point defects because they mediate mass-transport properties, as is discussed in Chapter 5. Although the influence of point defects on physical and chemical properties is well founded, relatively little has been reported concerning their affect on the mechanical behavior of solids. A mechanical force applied over a macroscopic area induces the displacement of a large numbers of atoms in a cooperative motion, mediated by dislocations and other extended defects. Recently, however, variations in the vacancy concentrations of a series of binary intermetallics have been associated with the differing hardening behavior of those phases. For example, the hardening of FeAl (CsCl structure) is extremely sensitive to thermal history, while that of the isostructural NiAl, is not. Interestingly, this has been attributed to a high vacancy concentration in FeAl, which is indicative of structural instability (Chang, 2000).

2.5.2 Extrinsic Point Defects

Nonequilibrium concentrations of point defects can be introduced by materials processing (e.g., rapid quenching or irradiation treatment), in which case they are classified as *extrinsic*. Extrinsic defects can also be introduced chemically. Often times, *nonstoichiometry* results from extrinsic point defects, and its extent may be measured by the defect concentration. Many transition-metal compounds are nonstoichiometric because the transition metal is present in more than one oxidation state. For example, some of the metal ions can be oxidized to a higher valence state. This requires either the introduction of cation vacancies or the creation of anion interstitials in order to maintain charge neutrality. The possibility for mixed-valency is not a prerequisite for nonstoichiometry, however. In the alkali halides, extra alkali metal atoms can diffused into the lattice, giving $M_{1+\delta}X(\delta \ll 1)$. The extra alkali metal atoms ionize and force an equal number of anions to vacate their lattice sites. As a result, there is an equal number of cations and anions residing on the surface of the crystal, with the freed electrons being bound to the a nion vacancies. Interestingly, the localized electrons (known as *F*-centers) have available to them discrete quantized energy levels, which fall within the visible region of the electromagnetic spectrum. The nonstoichiometric crystals thus become colored.

Another type of extrinsic point defect is the *impurity*, which may substitute for an atom in the lattice or reside in an interstitial site. The impurity need not be of the same valency as the host, but overall charge neutrality must be maintained in the crystal. This type of extrinsic point defect can greatly influence electrical transport properties (electrical conductivity). If the impurities or dopants are electrically active, they can vary the charge-carrier concentration and, hence, the electrical conductivity of the host. This topic is discussed in ample detail later but, for now, a brief explanation is appropriate. In a semiconductor, the magnitude of the energy gap separating the filled states (valence band) from the empty states (conduction band) is small enough that electrons may be thermally excited from the top of the valence band to the bottom of the conduction band. In an intrinsic semiconductor, the number of electrons that have been thermally excited to the conduction band equals the number of "holes" left behind in the valence band. Both the electrons and holes are charge carriers that contribute to electrical conduction. The concentration of either one of these carrier types can be increased by the addition of electrically active *dopants* (intentionally added impurities), to give an extrinsic semiconductor.

Dopants that donate electrons are termed *donors* or *n*-type dopants (e.g., phosphorus atoms in silicon), since the negatively charged electrons become the majority carrier. By comparison, impurities that accept electrons (boron atoms in silicon), creating holes, are termed *acceptors* or *p*-type dopants, since the positively charged holes become the majority carrier. In silicon, and other semiconductors, the energy levels associated with the dopants are located *within* the band gap. However, *n*-type dopant levels are very close to the conduction band edge, while *p*-type dopant levels are very close to the valence-band edge. Such dopants are called *shallow impurities*. Thermally excitation of charge carriers from these shallow impurities to one or other of the bands of the host is possible. Behavior of this type was discussed early on in semiconductor physics (see, for example, Wilson, 1931).

Vacancies can also make possible the low-temperature manipulation of a solid's structure. For example, aliovalent ion exchange (e.g., Ca^{2+} for Na^+) has been used to create extrinsic vacancies in some transition metal oxides at relatively low temperatures (McIntyre et al., 1998; Lalena et al., 1999). These vacancies were then used to intercalate new species into the lattice, again at relatively low temperatures. Intercalation refers to the insertion of an extrinsic species into a host without a major rearrangement of the crystal structure. If the new species is a strong reducing agent, mixed valency of the transition metal may be introduced into the host.

Walter Schottky (Courtesy of Siemens Archives—Munich © Siemens. Reproduced with permission.)

WALTER HANS SCHOTTKY

(1886–1976) received his doctorate in physics from the Humboldt University in Berlin in 1912, where he studied under Max Planck. Although his thesis was on the special theory of relativity, Schottky spent his life working in the area of semiconductor physics. He alternated between industrial and academic positions in Germany for several years. He was with Siemens AG until 1919 and the University of Wurzburg from 1920 to 1923. From 1923 to 1927, Schottky was Professor of Theoretical Physics at the University of Rostock. He rejoined Siemens in 1927, where he finished out his career. Schottky's inventions include: the ribbon microphone, the superheterodyne radio receiver, and the tetrode vacuum tube. In 1929, he published *Thermodynamik*, a book on the thermo-dynamics of solids. Schottky and Wagner studied the statistical thermodynamics of point defect formation. The cation/anion vacancy pair in ionic solids is named the Schottky defect. In 1938, he produced a barrier layer theory to explain the rectifying behavior of metal-semiconductor contacts. Metal-semiconductor diodes are now called Schottky barrier diodes.

2.5.3 Structural Distortions

The types of structural distortions observed in solids include bond-length altera-tions and bond-angle bending. Localized structural distortions occur around point defects such as vacancies and substitutional impurities due to their differing mass and force constants. Such bond distortions occur when they can lower the overall energy of the system. For example, we have seen how Pauling's third rule predicts that edge or face-sharing tetrahedra will be distorted due to increased repulsion between the neighboring cations. Octahedral rotation and tilting, commonly observed in the perovskites and related phases (e.g., brownmillerites), are similar low-energy distortion modes driven primarily by ion–ion repulsion (electrostatic) forces. Tilting occurs when the A-site cation is too small for its 12-coordinate site in the perovskite structure. The distortion effectively lowers the coordination number from 12 to 8. Goldschmidt recognized many years ago that octahedral tilting in perovskites results from a need to optimize the coordination around the A-site cation (Goldschmidt, 1926). He quantified the goodness-of-fit of the A cation with a tolerance factor, t:

$$t = (R_A + R_O)/[2^{1/2}(R_B + R_O)] \qquad (2.20)$$

where R_A, R_B, and R_O are the ionic radii of the A and B cations and oxygen, respectively. Using the radii by Shannon and Prewitt, perovskites are found for values ranging from 0.8 to 1.1 (Shannon and Prewitt, 1969).

More recently, Glazer developed a very influential notation for describing the 23 different tilt systems that have been observed (Glazer, 1972). The description is based on the rotations of the octahedra about each of the three Cartesian axes. The direction of the rotation about any one axis, relative to the rotations about the other two axes, is specified by a letter with a superscript that indicates whether the rotations in adjacent layers, which are coupled to the other layers via the vertex-sharing octahedra, are in the same (+) or opposite (−) directions. For example, $a^+a^+a^+$ signifies an equivalent rotation about the three axes in each of the adjacent layers. Similarly, in the $a^0a^0c^+$ tilt system, the octahedra are rotated only about the axes parallel to the z-axis, with the same direction in both layers.

Woodward has shown that tilt systems in which all of the A-site cations remain crystallographically equivalent are strongly favored when there is a single ion on the A site (Woodward, 1997). Of these systems, the lowest energy configuration depends on the competition between ionic bonding (A–O) and covalent (A–O σ, B–O σ, B–O π) bonding. The orthorhombic GdFeO$_3$ structure ($a^+b^-b^-$) is the one most frequently found when the Goldschmidt tolerance factor is smaller than 0.975, while the rhombohedral structure ($a^-a^-a^-$) is most commonly observed with $0.975 \le t \le 1.02$, and highly charged A cations. Undistorted cubic structures ($a^0a^0a^0$) are only found with oversized A cations and/or B–O π bonding. Tilt systems with nonequivalent A-site environments are favored when more than one type of A cation (with different sizes and/or bonding preferences) are present, with the ratio of large to small cations dictating the most stable tilt system.

Vertex-sharing polyhedra can also exhibit asymmetric M–X–M linkages (cation displacements), which arise from electronic driving forces, rather than electrostatic forces. For example, the *first-order* Jahn–Teller theorem states that nonlinear symmetrical geometries, which have degenerate electronic states, are unstable with respect to distortion. That is, such systems can lower their electronic energy by becoming less symmetrical and removing the degeneracy.

The first-order Jahn–Teller effect is important in *complexes* of transition-metal cations that contain nonuniformly filled degenerate orbitals, *if* the mechanism is not quenched by spin-orbit (Russell–Saunders) coupling. Thus, the Jahn–Teller effect can be expected with octahedrally coordinated d^9 and high spin d^4 cations and tetrahedrally coordinated d^1 and d^6 cations. The low spin state is not observed in tetrahedral geometry because of the small crystal-field splitting. Also, spin–orbit coupling is usually the dominant effect in T states so that the Jahn–Teller effect is not observed with tetrahedrally coordinated d^3, d^4, d^8, and d^9 ions.

In solids, the observance of Jahn–Teller distortions normally requires that the d electrons be localized. For example, the FeO_6 octahedra in the cubic perovskite $SrFeO_3$ might be expected to exhibit Jahn–Teller distortions. The oxide ion can serve as a good pi donor, so an Fe^{4+} cation octahedrally coordinated by six oxygens should have a high spin d^4 ($t_{2g}^3 e_g^1$) electron configuration. However, $SrFeO_3$ is metallic. The e_g d electrons are delocalized in a Bloch functions and not localized Jahn–Teller distortion is observed. However, a so-called *band Jahn–Teller effect* has been confirmed as the cause of the structural distortions observed in some cases, such as the intermetallic phase Ni_2MnGa, a *Heusler* alloy (Brown et al., 1999).

An increase in the extent of valence d electron localization is expected for smaller principal quantum numbers and as one moves to the right in a period because of a contraction in the size of the d orbitals. For example, with compounds of the late $3d$ metals, a mixture of $4s$ bands and more-or-less localized $3d$ atomic orbitals may coexist, in which case it becomes possible for cubic crystal fields to split the degenerate d orbitals and give rise to a localized Jahn–Teller distortion (e.g., a single octahedra), or *small polaron*, in physics terminology. High concentrations of Jahn–Teller ions, where the polyhedra share structural elements, are subject to a *cooperative Jahn–Teller effect*, which can cause distortion to a lower crystalline symmetry.

The case of *nearly* degenerate states is treated by a second-order correction in perturbation theory and gives rise to the *second-order*, or *pseudo*-Jahn–Teller effect (Öpik and Pryce, 1957). This theorem predicts that, in systems with a small HOMO–LUMO gap (say, <4 eV), the near-degeneracy can be removed, and the HOMO stabilized by distortion to a lower symmetry structure that brings about electron occupancy of the LUMO (Pearson, 1969). One important example is the out-of-center displacement frequently found with transition metals with low d electron counts, particularly d^0 transition-metal ions such as Ti^{4+}, V^{5+}, and Mo^{6+}. The distortion has been found to increase with increasing formal charge and decrease with increasing size of the cation (Kunz and Brown, 1995). The second-order Jahn–Teller effect becomes insignificant for high d electron counts and, for perovskites, d^n metals with $n \geq 1$ are calculated to be symmetric (Wheeler et al., 1986). Out-of-center distortion plays an important role in ferroelectric behavior (e.g., $BaTi^{IV}O_3$), in which the absence of a center of symmetry is required.

Another force that can result in distorted coordination polyhedra is the inert (lone) pair effect. The inert-pair effect refers to the reluctance of the heavy posttransition elements from groups 13–15 to exhibit the highest possible oxidation state, by retaining their pair of valence s electrons. The "lone" pair of electrons on these elements can be stereochemically active and take the place of an anion in the coordination sphere of a cation, or "squeeze" between the anions and the metal causing distortion of the polyhedra. This is because a nonbonding pair of electrons occupies a larger volume of space than a bonding pair, the volume being about the size of an oxide or fluoride anion (Hyde and Andersson, 1989).

2.5.4 Bond-Valence Sum Calculations

As just enumerated, distorted polyhedra can be expected in solids under many circumstances. The *expected* metal–ligand bond distances can be determined by what are known as *bond-valence sum* (BVS) calculations. Bond valences are defined empirically, using formal charges and experimental bond lengths. Although we know that formal charges are not an indication of the true charge on an ion, we can assume the bond lengths in two substances will be very similar if each contains the metal in similar environments and with the same formal charge. In a perfect nonstrained crystal, the formal charge on a cation or anion must equal the sum of the bond valences for all the bonds that it forms:

$$\sum_j v_{ij} = V_i \qquad (2.21)$$

The most commonly adopted empirical expression for the variation of bond length with valence is

$$v_{ij} = \exp[(R_{ij} - d_{ij})/b] \qquad (2.22)$$

where b is taken to be a universal constant equal to 0.37 Å. O'Keeffe and Brese determined bond-valence parameters, R_{ij}, for a very large number of bonds (O'Keefe and Brese, 1991), many of which are listed in Table 2.8.

If the BVS rule is not satisfied (i.e., when the bond valence sums are not very near to the formal charges for the ions), this may indicate metastability. In $LaNiO_{2.5}$, for example, BVS calculations give a lanthanum valence of $+2.63$ and valences of $+2.20$ and $+2.13$ for the octahedral and square-planar nickel cations, respectively. Although the Ni^{2+} cation prefers square-planar coordination, this oxide readily takes up oxygen upon heating in undergoing a structural transition to the perovskite $LaNiO_3$, where the bond-valence sums are $+3.05$ and $+3.01$ for lanthanum and nickel, respectively (Alonso et al., 1997). The following worked example illustrates how to use Eqs. 2.21 and 2.22 with Table 2.8.

Example 2.2 Two sets of crystallographic data are given below for the bond lengths between Ti^{4+} and O^{2-} in an oxide where the formal charge on titanium is $+4$. Each titanium ion is octahedrally coordinated by oxygen and the bond lengths given are those for the two axial, Ti–O(2), and four equatorial, Ti–O(1), distances. Use BVS calculations to predict which data are the most plausible.

X-Ray Diffraction Data	Neutron Diffraction Data
Ti–O(1) = 1.918	Ti–O(1) = 1.969
Ti–O(2) = 1.848	Ti–O(2) = 1.924

TABLE 2.8 Bond Valence Parameters for Some Halides, Nitrides, Oxides, Phosphides, and Sulfides

Cation	F	Cl	Br	I	N	O	P	S
Ag^I	1.80	2.09	2.22	2.38	1.85	1.805	2.22	2.15
Al^{III}	1.545	2.03	2.20	2.41	1.79	1.651	2.24	2.13
As^{III}	1.70	2.16	2.32	2.54	1.93	1.789	2.34	2.54
As^V	1.62	2.14				1.767		
Au^{III}	1.81	2.17				1.833		
Ba^{II}	2.19	2.69	2.88	3.13	2.47	2.29	2.88	2.77
Be^{II}	1.28	1.76	1.90	2.10	1.50	1.381	1.95	1.83
Bi^{III}	1.99	2.48	2.62	2.84	2.24	2.09	2.63	2.55
Bi^V	1.97	2.44				2.06		
C^{IV}	1.32	1.76	1.90	2.12	1.47	1.39	1.89	1.82
Ca^{II}	1.842	2.37	2.49	2.72	2.14	1.967	2.55	2.45
Cd^{II}	1.811	2.23	2.35	2.57	1.96	1.904	2.34	2.29
Ce^{III}	2.036	2.52	2.69	2.92	2.34	2.151	2.70	2.62
Ce^{IV}	1.995	2.41				2.028		
Co^{II}	1.64	2.01	2.18	2.37	1.84	1.692	2.21	2.06
Co^{III}	1.62	2.05				1.70		
Cr^{II}	1.67	2.09	2.26	2.45		1.73		
Cr^{III}	1.64	2.08			1.85	1.724	2.27	2.18
Cr^{VI}	1.74	2.12				1.794		
Cs^I	2.33	2.79	2.95	3.18	2.53	2.42	2.93	2.89
Cu^I	1.6	1.85	1.99	2.16	1.61	1.593	1.97	1.86
Cu^{II}	1.6	2.00				1.679		
Fe^{II}	1.65	2.06	2.26	2.47	1.86	1.734	2.27	2.16
Fe^{III}	1.67	2.09				1.759		
Ga^{III}	1.62	2.07	2.24	2.45	1.84	1.730	2.26	2.17
Gd^{III}	1.62	2.07	2.60	2.82	2.22	2.065	2.61	2.53
Ge^{IV}	1.66	2.14	2.30	2.50	1.88	1.748	2.32	2.22
Hf^{IV}	1.85	2.30	2.47	2.68	2.09	1.923	2.48	2.39
Hg^I	1.81	2.28	2.40	2.59	2.02	1.90	2.42	2.32
Hg^{II}	1.90	2.25				1.93		
Ho^{III}	1.908	2.401	2.55	2.77	2.18	2.023	2.56	2.48
In^{III}	1.79	2.28	2.41	2.63	2.03	1.902	2.43	2.36
La^{III}	2.057	2.545	2.72	2.93	2.34	2.172	2.73	2.64
Mn^{II}	1.66	2.14	2.26	2.49	1.87	1.760	2.24	2.20
Mn^{IV}	1.71	2.13				1.753		
Mn^{VII}	1.72	2.17				1.79		
Mo^{VI}	1.81	2.28	2.43	2.64	2.04	1.907	2.44	2.35
Nb^V	1.87	2.27	2.45	2.68	2.06	1.911	2.66	2.37
Ni^{II}	1.599	2.02	2.16	2.34	1.75	1.654	2.17	2.04
Os^{IV}	1.72	2.19				1.811		
Pb^{II}	2.03	2.53	2.64	2.78	2.22	2.112	2.64	2.55
Pb^{IV}	1.94	2.43				2.042		
Pd^{II}	1.74	2.05	2.19	2.38	1.81	1.792	2.22	2.10
Pt^{II}	1.68	2.05	2.18	2.37	1.77	1.768	2.19	2.08
Pt^{IV}	1.759	2.17				1.879		

TABLE 2.8   (*Continued*)

Cation	F	Cl	Br	I	N	O	P	S
Re^{VII}	1.86	2.23				1.97		
Sb^{III}	1.90	2.35	2.50	2.72	2.12	1.973	2.52	2.45
Sb^{V}	1.80	2.30				1.942		
Sc^{III}	1.76	2.23	2.38	2.59	1.98	1.849	2.40	2.32
Se^{IV}	1.73	2.22	2.33	2.54	1.93	1.811	2.34	2.25
Si^{IV}	1.58	2.03	2.20	2.41	1.77	1.624	2.23	2.13
Sn^{II}	1.926	2.36	2.55	2.76	2.14	1.984	2.45	2.45
Sn^{IV}	1.84	2.28				1.905		
Ta^{V}	1.88	2.30	2.45	2.66	2.01	1.920	2.47	2.39
Ti^{III}	1.723	2.17	2.32	2.54	1.93	1.791	2.36	2.24
Ti^{IV}	1.76	2.19				1.815		
V^{III}	1.702	2.19	2.30	2.51	1.86	1.743	2.31	2.23
V^{IV}	1.70	2.16				1.784		
V^{V}	1.71	2.16				1.803		
W^{VI}	1.83	2.27				1.921		
Y^{III}	1.904	2.40	2.55	2.77	2.17	2.014	2.57	2.48
Yb^{III}	1.875	2.371	2.51	2.72	2.12	1.985	2.53	2.43
Zn^{II}	1.62	2.01	2.15	2.36	1.77	1.704	2.15	2.09
Zr^{IV}	1.854	2.33	2.48	2.69	2.11	1.937	2.52	2.41

Source: After Brese, N. E.; O'Keeffe, M. 1991, *Acta. Cryst.*, *B47*, 192. Copyright © International Union of Crystallography. Reproduced with permission.

Solution   From Table 2.8, the bond-valence parameter, R_{ij}, for the Ti^{IV}–O bond is 1.815. Substituting the values for the various parameters in Eqs. 2.21 and 2.22 yields the valence on titanium calculated from each data set:

$$V_i(1) = \Sigma\{\exp[(R_{ij} - d_{ij})/b]\}_{axial} + \Sigma\{\exp[(R_{ij} - d_{ij})/b]\}_{equatorial}$$
$$= 2 \times \{\exp[(1.815 - 1.848)/0.37]\} + 4 \times \{\exp[(1.815 - 1.918)/0.37]\}$$
$$= 4.86$$
$$V_i(2) = \Sigma\{\exp[(R_{ij} - d_{ij})/b]\}_{axial} + \Sigma\{\exp[(R_{ij} - d_{ij})/b]\}_{equatorial}$$
$$= 2 \times \{\exp[(1.815 - 1.924)/0.37]\} + 4 \times \{\exp[(1.815 - 1.969)/0.37]\}$$
$$= 4.13$$

The X-ray diffraction data indicate overbonding on the titanium cations, or that the shortened Ti–O bonds are under compressive stress. The neutron diffraction data, on the other hand, give a titanium valence closer to the formal charge of 4+. The structure from which these bond distances were taken, presumably, should be the least strained.

2.6 STRUCTURE CONTROL AND SYNTHETIC STRATEGIES

In certain respects, synthetic reactions involving nonmolecular inorganic solids are not so different from synthetic reactions in other fields of chemistry. Specific phases, often with crystal structures close to that of the desired products, are chosen as the starting materials. Structural transformations are then carried out that modify the parent phases in some preconceived way.

We can very broadly distinguish between two types of transformations. The first is the formation of a solid solution, in which solute atoms are inserted into vacancies (lattice sites or interstitial sites) or substitute for a solvent atom on a particular sublattice. Many types of synthetic processes can result in this type of transforma tion, including ion-exchange reactions, intercalation reactions, alloy solidification processes, and the high-temperature ceramic method. Of these, ion exchange, intercalation, and other so-called *soft chemical* (*chimie douce*) *reactions* produce no structural changes except, perhaps, an expansion or contraction of the lattice to accommodate the new species. They are said to be under *topotactic* (or topochemical) control.

The second type of transformation includes all those reactions resulting in significant structural changes from those of the starting materials. Crystallization from aqueous solutions, gels, glasses, and melts may produce this type of transformation, but so do many solid-state reactions, such as combustion synthesis [also called self-propagating high-temperature synthesis SHS] and the ceramic method. Given only the chemical composition of a product phase, and no information about the starting materials or synthetic route, might one be able to predict its crystal structure (or properties) unambiguously? The answer, unfortunately, is that this is generally not possible. Again, this is not so different from other fields of chemistry—what is the structure of $C_{15}H_{12}N_6O_4$?

Although reaction schemes for carrying out specific structural transformations are commonplace in organic chemistry, solid-state chemists have only a limited number of predictive synthetic strategies at their disposal. The most commonly employed preparative technique for nonmetallic materials is the high-temperature ceramic route ($T \geq 500°C$). These reactions are kinetically slow even at very high temperatures, since they are controlled by the solid-state diffusion of ions and atoms through both the reactants and products. Hence, they are under thermo-dynamic control; the products formed are simply the ones that are the most thermodynamically stable. Even so, it is often possible to utilize simple concepts from basic inorganic chemistry [e.g., radius ratio rules, crystal field stabilization energy (CFSE)] to establish coordination preferences. This helps to minimize our extent of reliance on trial-and-error synthesis.

Consider the brownmillerite ($A_2B_2O_5$) structure examined earlier. This structure consists of alternating layers of vertex-sharing BO_3 octahedra and BO_2 tetrahedra. It is immediately obvious to the synthetic strategist that one has the opportunity to place cations with specific coordination preferences exclusively in one type of layer to give $A_2(BO_3)(B'O_2)$, as was previously discussed. For example, in Ca_2MnGaO_5,

Mn^{3+} exclusively occupies the BO_3 layers. In this case, that would have been predicted by a consideration of the CFSE.

It is well known from inorganic chemistry that a cubic crystal field splits the d orbitals on a transition-metal cation into a doubly degenerate set (e_g) and a triply degenerate set (t_{2g}). The splitting is about $10Dq$ for the octahedral case and $4.45Dq$ for the tetrahedral case. The quantity Dq is the crystal-field splitting energy per electron, which is proportional to the anion charge divided by the metal–anion bond length. With the O^{2-} anion, the crystal-field splitting is small for both the octahedral and tetrahedral cases and is, in fact, less than the repulsion felt between electrons in doubly occupied orbitals. Hence, the Mn^{3+} cation has a *high-spin* d^4 electron configuration in either an octahedral ($t_{2g}^3 e_g^1$) or tetrahedral ($e_g^2 t_{2g}^2$) field of O^{2-} anions. The *total* CFSE for the ground-state Mn^{3+} ion is established by summing the energies for each of the four d electrons. In an octahedral field, the t_{2g} orbitals are approximately $-4Dq$ each and the e_g orbitals are $+6Dq$ each. The CFSE is thus $-6Dq$. In the tetrahedral case, the t_{2g} orbitals are approximately $+1.78Dq$ each and the e_g orbitals are $-2.67Dq$ each (note the reversal in sign). The CFSE is thus $-1.78Dq$. Because the CFSE is greater in the octahedral rather than tetrahedral case, the Mn^{3+} cation in Ca_2MnGaO_5 would have been predicted to prefer the BO_3 layer.

In some cases, secondary forces may exert a sizable influence on the coordination environment as well. For example, we saw earlier how the second-order Jahn–Teller effect frequently manifests itself as a displacement of transition metals from the center of an octahedron. The phenomenon is only observed for metals with low d electron counts. This could be used advantageously in synthetic strategies where the goal is to selectively place cations in specific sites within a structure.

In $Na_2La_2Ti_{3-x}Ru_xO_{10}$, which is prepared by the ceramic route, for example, the transition-metal cations in the two outer octahedral layers of the triple-layer slab are electrostatically displaced toward the $(NaO)^-$ layers. However, the distribution of the nearly equal-sized Ti^{4+} and Ru^{4+} cations within the octahedra is nonrandom. It is found that the d^4 Ru^{4+} cations seem to have a definite, though not exclusive, preference for the middle, undistorted octahedral layer (Figure 2.28), while mostly d^0 Ti^{4+} cations are found in the out-of-center distorted outer octahedral layers (Lalena et al., 2000). For $x = 1.0, 0.75, 0.50$, and 0.25, the percentage of inner-layer octahedral sites occupied by Ru^{4+} are, respectively, 63.7, 45.4, 34.2, and 16. The remaining Ru^{4+} cations are distributed among the octahedral sites of the two outer layers. Thus, the percentage of inner-layer octahedral sites occupied by Ru^{4+} is about linearly proportional to x. Furthermore, compositions with $x > 1.0$ cannot be prepared, even though the isostructural triple-layer oxide $Sr_4Ru_3O_{10}$ is known. In $Sr_4Ru_3O_{10}$, there are no electrostatic or electronic driving forces for out-of-center distortion in the outer octahedral layers (Cao et al., 1997). Hence, although the origin for the cation displacement in the outer octahedral layers of $Na_2La_2Ti_{3-x}Ru_xO_{10}$ is electrostatic in nature, the electronic second-order Jahn–Teller effect places an upper limit on the ruthenium doping level that can be tolerated in these sites.

In custom designing materials with tailored properties, it is often necessary to synthesize metastable phases that will be kinetically stable under the temperature

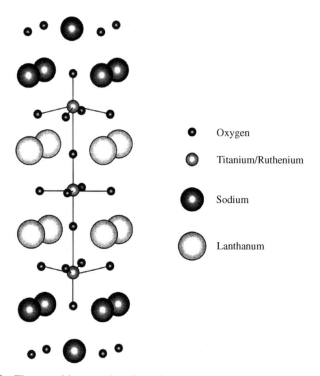

●	Oxygen
◉	Titanium/Ruthenium
⬤	Sodium
◯	Lanthanum

Figure 2.28   The transition-metal cations in the outer octahedral layers of Na_2La_2-$Ti_{3-x}Ru_xO_{10}$ are displaced from the centers of their octahedra. Only transition metal cations with a low d electron count (i.e., Ti^{4+}, d^0) can readily accommodate this disortion. Hence, the Ru^{4+} cations (d^4) are found mostly in the central undistorted layer.

and conditions of use. These phases are obtainable only through kinetic (chemical) control. In many cases, kinetic control has been achieved via the soft chemical low-temperature (e.g., electrochemical synthesis, sol-gel method) and/or topochemical routes (e.g., intercalation, ion exchange, dehydration reactions), since these routes use mild synthetic conditions. It should be noted that not all soft chemical routes are topochemical. A reaction is said to be under topochemical control only if it follows the pathway of minimum atomic or molecular movement (Elizabé et al., 1997). Accordingly, topochemical reactions are those in which the lattice of the solid product shows one or a small number of crystallographically equivalent definite orientations relative to the lattice of the initial crystal, and if the reaction has proceeded throughout the entire volume of the initial crystal (Günter, 1972).

Under the proper circumstances, most soft chemical processes can allow ready manipulation of the ionic component of many nonmolecular materials. Indeed, the solid-state literature has seen an enormous growth in the number of reports, wherein the utility of these synthetic strategies are exploited. However, these methods typically leave the covalent framework of ionocovalent structures intact. It would be extremely desirable to exercise kinetic control over the *entire* structure of a solid, thereby maximizing the ability to tune its properties.

Probably the most challenging task is predicting, a priori, the *extended* structure—not only the coordination preferences of all the atoms or ions, but the polyhedra connectivity as well. In this area, Tulsky and Long have taken a major step forward by formalizing the application of *dimensional reduction* to treat the formation of ternary phases from binary solids (Tulsky and Long, 2001):

$$MX_x + nA_aX \rightarrow A_{na}MX_{x+n} \qquad (2.23)$$

In this reaction, the first reactant on the left is the binary "parent" phase, and the product to the right of the arrow is a ternary "child." The second reactant on the left is an ionic reagent termed the *dimensional reduction agent*, in which X is a halide, oxide, or chalcogenide. The added X anions terminate M–X–M bridges in the parent, yielding a framework in the child that retains the metal coordination geometry and polyhedra connectivity mode (i.e., corner-, edge-, or face-sharing), but that has a *lower* connectivity (e.g., fewer polyhedra corners shared). The A cations serve to balance charge, ideally, without influencing the M–X covalent framework.

The utility of this approach lies in its ability to predict the framework connectivity of the child compound. Connectivity is defined as the average number of distinct M–X–M linkages around the metal centers or, alternatively, the average number of bonds that must be broken to liberate a discrete polyhedron. For frameworks with only one kind of metal and one kind of anion, the connectivity of the parent is given by

$$CN_M \times (CN_X - 1) \qquad (2.24)$$

where CN_M is the coordination number of the metal, and CN_X denotes the number of metal atoms coordinated to X. The framework connectivity of an $A_{na}MX_{x+n}$ child compound is predicted to be

$$2 \times [CN_M - (x + n)] \qquad (2.25)$$

The reaction schemes of Eqs. 2.25 and 2.26 illustrate how, as Eq. 2.25 predicts, connectivity is progressively lowered by reaction of a binary parent with increasing amounts of a dimensional reduction agent:

$$VF_3 \xrightarrow{A_aF} A_a VF_4 \xrightarrow{A_aF} A_{2a}VF_5 \xrightarrow{A_aF} A_{3a}VF_6 \qquad (2.26)$$

$$\underset{\text{6-connected}}{} \quad \underset{\text{4-connected}}{} \quad \underset{\text{2-connected}}{\phantom{A_{2a}VF_5}} \quad \underset{\text{0-connected}}{\phantom{A_{3a}VF_6}}$$

where A = alkali metal or Tl; $a = 1$

$$MnCl_2 \xrightarrow{2/3NaCl} Na_2Mn_3Cl_8 \xrightarrow{1/3NaCl} NaMnCl_3 \xrightarrow{NaCl} Na_2MnCl_4 \qquad (2.27)$$

$$\underset{\text{12-connected}}{} \quad \underset{\text{8-connected}}{} \quad \underset{\text{6-connected}}{} \quad \underset{\text{4-connected}}{}$$

Thus, solid-state reaction between VF_3, with corner-sharing VF_6 octahedra in three dimensions, and A_aF in a 1:3 molar ratio yields A_3VF_6, containing discrete octahedral anions. Similarly, reaction of $MnCl_2$, which consists of $MnCl_6$ octahedra sharing six edges, with NaCl in a 1 : 1 molar ratio yields a 2D framework with octahedra sharing only three edges.

Solids with octahedral, tetrahedral, square-planar, and linear metal coordination geometries, including many different types of polyhedra connectivity modes, are amenable to dimensional reduction. Tulsky and Long compiled an enormous database of over 300 different allowed combinations of M and X and over 10,000 combinations of A, M, and X. The formalism may be extendable to quaternary phases as well. However, frameworks containing anion–anion linkages, anions other than halides, oxide, or chalcogenides, nonstoichiometric phases, and mixed-valence compounds were excluded from their initial study.

Other limitations of the approach include its inability to predict which of many possible isomers, each satisfying the criteria set by Eq. 2.25, will result. Furthermore, Eq. 2.25 is only valid for compounds derived from parents with one- and two-coordinate anions, albeit this is the majority of cases. Finally, the effect (most probably size and polarizing power) of the A counter ion may be significant enough in some cases to destabilize the covalent MX_x framework, diminishing the success of dimensional reduction. Despite these limitations, Tulsky and Long's formalism is an exciting development in the evolution of what has come to be known as *rational materials design*. It is hoped that dimensional reduction can be coupled with other synthetic strategies to provide access to a wider range of tailored materials.

REFERENCES

Alberti, G.; Costantino, U.; Allulli, S.; Tomassini, J. 1978, *J. Inorg. Nucl. Chem.*, *40*, 1113.

Alonso, J. A.; Martínez-Lope, M. J.; García-Muñoz, J. L; Fernández-Díaz, M. T. 1997, *J. Phys.: Condens. Mater.*, *9*, 6417.

Bertaut, E. F. 1952, *J. Phys. Radium*, *13*, 499.

Bethe, H. 1929, *Ann. Phys.*, *3*, 133.

Bloch, F. 1928, *Z. Physik.*, *52*, 555.

Born, M.; Landé, A. 1918, *Sitzungsber Deut. Akad. Wiss. Berlin*, *45*, 1048.

Brese, N. E.; O'Keefe, M. 1991, *Acta Cryst.*, *B47*, 192.

Brown, P. J.; Bargawi, A. Y.; Crangle, J.; Neumann, K-U.; Ziebeck, K. R. A. 1999, *J. Phys. Condens. Mater.*, *11(24)*, 4715.

Cao, G.; McCall, S.; Crow, J. E. 1997, *Phys. Rev. B*, *56*, R5740.

Car, R.; Parrinello, M. 1985, *Phys. Rev. Lett.*, *55*, 2471.

Chang, Y. A. 2000, In: Turchi, P. E. A., Shull, R. D., editors, *The Science of Alloys for the 21st Century: A Hume-Rothery Symposium Celebration*, The Minerals, Metals, and Materials Society, Warrendale, PA.

Corbett, J. D.; Zhao, J.-T. 1995, *Inorg. Chem.*, *34*, 378.

Cotton, F. A. 1963, *Chemical Applications of Group Theory*, 3rd ed., John Wiley & Sons, New York. Reprinted in 1971 and 1990.

Delmas, C.; Fouassier, C.; Hagenmuller, P. 1975, *J. Solid State Chem.*, *13*, 165.

Elizabé, L.; Kariuke, B. M.; Harris, K. D. M.; Tremayne, M.; Epple, M.; Thomas, J. M. 1997, *J. Phys. Chem. B*, *101*, 8827.

Evarestov, R. A.; Leko, A. V.; Veryazov, V. A. 1997, *Phys. Status Solidi B*, *203(1)*, R3.

Ewald, P. P. 1913, *Phys. Z.*, *14*, 465.

Ewald, P. P. 1921, *Ann. Phys. (Leipzig)*, *64*, 253.

Fabris, S.; Paxton, A. T.; Finnis, M. W. 2002, *Acta Mater.*, *50(20)*, 5171.

Fajans, K. 1915a, *Umschau*, *19*, 661.

Fajans, K. 1915b, *Z. Physik*, *16*, 456.

Fajans, K. 1924, *Z. Physik*, *25*, 596.

Férey, G. 2001, *Chem. Mater.*, *13*, 3084.

Finkelstein, B. N.; Horowitz, G. E. 1928, *Z. Physik*, *48*, 118.

Fisher, C. A. J.; Islam, M. S. 1999, *Solid State Ionics*, *118*, 355.

Frank, F. C.; Kasper, J. S. 1958a, *Acta Cryst.*, *11*, 184.

Frank, F. C.; Kasper, J. S. 1958b, *Acta Cryst.*, *12*, 483.

Frenkel, J. 1926, *Z. Phys.*, *35*, 652.

Friauf, J. B. 1927, *Phys. Rev.*, *29*, 34.

Fouassier, C.; Delmas, C.; Hagenmuller, P. 1975, *Mat. Res. Bull.*, *10*, 443.

Gardner, J. S.; Gaulin, B. D.; Berlinsky, A. J.; Waldron, P.; Dunsiger, S. R.; Raju, N. P.; Greedan, J. E. 2001, *Phys. Rev. B*, *64*, 224416.

Glazer, A. M. 1972, *Acta Cryst.*, *B28*, 3384.

Goldschmidt, V. M., Ulrich, F., Barth, T. 1925, *Norske Vidensk. Akad. Skrifter I Mat. Naturv. Kl.*, *5*, 1.

Goldschmidt, V. M., Barth, T., Holmsen, P., Lunde, G., Zachariasen, W. 1926a, *Norske Vidensk. Akad. Skrifter I Mat. Naturv. Kl.*, *1*, 1.

Goldschmidt, V. M., Barth, T., Lunde, G., Zachariasen, W. 1926b, *Norske Vidensk. Akad. Skrifter I Mat. Naturv. Kl.*, *2*, 1.

Goldschmidt, V. M., Barth, T., Lunde, G., Zachariasen, W. 1926c, *Norske Vidensk. Akad. Skrifter I Mat. Naturv. Kl.*, *8*, 1.

Goldschmidt, V. M. 1926, *Naturwissenchaften*, *14*, 477.

Goodenough, J. B. 1965, *Bull. Soc. Chim. Fr.*, 1200.

Goodenough, J. B. 1966., *Magnetism and the Chemical Bond*, 2nd print., Interscience Publishers, New York.

Goodenough, J. B. 1967, *Mat. Res. Bull.*, *2*, 37.

Goodenough, J. B.; Ruiz-Diaz, J. E.; Zhen, Y. S. 1990, *Solid State Ionics*, *44*, 21.

Greenblatt, M. 1996, *Acc. Chem. Res.*, *29*, 219.

Greenwood, N. N.; Earnshaw, A. 1997, *Chemistry of the Elements*, 2nd ed., Butterworth-Heinemann, Oxford.

Gschneidner, K. A.; Pecharsky, V. K.; Pecharsky, A. O. 2000, In: Turchi, P. E. A.; Shull, R. D.; editors, *The Science of Alloys for the 21st Century: A Hume-Rothery Symposium Celebration*, The Minerals, Metals, and Materials Society, Warrendale, PA.

Günter, J. R. 1972, *J. Solid State Chem.*, *5*, 354, and references therein.

Hansen, W. C.; Brownmiller, L. T.; Bogue, R. H. 1928, *J. Am. Chem. Soc.*, *50*, 396.

Holleman, A. F.; Wiberg, E. 2001, *Inorganic Chemistry*, Academic Press, San Diego.

Holt, A.; Norby, T.; Glenne, R. 1999, *Ionics*, *5*, 434.

Heitler, W.; London, F. 1927, *Z. Phys.*, *44*, 455.

Heusler, F. 1903, *Verh. Dtsch. Ges.*, *5*, 219.

Hume-Rothery, W. 1930, *Philos. Mag.*, *9*, 65.

Hume-Rothery, W. 1931, *Philos. Mag.*, *11*, 649.

Hume-Rothery, W. 1934, *Philos. Trans. R. Soc.*, *A244*, 1.

Hund, F. 1926, *Z. Physik*, *36*, 657.

Hund, F. 1927a, *Z. Physik*, *40*, 742.

Hund, F. 1927b, *Z. Physik*, *43*, 805.

Hyde, B. G.; Andersson, S. 1989, *Inorganic Crystal Structures*, John Wiley & Sons, New York.

Imada, M.; Fujimori, A.; Tohura, Y. 1998, *Rev. Mod. Phys.*, *70*, 1039.

Johnson, Q. C.; Templeton, D. H. 1962, *J. Chem. Phys.*, *34*, 2004.

Kohn, W.; Sham, L. J. 1965, *Phys. Rev.*, *140*, 1133.

Kuang, X.; Jing, X.; Loong, C.-K.; Lachowski, E. E.; Skakle, J. M. S.; West, A. R. 2002, *Chem. Mater.*, *14*, 4359.

Kunz, M.; Brown, D. 1995, *J. Solid State Chem.*, *115*, 395.

Lalena, J. N.; Falster, A. U.; Simmons, W. B.; Carpenter, E. E.; Wiggins, J.; Hariharan, S.; Wiley, J. B. 1998, *Inorg. Chem.*, *37*, 4484.

Lalena, J. N.; Falster, A. U.; Simmons, W. B.; Carpenter, E. E.; Wiggins, J.; Hariharan, S.; Wiley, J. B. 2000, *Chem. Mater.*, *12*, 2418.

Lennard-Jones, J. E. 1929, *Trans. Faraday Soc.*, *25*, 668.

Madelung, E. 1918, *Z. Phys.*, *19*, 524.

Mackay, A. L.; Kramer, P. 1985, *Nature*, *316*, 17.

McIntyre, R. A.; Falster, A. U.; Li, S.; Simmons, W. B.; O'Connor, C. J.; Wiley, J. B. 1998, *J. Am. Chem. Soc.*, *120*, 217.

Mendiboure, A.; Delmas, C.; Hagenmuller, P. 1985, *J. Solid State Chem.*, *57*, 323.

Moller, K.; Bein, T. 1998, *Chem. Mater*, *10*, 2950.

Morellon, P. A.; Algarabel, P. A.; Ibarra, M. R.; Blasco, J.; García-Landa, B.; Arnold, Z.; Albertini, F. 1998, *Phys. Rev. B*, *58*, R14721.

Mott, N. F. 1958, *Nuovo Cimento*, *7* (*10*) *Suppl.*, 312.

Mulliken, R. S. 1926, *Proc. Natl. Acad. Sci. USA*, *112*, 144.

Mulliken, R. S. 1928a, *Phys. Rev.*, *32*, 186.

Mulliken, R. S. 1928b, *Phys. Rev.*, *32*, 761.

Mulliken, R. S. 1932, *Phys. Rev.*, *40*, 55.

Norby, T. 2001, *J. Mater. Chem.*, *11*, 11.

Ohno, K.; Esfarjani, K.; Kawazoe, Y. 1999, *Computational Materials Science From Ab Initio to Monte Carlo Methods*, Springer-Verlag, Berlin.

Olazcuaga, R.; Reau, J-M.; Devalette, M.; Le Flem, G.; Hagenmuller, P. 1975, *J. Solid State Chem.*, *13*, 275.

Öpik, U.; Pryce, M. L. H. 1957, *Proc. Roy. Soc. London*, *A238*, 425.

Parr, R. G.; Donnelly, M.; Levy, M.; Palke, W. E. 1978, *J. Chem. Phys.*, *68*, 3801.

Parthé, E. 2000, In: Turchi, P. E. A., Shull, R. D., editors, *The Science of Alloys for the 21st Century: A Hume-Rothery Symposium Celebration*; The Minerals, Metals, and Materials Society, Warrendale, PA.

Pauling, L. 1927, *Proc. R. Soc. London*, *A114*, 181.

Pauling, L. 1928, *Z. Kristallogr.*, *67*, 377.

Pauling, L. 1929, *J. Am. Chem. Soc.*, *51*, 1010.

Pauling, L. 1931, *J. Am. Chem. Soc.*, *53*, 1367.

Pauling, L. 1932, *J. Am. Chem. Soc.*, *54*, 988.

Pearson, R. G. 1969, *J. Am. Chem. Soc.*, *91*, 4947.

Pearson, W. B. 1967, *A Handbook of Lattice Spacings and Structures of Metals and Alloys*, Vol. 2, Pergamon Press, Oxford.

Pearson, W. B. 1964, *Acta Crsyt.*, *17*, 1.

Pecharsky, V. K.; Gschneidner, K. A. 1997a, *Phys. Rev. Lett.*, *78*, 4494.

Pecharsky, V. K.; Gschneidner, K. A. 1997b, *Appl. Phys. Lett.*, *70*, 3299.

Porterfield, W. W. 1993, *Inorganic Chemistry A Unified Approach*, 2nd ed., Academic Press, San Diego, CA.

Prewitt, C. T.; Shannon, R. D.; Rogers, D. B. 1971, *Inorg. Chem.*, *10*, 719.

Rao, C. N. R.; Raveau, B. 1998. *Transition Metal Oxides: Structure, Properties, and Synthesis of Ceramic Oxides* 2nd ed., Wiley-VCH, New York.

Reynolds, P. W.; Hume-Rothery, W. 1937, *J. Inst. Metals*, *60*, 365.

Rogers, D. B.; Shannon, R. D.; Prewitt, C. T. 1971, *Inorg. Chem.*, *10*, 723.

Saitta, A. M. 1997, Doctoral Thesis, Scuola Internazionale Superiore DiStudi Avanzati–International School for Advanced Studies, Trieste, Italy.

Sanderson, R. T. 1951, *Science*, *114*, 670.

Seitz, F. 1940, *The Modern Theory of Solids*, McGraw-Hill Book Company, New York.

Seshadri, R.; Felser, C.; Thieme, K.; Tremel, W. 1998, *Chem. Mater.*, *10*, 2189.

Shannon, R. D.; Prewitt, C. T. 1969, *Acta Cryst.*, *B25*, 925.

Shannon, R. D.; Rogers, D. B.; Prewitt, C. T. 1971, *Inorg. Chem.*, *10*, 713.

Shechtman, D.; Blech, I.; Gratias, D.; Cahn, J. W. 1984, *Phys. Rev. Lett.*, *53*, 1951.

Slater, J. C. 1931, *Phys. Rev.*, *37*, 481.

Thackery, M. M.; David, W. I. F.; Goodenough, J. B. 1982, *Mater. Res. Bull.*, *17*, 785.

Thiên-Nga, T.; Paxton, A. T. 1998, *Phys. Rev. B*, *58*, 13233.

Thomson, J. J. 1897, *Philos. Mag.*, *44*, 293.

Thomson, J. J. 1904, *Electricity and Matter*, Yale University Press, New Haven, CT.

Turnbull, D.; Aziz, M. J. 2000, In: Turchi, P. E. A., Shull, R. D., editors, *The Science of Alloys for the 21st Century: A Hume-Rothery Symposium Celebration*, The Minerals, Metals, and Materials Society, Warrendale, PA.

Tulsky, E. G.; Long, J. R. 2001, *Chem. Mater.*, *13*, 1149.

Vidyasagar, K.; Reller, A.; Gopalakrishnan, J.; Rao, C. N. R. 1985, *J. Chem. Soc. Chem. Commun.*, 7.

Wade, K. 1976, *Adv. Inorg. Chem. Radiochem.*, *18*, 1.

West, A. R. 1984, *Solid State Chemistry and its Applications*, John Wiley & Sons, Chichester, UK.

Wheeler, R. A.; Whangbo, M.-H.; Hughbanks, T.; Hoffmann, R.; Burdett, J. K.; Albright, T. A. 1986, *J. Am. Chem. Soc.*, *108*, 2222.

Wilson, A. H. 1931, *Proc. Roy. Soc.*, *133*, 458.

Woodward, P. M. 1997, *Acta Cryst.*, *B53*, 44.

Yamanak, S.; Horie, H.; Nakano, H.; Ishikawa, M. 1995, *Fullerene Sci. Tech.*, *3*, 21.

Zintl, E. 1939, *Angew Chem.*, *51*, 1.

CHAPTER 3

The Electronic Level, I: An Overview of Band Theory

In the present chapter, we lay the foundation for understanding band structure in crystalline solids. The presumption is, of course, that the electronic structure is more appropriately described from the standpoint of an MO (or Bloch) type approach, rather than the Heitler–London valence-bond approach. We start with the many-body Schrödinger equation and the independent-electron (Hartree–Fock) approximation. This is followed with Bloch's theorem for wave functions in a periodic potential and an introduction to reciprocal space. Two general approaches are then described for solving the extended electronic structure problem: the free-electron model and the LCAO method, both of which rely on the independent-electron approximation. Finally, the consequences of the independent-electron approximation are examined. In Chapter 4, we study the tight-binding method in detail. Chapter 5 focuses on electron and atomic dynamics (i.e., transport proper-ties), and the metal–nonmetal transition is discussed in Chapter 6.

3.1 THE MANY-BODY SCHRÖDINGER EQUATION

The Schrödinger equation for a many-electron system is written as:

$$\mathcal{H}\Psi = E\Psi \tag{3.1}$$

where $\mathcal{H}$ is known as the Hamiltonian operator, E is the *eigenvalue* representing the total energy of the system, and Ψ is the *eigenfunction* (or *eigenstate*). For a system containing N electrons and K nuclei, the Hamiltonian operator is expressed, in atomic units, as:

$$-\frac{1}{2}\sum_{\mu=1}^{N}\nabla_\mu^2 - \frac{1}{2}\sum_{n=1}^{K}\nabla_n^2 + \sum_{\mu>\nu}^{N} 1/|r_\mu - r_\nu| - \sum_{n=1}^{K}\sum_{\mu=1}^{N} Z_n/|r_\mu - R_n|$$

$$+ \sum_{n>m}^{K} Z_n Z_m/|R_n - R_m| \tag{3.2}$$

Principles of Inorganic Materials Design By John N. Lalena and David A. Cleary
ISBN 0-471-43418-3 Copyright © 2005 John Wiley & Sons, Inc.

These terms represent, from left to right, electron kinetic energy, nuclear kinetic energy, electron–electron Coulomb repulsion, electron–nuclear Coulomb attraction, and nuclear–nuclear Coulomb repulsion. The eigenfunction of Eq. 3.1 thus depends on the positions of all the electrons, r_μ, and nuclei, R_n. The problem of solving for the eigenfunction can be simplified by separating the degrees of freedom connected with the motion of the nuclei from those of the electrons. The Born–Oppenheimer approximation asserts that because the nuclei are so much heavier than the electrons, their kinetic energy (motion) can be neglected (Born and Oppenheimer, 1927). The second term in Eq. 3.2 can then be set to zero, and the last term treated as a *parameter*.

We now write the Schrödinger equation for the electrons in the system as:

$$\left[-\frac{1}{2} \sum_{\mu=1}^{N} \nabla_\mu^2 + \sum_{\mu>\nu}^{N} 1/|r_\mu - r_\nu| - \sum_{n=1}^{K} \sum_{\mu=1}^{N} Z_n/|r_\mu - R_n| \right] \Psi = E\Psi \tag{3.3}$$

where Ψ is the many-body (*N*-electron) eigenfunction, which is, in general, a wave-like solution, and E is the electronic energy. The total energy may be obtained by adding in the electrostatic energy of the nuclei.

The simplest approach to approximating a solution to Eq. 3.3 is to assume that all the electrons move independently of one another. That is, we imagine they mutually interact only via an *averaged* potential energy. This is known as the Hartree approximation. It enables us to write the Hamiltonian for the *N*-electron system as a sum of *N* one-electron Hamiltonians, h_μ, and the many-body wave function as a product of *N* one-electron wave functions, ψ_μ (normally written as a single Slater determinant). Note that the one-electron functions may be of the atomic, molecular, or Bloch type, depending on the problem at hand. Finally, the total energy of the *N*-electron system is the sum of the eigenvalues for all of the one-electron eigenvalue equations, ε_μ. The relationships between the *N*-electron quantities and one-electron quantities can be represented mathematically as:

$$\mathcal{H} = \sum_\mu h_\mu \tag{3.4}$$

$$\Psi = 1/\sqrt{N!} \sum_P (-1)^P \prod_\mu P\psi_\mu(\mathbf{x}_\mu) = 1/\sqrt{N!} \det[\psi_\mu(\mathbf{x}_\nu)] \tag{3.5}$$

$$E = \sum_\mu \varepsilon_\mu \tag{3.6}$$

Each one-electron wave function is given by a spin-orbital. A spin-orbital is a function depending on both the spatial coordinates (position), r_μ, and spin coordinates, s_μ, of the electron. It can be denoted as $\psi(\mathbf{x})$, where $\mathbf{x}_\mu = (r_\mu, s_\mu)$. Using a single Slater determinant as a trial-ground state wave function for the *N*-electron system, it is found upon application of the variational principle that the one-electron wave functions themselves must satisfy the following equation, which we provide here without derivation:

$$\left[-\tfrac{1}{2}\nabla_\mu^2 + \left[\sum_{i=1}^{N}\sum_{s_\nu}\int \psi_i^*(\nu)\psi_i(\nu)/\left|r_\mu - r_\nu\right|dr_\nu\right] - \sum_{n}Z_n/\left|r_\mu - R_n\right|\right]\psi_j(\mu) = \epsilon_j\psi_j(\mu)$$

(3.7)

In Eq. 3.7, ψ_j is the jth one-electron orbital accommodating the μth electron with spatial coordinate r_μ and spin coordinate s_μ. Likewise, the νth electron resides in the ith orbital denoted by ψ_i. The Lagrange multiplier, ε_j, guarantees that the solutions to the equation form an orthonormal set and is the expectation value for the equation. Hence, it is the quantized one-electron orbital energy. The second term in brackets on the left-hand side of Eq. 3.7 is called the direct term (also known as the Coulomb term, or Hartree term). It represents the potential felt by the μth electron resulting from a charge distribution caused by all the electrons, including the μth electron itself (the so-called self-interaction). Equation 3.7 was derived by the British mathematician Douglas Rayner Hartree (1897–1958) in 1928 (Hartree, 1928).

Equation 3.7 has the form of a self-consistency problem. The solution to the equation is $\psi_j(\mu)$, but the exact form of the equation is determined by $\psi_j(\mu)$ itself. It must be solved by an iterative procedure [called the *self-consistent-field* (SCF) approach], in which convergence is taken to occur at the step where both ε_j and $\psi_j(\mu)$ do not differ appreciably from the prior step.

A little further discussion on electron spin is in order now. Spin-orbitals are necessary because an electron possess a spin quantum number ($+\tfrac{1}{2}$ or $-\tfrac{1}{2}$). In the absence of a magnetic field, the "up" and "down" spins are energetically degenerate, or indistinguishable. The Pauli exclusion principle says that electronic wave functions must be antisymmetric (they change sign) under the interchange of any two electrons. Because of this antisymmetry, two electrons are not allowed to occupy the same quantum state.

The Russian mathematician Vladimir Alexandrovich Fock (1898–1974) extended the Hartree equation by considering this antisymmetry (Fock, 1930), resulting in what is called the Fock, or Hartree–Fock, equation (for a derivation, see Thijssen, 1999). With the Hartree–Fock approximation, the following term is added to Eq. 3.7:

$$-\sum_{i=1}^{N}\sum_{s_\nu}\int \psi_i^*(\nu)\psi_j(\nu)/\left|r_\mu - r_\nu\right|dr_\nu\psi_i(\mu)$$

(3.8)

This new term is called the *exchange term*, or Fock term. It is similar to the Hartee term, but with the spin-orbitals interchanged. The minus sign results from the antisymmetry of the wave function with respect to two-particle exchange. The exchange term lowers the Coulomb interaction between electrons, since every electron is surrounded by an "exchange hole" in which other electrons with parallel spin are hardly found. Exchange introduces correlation by keeping electrons with parallel spin apart. However, the term "correlation" is normally reserved for electron correlation apart from that due to exchange, that is, Coulomb repulsion between antiparallel spin electrons, usually called *dynamic correlation*. These dynamic correlation effects are neglected in the Hartree–Fock theory.

The Hartree–Fock equation can be written as a generalized eigenvalue problem:

$$\mathcal{F}\psi_j(\mu) = \varepsilon_j\psi_j(\mu) \tag{3.9}$$

where the Fock operator, $\mathcal{F}$, is expressed as

$$\mathcal{F} = -\tfrac{1}{2}\nabla^2_\mu - \sum_n Z_n/(r_\mu - R_n) + J + K \tag{3.10}$$

in which J and K are, respectively, the Hartree (Coulomb) and Fock (exchange) terms. The Fock operator is sometimes referred to as the *effective one-electron Hamiltonian*. Like Eq. 3.7, Eq. 3.9 must also be solved by the SCF approach.

The Hartree–Fock approximation is much more tractable with systems containing a small number of atoms than with crystals. With crystals, it is necessary to include corrections such as the random-phase approximation (RPA). However, a discussion of these is beyond the scope of this book. But wait! Alas, not all of our problems are solved yet. The issue of how to computationally handle the near infinite number of one-electron wave functions in a solid has not been addressed. We now take this up.

3.2   BLOCH'S THEOREM

The potential felt by an electron is given by the sum of the last three terms on the right-hand side of Eq. 3.10. This potential has a perfect periodicity in a crystalline solid, and may be represented as

$$V(r + R) = V(r) \tag{3.11}$$

where r is an atomic position and R is a lattice vector. If Eq. 3.11 is true, a symmetry operation, such as translation, T_R, that transforms a crystal into itself, does not change the Hamiltonian, $\mathcal{H}$ (or the Fock operator, $\mathcal{F}$), that is, the Hamiltonian is translational invariant. Hence, T_R commutes with $\mathcal{H}$ (and $\mathcal{F}$): $T_R\mathcal{H} - \mathcal{H}T_R = 0$. Therefore, a one-electron wave function satisfying the Schrödinger equation (or Hartree–Fock equation) is also an eigenstate of T_R, or:

$$T_R\psi_j(r) = \psi_j(r + R) = f\psi_j(r) \tag{3.12}$$

where, for clarity, we have emphasized just the spatial coordinate of the electron, r, and dropped the subscript μ for generality.

If we now impose what is known as "periodic" boundary conditions on the entire crystal by joining its faces, it restricts the number of wavelengths that fit into the crystal to an integer, which corresponds to running wave solutions of the Schrödinger equation. This method was first introduced in Born and von Kármán's treatment of the surface atomic dynamics in the theory of specific heat (Born and von Kármán, 1912). Essentially, the crystal is taken to be surfaceless. It is easy to envision a 1D chain with its two ends joined together to form a ring, but not so easy to picture the analogous situation for 3D crystals. The eigenvalue satisfying this requirement and Eq. 3.12 is:

$$f = \exp(i\boldsymbol{k} \cdot \boldsymbol{R}) \tag{3.13}$$

where $\boldsymbol{R}$ is a "real-space" lattice vector equal to $n_1\boldsymbol{a}_1 + n_2\boldsymbol{a}_2 + n_3\boldsymbol{a}_3$ (n_1, n_2, and n_3 are integers, and $\boldsymbol{a}_1$, $\boldsymbol{a}_2$, and $\boldsymbol{a}_3$ are primitive translation vectors). The quantity $\boldsymbol{k}$ is the "reciprocal-space" *wave vector* that can only take on the values of 0 or $\pm 2\pi n/L$, where n is an integer and L is the sample dimension. Also, note that i in Eq. 3.13 is the square root of -1. An important theorem, called Bloch's theorem, can be derived from Eqs. 3.12 and 3.13. Bloch's theorem states that, owing to the translational invariance of the Hamiltonian, any one-electron wave function can be represented by a modulated plane wave (Bloch, 1928):

$$\psi_{kj}(\boldsymbol{r}) = \exp(i\boldsymbol{k} \cdot \boldsymbol{r})u_{kj}(\boldsymbol{r}) \tag{3.14}$$

where $\exp(i\boldsymbol{k} \cdot \boldsymbol{r})$ is the plane wave and $u_{kj}(\boldsymbol{r})$ is a function with the periodicity of the real-space Bravais lattice, satisfying the following equation:

$$u_{kj}(\boldsymbol{r}) = u_{kj}(\boldsymbol{r} + \boldsymbol{R}) = \exp(-i\boldsymbol{k} \cdot \boldsymbol{r})\psi_{kj}(\boldsymbol{r}) \tag{3.15}$$

Note that $\exp(i\boldsymbol{k} \cdot \boldsymbol{r})\exp(-i\boldsymbol{k} \cdot \boldsymbol{r})\psi_{kj}(\boldsymbol{r}) = \psi_{kj}(\boldsymbol{r})$. Equation 3.15 implies that the following relation also holds:

$$\psi_{kj}(\boldsymbol{r} + \boldsymbol{R}) = \exp(i\boldsymbol{k} \cdot \boldsymbol{R})\psi_{kj}(\boldsymbol{r}) \tag{3.16}$$

The wave function has the same amplitude at equivalent positions in each unit cell. Thus, the full electronic structure problem is reduced to a consideration of just the number of electrons in the unit cell (or half that number if the electronic orbitals are assumed to be doubly occupied) and applying boundary conditions to the cell as dictated by Bloch's theorem (Eq. 3.14). Each unit cell face has a "partner" face that is found by translating the face over a lattice vector $\boldsymbol{R}$. The solutions to the Schrödinger equation on both faces are equal up to the phase factor $\exp(i\boldsymbol{k} \cdot \boldsymbol{R})$, determining the solutions inside the cell completely.

Since the wave functions are subject to boundary conditions, the energy eigenvalues are quantized. There are N one-electron eigenstates, Ψ_j, with corresponding eigenvalues, ε_j (but there may be sets of energetically degenerate eigenstates for some values of $\boldsymbol{k}$). Each of the unique eigenvalues is termed an *energy level*. Because N is so large in a solid, quasi-continuous energy bands form in the density of states, with an infinitesimal separation between the different energy levels. The highest energy level occupied by electrons in the density of states is termed the *Fermi level*, and the eigenvalue energy at this level is termed the *Fermi energy* after the Italian physicist Enrico Fermi (1901–1954). All levels below the Fermi level are occupied with electrons and all levels above it are empty. The eigenvalues change smoothly as k changes, forming curves as they move between one k-points in ε–k space. The dispersions of the various curves are displayed in *band structure*, or *band dispersion*, diagrams.

Felix Bloch (Courtesy Stanford University Archives, Green Library. © Stanford University. Reproduced with permission.)

FELIX BLOCH

(1905–1983) received his Ph.D. in physics in 1928 from the University of Leipzig, where he studied under Schrödinger and, later, Heisenberg. In his Ph.D. thesis, Bloch introduced the theorem that allows us to write the electron wave function in a periodic lattice as a modulated plane wave. This concept is fundamental to electronic structure calculations on crystalline solids. From 1928 to 1929, Bloch was an assistant to Wolfgang Pauli at the Swiss Federal Institute of Technology in Zurich. Bloch is no less famous for his contributions as an experimentalist. Upon Hitler's ascent to power, he left Europe and accepted a position at Stanford University, where he carried out research on nuclear magnetic measurements. In 1939, he determined the magnetic moment of the neutron with an accuracy of about 1%. Bloch and E. M. Purcell were awarded the 1952 Nobel Prize in Physics for the first successful nuclear magnetic resonance experiments, performed independently in 1946. Purcell measured the nuclear magnetic absorption, while Bloch's method utilized nuclear induction, the well-known Bloch equations describing the behavior of a nuclear spin in a magnetic field under the influence of radio-frequency pulses. Nuclear magnetic resonance has since been applied to the investigation of molecular structure and medical diagnostics, via magnetic resonance imaging. Bloch was elected to the U.S. National Academy of Sciences in 1948.

(*Primary source*: R. Hofstadter *Biographical Memoirs of the U.S. National Academy of Sciences*, Vol. 64, **1994**, pp. 34–71.)

3.3 RECIPROCAL SPACE

We have just stated that a band-structure diagram is a plot of the energies of the various bands in a periodic solid versus the value of the reciprocal-space wave vector k. It is now necessary to discuss the concept of the reciprocal-space lattice and its relation to the real-space lattice, which was introduced in Section 2.1.3. The crystal structure of a solid is ordinarily presented in terms of the real-space lattice comprised of *lattice points*, which have an atom or group of atoms associated with them whose positions can be referred to them. Two real-space lattice points are connected by a primitive translation vector, R:

$$R = u a_1 + v a_2 + w a_3 \tag{3.17}$$

where u, v, w are integers and a_1, a_2, a_3 are called *basis vectors*. The primitive vectors for the cubic lattices, for example, are given in Table 3.1.

It may be recalled from Section 2.1.3 that an alternative description for a crystal structure can be made in terms of sets of *lattice planes*, which intersect the unit cell axes at $u' a_1$, $v' a_2$, and $w' a_3$. The reciprocals of the coefficients are transformed to the smallest three integers having the same ratios, h, k, and l, which are used to denote the plane (hkl). Of course, the lattice planes may or may not coincide with the layers of atoms. Any such set of planes is completely specified by the interplanar spacing, d_{hkl}, and the unit-vector normal to the set, n_{hkl}, since the former is given by the projection of, for example, $u' a_1$ onto n_{hkl}, that is, $d_{hkl} = u' a_1 \cdot n_{hkl}$. The reciprocal-lattice vector is defined as:

$$G_{hkl} = 2\pi n_{hkl} / d_{hkl} \tag{3.18}$$

TABLE 3.1 Primitive Translation Vectors of the Real-Space Cubic Lattices:
$R = u a_1 + v a_2 + w a_3$

Lattice	Primitive Vectors		
Simple cubic (sc)	$a_1 = ax$	$a_2 = ay$	$a_3 = az$
Face-centered cubic (fcc)	$a_1 = a/2(y+z)$	$a_2 = a/2(x+z)$	$a_3 = a/2(x+y)$
Body-centered cubic (bcc)	$a_1 = a/2(-x+y+z)$	$a_2 = a/2(x-y+z)$	$a_3 = a/2(x+y-z)$

The factor 2π is usually omitted from the definition of a reciprocal-lattice vector in crystallography. This is because Bragg's law defines the deviation of a diffracted ray from the direct ray in terms of the half-wavelength of the radiation and the quantity $1/d$, which, in crystallography, is taken as the reciprocal-lattice vector:

$$\sin \theta = n(\lambda/2)(1/d) \qquad (3.19)$$

The factor 2π arises, however, when we make use of the relation $\lambda = 2\pi/k$ to express the periodicity of the incident radiation. Each vector G_{hkl} of the reciprocal lattice corresponds to a set of planes in the real-space lattice with Miller indices (hkl). The coordinates of the endpoint of each vector are h, k, and l. The vectors are perpendicular to the (hkl) planes, and the lengths of the vectors are equal to the reciprocal of the plane spacing. Hence, each crystal structure has associated with it a real-space lattice and a reciprocal-space lattice.

A reciprocal-lattice vector can also be defined in terms of basis vectors:

$$G = hb_1 + kb_2 + lb_3 \qquad (3.20)$$

(Compare Eq. 3.20 to Eq. 3.17.) The basis vectors of the reciprocal lattice (b_1, b_2, b_3) can be obtained from those of the real-space lattice (a_1, a_2, a_3) by a relation given by J. W. Gibbs (Wilson, 1907; Margenau and Murphy, 1956):

$$b_1 = [2\pi/(a_2 \times a_3)]/a_1(a_2 \times a_3) \qquad (3.21)$$
$$b_2 = [2\pi/(a_1 \times a_3)]/a_1(a_2 \times a_3) \qquad (3.22)$$
$$b_3 = [2\pi/(a_1 \times a_2)]/a_1(a_2 \times a_3) \qquad (3.23)$$

In Eqs. 3.21–3.23, the numerators contain cross products and the denominators are scalar triple products equal to the volume of the real-space unit cell.

Taking the primitive translation vectors for one of the real-space cubic lattices from Table 3.1, Eqs. 3.21–3.23 can be used to obtain the primitive translation vectors for the corresponding reciprocal lattice, which are given in Table 3.2. By comparing Tables 3.1 and 3.2, it is seen that the primitive vectors of the reciprocal lattice for the real-space fcc lattice, for example, are the primitive vectors for a bcc lattice. In other words, the fcc real-space lattice has a bcc reciprocal lattice.

We now relabel the basis vectors: $a_1 = a$, $a_2 = b$, $a_3 = c$; $b_1 = a^*$, $b_2 = b^*$, $b_3 = c^*$, and introduce a general vector, k, in reciprocal space:

$$k = k_x a^* + k_y b^* + k_z c^* \qquad (3.24)$$

TABLE 3.2   Primitive Translation Vectors of the Reciprocal Lattices

Real-Space Lattice	Corresponding Reciprocal Lattice Primitive Vectors		
sc	$b_1 = (2\pi/a)x$	$b_2 = (2\pi/a)y$	$b_3 = (2\pi/a)z$
bcc	$b_1 = (2\pi/a)(y + z)$	$b_2 = (2\pi/a)(x + z)$	$b_3 = (2\pi/a)(x + y)$
fcc	$b_1 = (2\pi/a)(-x + y + z)$	$b_2 = (2\pi/a)(x - y + z)$	$b_3 = (2\pi/a)(x + y - z)$

Note that this vector appears in the expression for the electronic wave function (e.g., Eq. 3.14). It is used to define a unit cell in reciprocal space, which may be derived in the following manner. First, reciprocal-lattice vectors G are drawn between a given reciprocal-lattice point (the origin) and all other points. Perpendicular planes are added at the *midpoints* to these G vectors (giving new vectors, k, originating from the origin and terminating on the bisecting planes). The smallest volume enclosed by the polyhedra defined by the intersection of these planes about the central lattice point is termed the *Wigner–Seitz cell*. This is named after Hungarian-born American physicist and Nobel laureate Eugene Paul Wigner (1902–1995) and his first graduate student at Princeton, Frederick Seitz (b. 1911), who studied the symmetry properties of wave functions in crystals by group theoretical methods (Seitz, 1936, Bouckaert et al., 1936).

The Wigner–Seitz cell actually is a space-filling primitive cell that may be constructed to represent a real-space lattice or reciprocal-space lattice. The Wigner–Seitz cell of the reciprocal lattice, however, is conventionally referred to as the *first Brillouin zone* (BZ), after French physicist Léon Brillouin (1889–1969) who had earlier shown that the surfaces of discontinuity in reciprocal-space-form polyhedra (Brillouin, 1930). The first BZ contains all the symmetry of the reciprocal lattice, that is, all of reciprocal space is covered by periodic translation of this unit cell. There are higher-order zones that can be constructed similarly to the first BZ, by drawing vectors to the next nearest neighbors. The result is that the higher-order zones are fragmented pieces separated from each other by the lower zones. Each zone occupies an equal volume of k-space.

The first BZs for the simple cubic (sc), bcc, fcc lattices are shown in Figure 3.1a–1c. The *inner* symmetry elements for each BZ are: the center, Γ; the threefold axis, Λ; the fourfold axis, Δ; and the twofold axis, Σ. The symmetry points on the BZ boundary (faces) (X, M, R, etc.) depend on the type of polyhedron. The reciprocal lattice of a real-space sc lattice is itself a sc lattice. The Wigner–Seitz cell is the cube shown in Figure 3.1a. Thus, the first Brillouin zone for the sc real-space lattice is a cube with the following high-symmetry points:

k-Point Label	Cartesian Coordinates
Γ	$(0, 0, 0)$
X	$(\pi/a, 0, 0)$
M	$(\pi/a, \pi/a, 0)$
R	$(\pi/a, \pi/a, \pi/a)$

The reciprocal lattice of a bcc real-space lattice is an fcc lattice. The Wigner–Seitz cell of the fcc lattice is the rhombic dodecahedron in Figure 3.1b. The volume enclosed by this polyhedron is the first BZ for the bcc real-space lattice. The high-symmetry points are

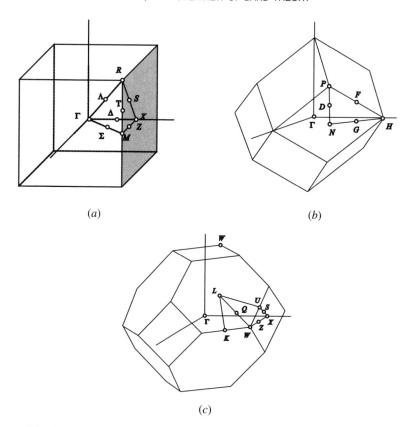

(a)

(b)

(c)

Figure 3.1 (a) The Wigner–Seitz cell of reciprocal space (the first Brillouin zone) for the simple cubic real-space lattice is itself a simple cubic lattice. (b) For the bcc real-space lattice, the first BZ is a rhombododecahedron. (c) For the fcc real-space lattice, the first BZ is a truncated octahedron. (After Bouckaert, Smoluchowski, and Wigner, 1936. Reproduced by permission of the American Physical Society.)

k-Point Label	Cartesian Coordinates
Γ	$(0, 0, 0)$
H	$(2\pi/a, 0, 0)$
N	$(\pi/a, \pi/a, 0)$
P	$(\pi/a, \pi/a, \pi/a)$

The reciprocal lattice for the fcc real-space lattice is a bcc lattice. The Wigner–Seitz cell is a truncated octahedron (Figure 3.1c). The shapes of the BZs for the sc and bcc real-space lattices are completely determined by the condition that each inner vector, k, go over into another by all the symmetry operations. This is not the case for the truncated octahedron. The surface of the Wigner–Seitz cell is only fixed at the truncating planes, not the octahedral planes. Nonetheless, the volume enclosed by the truncated octahedron is taken to be the first BZ for the fcc real-space lattice (Bouckaert et al., 1936). The special high-symmetry points are:

k-Point Label	Cartesian Coordinates
Γ	$(0, 0, 0)$
X	$(2\pi/a, 0, 0)$
W	$(2\pi/a, \pi/a, 0)$
K	$(3\pi/2a, 3\pi/2a, 0)$
L	$(\pi/a, \pi/a, \pi/a)$

3.4 A CHOICE OF BASIS SETS

The one-electron wave function in an extended solid can be represented with different basis sets. We discuss only two types, representing opposite extremes: the plane-wave basis set (free-electron and nearly free-electron models) and the Bloch sum of atomic orbitals basis set (LCAO method). A periodic solid can be considered constructed by the coalescence of these isolated atoms into extended Bloch wave functions. On the other hand, within the free-electron framework, in the limit of an infinitesimal periodic potential ($V = 0$), a Bloch wave function becomes a simple unmodulated plane wave. We focus on the LCAO approach in this book, in fact, devoting all of Chapter 4 to it. However, a brief presentation of the free-electron model is warranted due to its historical significance and for certain insights it provides, particularly with metals. In the early years, the plane-wave expansion method came into favor with physicists and metallurgists, mainly because of the substantial efforts at the time to understand the conductivity of metals.

3.4.1 Plane-Wave Expansion: The Free-Electron Models

The free-electron model originated from the work of Paul Karl Ludwig Drude (1863–1906), who utilized the kinetic theory of gases to treat electrical conduction in metals (Drude, 1900a, 1900b, 1902). In Drude's model, a metal is regarded as a "gas" of free-valence electrons immersed in a sea of metal ions. Because the ions are homogeneously distributed in the solid, their net positive charge is considered uniformly smeared out, forming what is referred to as a *jellium*. In this model, the free electrons experience a constant electrostatic potential everywhere in the metal and all of them have the same average kinetic energy, $3/2\ k_BT$, which corresponds to the $1/2mv^2$ in Newton's second law. Soon afterward, Lorentz treated the free electrons as classic distinguishable particles that obey the Maxwell–Boltzmann distribution laws (Lorentz, 1904–1905).

After the discovery of the Pauli principle, Arnold Sonnerfeld (1868–1951) regarded the free electrons of a metal as a degenerate Fermi gas, with the free electrons subject to Fermi–Direc statistics, thereby transforming the classic Drude–Lorentz model into the realm of quantum theory (Sonnerfeld, 1928). Whereas every electron is considered to have the same average kinetic energy in a classic electron gas, the condition that electron states be solutions of a wave equation subject to boundary conditions naturally gives rise to a spread of quantized energy levels in Sonnerfeld's theory of metals. The allowed electronic states are running-wave solutions to the time-independent Schrödinger equation. The appropriate boundary conditions correspond to an integral number of wavelengths of the running wave along each crystal dimension.

Further development of Sonnerfeld's theory of metals would take us well outside the intended scope of this textbook. The interested reader may refer to any of several books for this (e.g., Seitz, 1940). Rather, we wish to discuss the band approximation based upon the Bloch scheme. In the Bloch scheme, Sonnerfeld's model corresponds to an "empty lattice," in which the electronic Hamiltonian contains only the electron kinetic energy term. The lattice potential is assumed constant, and taken to be zero, without any loss of generality. The solutions of the time-independent Schrödinger equation in this case can be written as simple plane waves, $\psi_k(r) = \exp[ik \cdot r]$. Because the wave function does not change if one adds an arbitrary reciprocal-lattice vector G to the wave vector k, BZ symmetry may be superimposed on the plane waves to reduce the number of wave vectors that must be considered:

$$\psi_k(r) = \psi_{k+G}(r) = \exp[i(k + G) \cdot r] \tag{3.25}$$

Note that the periodic $u_k(r)$ function that appears in Eq. 3.14 is absent, since the periodic potential $V(r)$ is assumed to be infinitesimally small (i.e., in the limit $V(r) \to 0$). In the free-electron model, the conduction electrons are regarded as free to move throughout the crystal unimpeded by the ions they left behind. The energy eigenvalues, in atomic units, are given by

$$\varepsilon = (k + G)^2/2 \tag{3.26}$$

From the form of Eq. 3.26, it is seen that ε is a simple parabolic function of $k[\varepsilon(k) \equiv \varepsilon(k + G)]$. The band structure describes this dependence of $\varepsilon(k)$ on k, and it is an $(n + 1)$-dimensional quantity, where n is the number dimensionality of the crystal. To visualize it, $\varepsilon(k)$ is plotted along particular projectories between high-symmetry points. Hence, for a 1D crystal, the band structure will consist of a single parabola in the free-electron approximation. The parabola shows all the degenerate (positive and negative) values for k. For 3D crystals, a single paraboloid is obtained.

The electronic properties of most main group s- and p-block elements are better described by introducing a periodic potential as a small perturbation. In the context of the present model, this approach is known as the *nearly free-electron* (NFE) *model*. In 1930, Peierls showed that, in the NFE limit, band gaps arise from electron diffraction, a natural consequence of wave propagation in a periodic structure (Peierls, 1930). Brillouin generalized the result and showed that, in three dimensions, the surfaces of discontinuity form polyhedra in reciprocal space—BZs (Brillouin, 1930).

The diffraction condition for electrons in a solid, with periodicity a, is satisfied when the wave vector lies on the bisector plane of a reciprocal lattice vector, that is, at the BZ boundaries. This is given by:

$$k = \pm G/2 = \pm n\pi/a \qquad (3.27)$$

The wave functions with these wave vectors correspond to standing-wave solutions, rather than running-wave solutions, to the Schrödinger equation. There must now be an integral number of half-wavelengths along each dimension of the crystal. The two wave functions $\psi_{-G/2}(r)$ and $\psi_{+G/2}(r)$ at the opposite zone boundaries are degenerate in the free-electron limit. However, it can be shown by perturbation theory that these levels split into nondegenerate levels by the introduction of a small periodic potential (see Elliot, 1998). Perturbation theory gives the energies, in first order, as:

$$\varepsilon_{\pm} = (k + G)^2/2 \pm V_G \qquad (3.28)$$

Hence, ψ_{+} is lower than the free-electron value by V_G, and ψ_{-} is higher than the free-electron value by V_G, thereby opening up a band gap of magnitude $2V_G$, as shown pictorially for a 1D crystal in Figure 3.2.

3.4.2   The Fermi Surface and Phase Stability

In Figure 3.2, the eigenvalues are concentrated in intervals separated by forbidden regions. In three-dimensions, this corresponds to discontinuities in the energy contours at the zone boundaries. Electrons deep in the valence band possess low

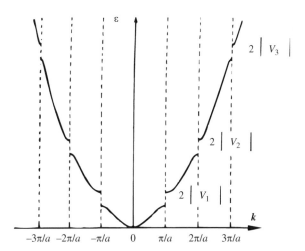

Figure 3.2   The electronic structure of a 1D crystal in the nearly free-electron approximation is a single parabola with energy gaps of magnitude $2V$ occurring at the BZ boundaries. (After Elliot, 1998, *The Physics and Chemistry of Solids*. Copyright © John Wiley & Sons, Inc. Reproduced with permission.)

energies and thus have k vectors that terminate well short of the first BZ boundary. The energy contours for these electrons are thus spheres contained within the first BZ. As their energy increases, electrons have k vectors that gradually lengthen to fill the first BZ. The most important contour to consider is, of course, the Fermi surface, which match the electrons with the Fermi energy. An electron with the Fermi energy has a wave vector (called the Fermi wave vector) that is proportional to the square root of the Fermi energy, which, in turn, is proportional to the electron density (N/V). When the Fermi wave vector does not touch the BZ boundary, but rather is of a length appreciably less than the distance to the first BZ boundary (i.e., when the Fermi wave vector lies well *within* the first BZ), the Fermi surface is spherical. This is the case for the monovalent (one conduction electron) fcc alkali metals (Figure 3.3a).

Recall that electrons are diffracted at a zone boundary into the next zone. This means that k vectors cannot *terminate* on a zone boundary because the associated energy value is forbidden, that is, the first BZ is a polyhedron whose faces satisfy the Laue condition for diffraction in reciprocal space. Actually, when a k vector terminates very near a BZ boundary the Fermi surface topology is perturbed by NFE effects. For k-values just below a face on a zone boundary, the electron energy is lowered so that the Fermi sphere necks outward towards the face. This happens in monovalent fcc copper, where the Fermi surface necks toward the L-point on the first BZ boundary (Figure 3.3b). For k-values just above the zone boundary, the electron energy is increased and the Fermi surface necks down toward the face.

An interesting area still under debate in the field of metallurgy is the consequences of Fermi surface topology on the phase equilibria in alloy systems. The connection between these two seemingly unrelated features started with the work of

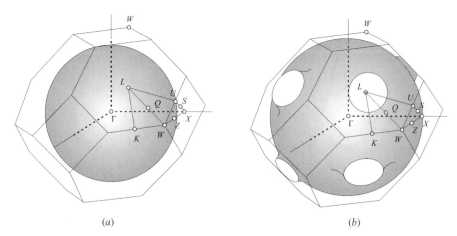

(a)   (b)

Figure 3.3 For the monovalent fcc alkali metals (a) with a low electron density, the Fermi wave vector (the radius of the Fermi sphere) lies well below the first BZ boundary. The Fermi surface is unperturbed. For monovalent fcc copper (b), the increased electron density forces the Fermi wave vector to terminate very near the L-point. The electron energy is lowered and the Fermi sphere necks outward toward that face of the BZ boundary.

William Hume-Rothery, who reported that the critical valence electron-to-atom (e/a) ratios corresponding to maximum solubility in the α phase and for the occurrence of the β and γ intermediate phases in some copper and silver alloys were 1.36, 1.48, and 1.54, respectively (Hume-Rothery et al., 1934). Mott and Jones later showed that the Fermi surface touches the BZ edge in those phases at precisely these same values, which strongly pointed to a connection between BZ touching and phase stability (Mott and Jones, 1936).

The connection follows the line of reasoning just presented. As a polyvalent metal is dissolved in a monovalent metal, the electron density increases, as does the Fermi energy and Fermi wave vector. Eventually, the Fermi sphere touches the BZ boundary and the crystal structure becomes unstable with respect to alternative structures (Raynor, 1947; Pettifor, 2000). Subsequent work has been carried out confirming that the structures of Hume-Rothery's alloys (alloys comprised of the noble metals with elements to the right on the periodic table) do indeed depend only on their (e/a) ratio (Stroud and Ashcroft, 1971; Pettifor and Ward, 1984; Pettifor, 2000). Unfortunately, the importance of the e/a ratio on phase equilibria is much less clear when it does not correspond precisely to BZ touching.

For low-dimensional (1D, 2D) metals, the topology of the Fermi surface is especially significant, as this can lead to a charge density wave state. When a portion of the Fermi surface can be translated by a vector q and superimposed on another portion of the Fermi surface, the Fermi surface is *nested*. Geometrical instabilities can result when large sections of the Fermi surface are nested. For one-dimensional metals, if $q = b/n$, then a distortion, known as the Peierls distortion, leading to a unit cell n times as large as the original one is predicted (Burdett, 1996). If the nesting is complete, the system will exhibit a metal–nonmetal transition after the modulation destroys the entire Fermi surface by opening a band gap (Canadell, 1998). When the nesting is less than complete, the driving force for distortion is reduced, the gap opening being only partial. Charge-density waves and the Peierls distortion are discussed in more detail in Section 6.5.

3.4.3 Bloch Sum Basis Set: The LCAO Method

Plane-wave expansion in a periodic potential requires a very large number of plane waves to achieve convergence. In general, finite plane-wave expansions are inadequate for describing the strong oscillations in the wave functions near the nuclei. To alleviate the problem, other approaches, such as the augmented plane-wave (APW) method, which treats the atomic core and interstitial regions differently, and the pseudopotential method, which neglects the core electrons completely in the calculation scheme, have been introduced. Finally, because the conduction electron kinetic energy is the dominant attribute of metallic systems, metals have been the chief application of the NFE approach to band theory. It is not as appropriate for systems where covalent energy is the dominant attribute, transition metal and rare-earth systems with tightly bound valence electrons (valence d and $4f$ orbitals do not extend as far from the nucleus as valence s and p orbitals), or for describing inner-shell core electrons in systems.

For solids with more localized electrons, the LCAO approach is perhaps more suitable. Here, the starting point is the isolated atoms (for which it is assumed that the electron wave functions are already known). In this respect, the approach is the extreme opposite of the free-electron picture. A periodic solid is constructed by bringing together a large number of isolated atoms in a manner entirely analogous to the way one builds molecules with the LCAO-MO theory. The basic assumption is that overlap between atomic orbitals is small enough that the extra potential experienced by an electron in a solid can be treated as a perturbation to the potential in an atom. The extended (Bloch) wave function is treated as a superposition of "localized orbitals," χ, centered at each atom:

$$\psi_{k\mu}(\mathbf{r}) = N^{-1/2} \sum_{\mathbf{R}} e^{i\mathbf{k}\cdot\mathbf{R}} \chi_{\mu}(\mathbf{r} - \mathbf{R}) \tag{3.29}$$

One defines a Bloch sum (or Bloch orbital) for each atomic orbital in the chemical-point group (or lattice point), and crystal orbitals are then formed by taking linear combinations of the Bloch sums.

First, let us take the simplest possible case, a monatomic solid with a primitive Bravais lattice containing one atomic orbital per lattice point. The crystal orbitals are then equivalent to simple Bloch sums. The wavelength, λ, of a Bloch sum is given by the following relation:

$$|\mathbf{k}| = 2\pi/\lambda \tag{3.30}$$

We can see how the phase of a Bloch sum changes in a periodic lattice by considering a simple 1D lattice of (σ-bonded) p atomic orbitals, with a repeat distance d. Figure 3.4 shows such a chain and the sign combinations of the atomic

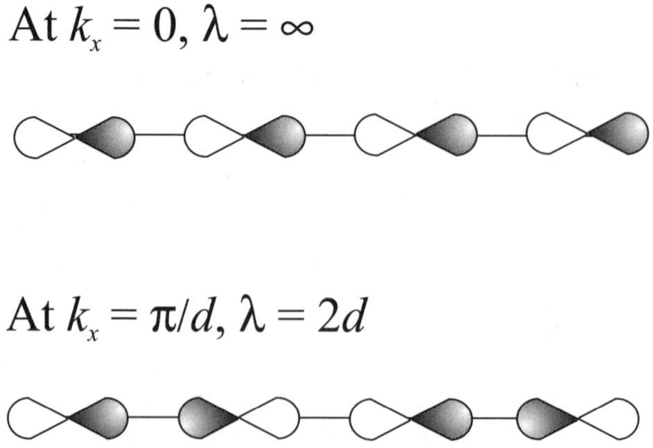

At $k_x = 0$, $\lambda = \infty$

At $k_x = \pi/d$, $\lambda = 2d$

Figure 3.4 A 1D periodic chain of p atomic orbitals. At top is shown the sign combinations corresponding to the $\mathbf{k}$-point Γ ($\mathbf{k} = 0$), where $\lambda = \infty$. At bottom are the sign combinations for the $\mathbf{k}$-point X ($\mathbf{k} = \pi/d$), where $\lambda = 2d$.

orbitals for three special values of k ($= k_x$ in the 1D case). Euler's relation, $e^{\pm i\theta} = \cos\theta \pm i\sin\theta$, can be used to evaluate the term $\exp(ik_x nd)$. At Γ, where $k_x = 0$ $\exp(ik_x nd) = (1)^n = 1$, so there is no phase change from one unit cell (atomic orbital) to the next and, from Eq. 3.30, $\lambda = 8$. At X, $k_x = \pi/d$ and $\exp(ik_x nd) = (-1)^n$, so the phase alternates as n, the signs of the atomic orbitals, and $\lambda = 2\pi/(\pi/d) = 2d$.

For lattices with more than one atom per lattice point, we have to consider combinations of Bloch sums. In general, the LCAO approach requires that we end up with the same number of molecular orbitals (crystal orbitals in solids) as the number of atomic orbitals (Bloch sums in solids) with which we start. Thus, expressing the electron wave functions in a crystalline solid as linear combinations of atomic orbitals (Bloch sums) is really the same approach used in the 1930s by Hund, Mulliken, Hückel and others to construct molecular orbitals for discrete molecules (the LCAO-MO theory).

The orbitals used to construct Bloch sums are usually Slater-type orbitals (STOs) or Gaussian-type orbitals (GTOs) since these types of orbitals well describe the electron density in molecules and solids, having the correct cusp behavior near the nucleus and the correct falloff behavior far from the nucleus. This is called a minimal basis. The use of a minimal basis together with a semiempirical two-center fixing of the Hamiltonian matrix elements is known as the *tight-binding method*. Because of its wide applicability, all of Chapter 4 is dedicated to the application of the tight-binding method to crystalline solids. We should note that it is also possible to write a tight-binding Hamiltonian for an amorphous substance with random atomic positions (off-diagonal configurational disorder), but the wave functions in such a system cannot be written as Bloch sums. To simplify the calculations, geometrical mean models have been introduced that reduce the off-diagonal disorder in the Hamiltonian into diagonal disorder (diagonal Hamiltonian matrix elements) (Kakehashi et al., 1993). Electronic-structure calculations on amorphous solids are not discussed in this book.

3.5 UNDERSTANDING BAND-STRUCTURE DIAGRAMS

We showed in Section 2.3.4 how a MO-like treatment of the smallest repeating chemical-point group, or lattice point, can be used to approximate the relative locations of the bands at the center of the BZ. To reinforce this idea, let us repeat the analogy with yet another example. Consider the energy-level diagram for a hypothetical octahedral ReO_6 molecule, shown in Figure 3.5, where we show the Re d–O p π-interactions and the Re d–O p σ-interactions. If we place electrons in the diagram in accordance with Hund's rule and the aufbau principle, there will be 36 electrons contributed by the six oxygens and one from the Re^{6+} (d^1) cation. Note that since the Re–O p π-mixing is weaker than the Re–O p σ-mixing, the bonding combinations resulting from the π-interactions will be higher in energy than the σ-interactions. Thus, the HOMOs have π-symmetry. Furthermore, the π^*-antibonding combinations are lower in energy than the σ^*-antibonding combinations on the

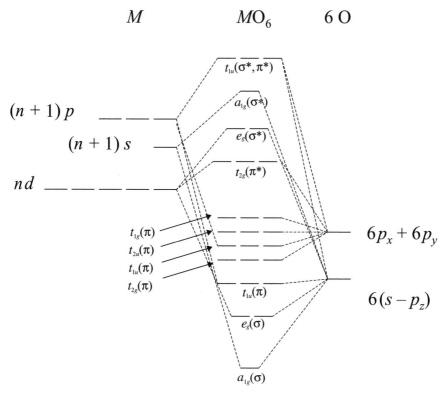

Figure 3.5 A MO energy-level diagram for octahedral transition metal complexes with metal-ligand σ- and π-bonding. (After Ballhausen and Gray, 1964, *Molecular Orbital Theory.* Copyright © Benjamin Cummings Publishing Company. Reproduced with permission.)

same grounds, so that the LUMO also has π symmetry. Shortening the Re–O bond distance will lower the bonding orbital energy and raise the antibonding orbital energy. The bonding and nonbonding MOs for our hypothetical ReO_6 are all filled with electrons, and one of the antibonding orbitals is half-filled (singly occupied). This is the condition, in a solid, that would be expected to lead to metallic behavior.

The particular splitting pattern of the *d*-orbitals in Figure 3.5 is characteristic of cubic (octahedral type) crystal fields. In the tetrahedral type, the e_g and t_{2g} ordering would be reversed. In other symmetries, the *d*-splitting is as shown in Figure 3.6.

There is much more information contained in a band-structure diagram than in a MO energy-level diagram, however. In the former, the band dispersion, or variation in electron energy, is plotted as one moves between high-symmetry *k*-points. The real solid most closely related to our hypothetical ReO_6 molecule is the perovskite ReO_3, which contains vertex-sharing ReO_6 octahedra, linked through Re *d*–O *p* π-bonds. The band structure and DOS for ReO_3, as calculated by Mattheiss (Mattheiss, 1969), is shown in Figure 3.7.

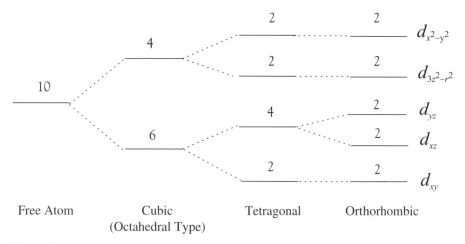

Figure 3.6 The crystal-field splitting of the d-orbitals under cubic, tetragonal, and orthorhombic symmitries.

When looking at any band structure diagram, some important points to remember are:

1. A band-structure diagram is a map of the variation in the energy, or dispersion, of the extended wave functions (called bands) for specific k-points within the first BZ (also called the Wigner–Seitz cell), which is the unit cell of k-space.

2. The total number of bands shown in a band-structure diagram is equal to the number of atomic orbitals contributed by the chemical point group, which constitutes a lattice point. Because the full crystal structure is generated by the repetition of the lattice point in space, it is also referred to as the basis of the structure.

3. There are $2N$ electrons in the BZ (where N is the number of unit cells in the crystal). Each crystal orbital can hold two electrons of opposite spin per two-center bonding site.

4. Band structure diagrams show k values of only *one* sign, positive or negative.

Now let us look at the band structure of ReO_3. On the left-hand side of Figure 3.7 is the band-dispersion diagram, and on the right-hand side the density-of-states curves are shown, which give the density of states per ΔE interval. Accounting for the way the atoms are shared between neighboring unit cells, we can see that, in ReO_3, there is one rhenium atom and three oxygen atoms contained within a single unit cell. In fact, the Bravais lattice for perovskite is simple cubic, and this same combination of atoms is associated with each lattice point. In his calculations, Mattheiss considered combinations made up of Bloch sums formed from the five rhenium $5d$ orbitals and the $2p$ orbitals of the three oxygen atoms. Because there are 14 atomic orbitals in his basis set of atomic orbitals, there are 14 different Bloch sums, which will

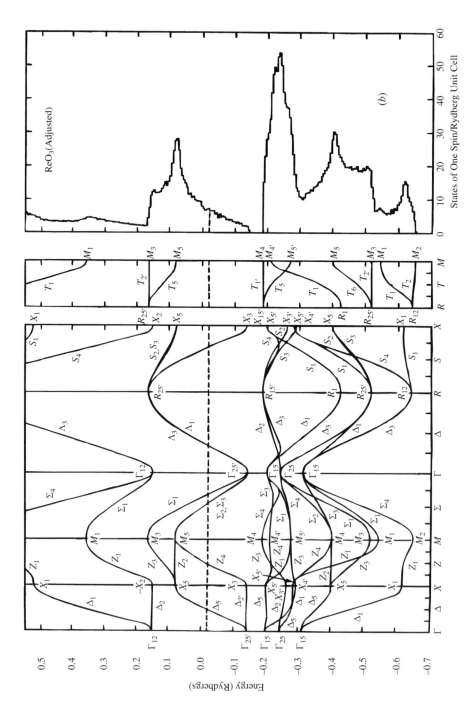

Figure 3.7 The LCAO (tight-binding) band structure for ReO_3. (After Mattheiss, 1969, *Physics Review.* Reproduced by permission of the American Physical Society.)

combine to give 14 crystal orbitals. We should thus expect to find 14 bands in the band structure diagram, which is the case.

Starting from the bottom up, the lowest group of nine bands in the diagram, with a bandwidth spanning from about -0.65 Ry up to -0.2 Ry, correspond to the oxygen $2p$ states. It is customary to refer to the entire set as the "p band," even though they are really crystal orbitals comprised of linear combinations of metal d and oxygen p Bloch sums. The reason for this is that these crystal orbitals have mostly p character, since they are closer in energy to the oxygen p Bloch sums. This set constitutes the top of the valence band in ReO_3 (the oxygen $2s$ states, which were neglected in the calculations, lie lower in energy). The next highest in energy, spanning from about -0.15 Ry to 0.15 Ry, correspond to the antibonding rhenium t_{2g}^* orbitals (xy, yz, zx) or, more precisely, the t_{2g}^* manifold of bands under the periodic potential. Finally, the two highest energy bands are the antibonding rhenium e_g^* manifold (the $3z^2 - r^2$, $x^2 - y^2$ bands). It is customary to refer to the two manifolds together as the "d-band," and it makes up the bottom of the conduction band (the rhenium $4s$ states lie higher in energy). We carry out the *band filling* in the same manner as in our earlier MO treatment using rule number three in the preceding list. Thus, the valence band is completely full and the lowest-lying t_{2g} band is half-filled. Because the Fermi level (HOMO) lies in a partially filled conduction band, ReO_3 is predicted to be metallic.

3.6 BREAKDOWN OF THE INDEPENDENT ELECTRON APPROXIMATION

Both the LCAO and NFE methods are complementary approaches to one-electron band theory, in which electrons are allowed to move independently of one another, through an averaged potential generated by all the other electrons. The true Hamiltonian is a function of the position of all the electrons in the solid and contains terms for all the interactions between these electrons, that is, all of the electron–electron Coulombic repulsions. Electronic motion is *correlated*; the electrons tend to stay away from one another because of Coulombic repulsion.

In the Hartree–Fock theory, correlation due to the exchange hole (involving parallel spin electrons at different sites) is included in the form of the aforementioned average potential, but the Coulomb repulsion between antiparallel spin electrons at the same site (intrasite, or dynamic correlation) is neglected. It is possible to account for dynamic correlation in order to improve on the results from Hartree–Fock calculations. One technique involves the introduction of the configuration interaction (CI), which represents many-electron wave functions with linear combinations of several Slater determinants. The first determinant is the Hartree–Fock ground state; the second one is the first excited state, and so on. Another method, popular among chemists, is the Møller–Plesset (MP) perturbation theory (Møller and Plesset, 1934). In this method, the difference between the Fock operator and the true Hamiltonian is treated as a perturbation. Unfortunately, it is difficult to implement all of these techniques on systems containing a large number of atoms.

What exactly is the cost of neglecting dynamic correlation effects in solids? The most important consequence is the possibility that the wrong type of electric transport behavior (i.e., metallic, semiconducting, or insulating) may be predicted. For example, one-electron band theory predicts metallic behavior whenever the bands are only half-filled, *regardless* of the interatomic separations. However, this is incorrect and even counterintuitive—isolated atoms are electrically insulating. The excitation energy, or electron transfer energy, for electronic conduction in a solid is essentially equal to the Coulomb repulsion between electrons at the same bonding site. In fact, the ratio of this on-site Coulomb repulsion to the one-electron bandwidth determines whether an electron is localized or itinerant. Unfortunately, as we have seen, the intrastate Coulomb repulsion is not accounted for in the Hartree–Fock approach.

Consider a *d*-electron system, such as a transition metal compound. The valence *d* atomic orbitals do not range far from the nucleus, so crystal orbitals comprised of Bloch sums of *d* orbitals and, say, O 2*p* orbitals, tend to be narrow. As the interatomic distance increases, the bandwidth of the crystal orbital decreases because of poorer overlap between the *d* and *p* Bloch orbitals. In general, when the interatomic distance is greater than a critical value, the bandwidth is so small that the electron transfer energy becomes prohibitively large. Thus, the condition for metallic behavior is not met—insulating behavior is observed.

The archetypal examples are the 3*d* transition metal monoxides NiO, CoO, FeO, MnO, VO, and TiO. All of these oxides possess the rock-salt structure (which makes both cation–cation and cation–anion–cation overlap important). One-electron band theory correctly predicts the metallic behavior observed in TiO, which is expected of a partially filled 3*d* band. However, except for VO, all the other monoxides are nonconducting. This can be explained by the presence of *localized* electrons. Because the radial extent of the *d* atomic orbitals increases as one moves to the left in a period, the extent of electron localization decreases for the lighter transition elements in a row (Harrison, 1989). By contrast, the heavier oxides in this group represent cases where the bandwidth is so narrow that dynamic correlation effects dominate. A material that is insulating because of this type of electron localization is referred to as a *Mott insulator* or, sometimes, a *Heitler–London-type insulator*, since it will be recalled that in the Heitler–London theory of chemical bonding, valence electrons are in orbitals centered on atoms, as opposed to delocalized molecular orbitals.

3.7 DENSITY FUNCTIONAL THEORY: AN ALTERNATIVE TO THE HARTREE–FOCK APPROACH

Most electronic-structure calculations on solids are now performed with the density functional theory (DFT), based on the work of Hohenberg, Kohn, and Sham (Hohenberg and Kohn, 1964; Kohn and Sham, 1965). In DFT, as in the Hartree–Fock approach, an effective independent-particle Hamiltonian is arrived at, and the electronic states are solved for self-consistently. The many-electron wave function is still written as a Slater determinant. However, the wave functions used to construct the Slater determinant are not the one-electron wave functions of the Hartree–Fock approximation. In the DFT, these wave functions have no individual meaning. They are merely used to construct the total electron charge density. The difference between the HF and DFT approaches lies in the dependence of the Hamiltonian in DFT on the exchange correlation potential, $V_{xc}[n](r)$, a functional derivative of the exchange correlation energy, E_{xc}, that, in turn, is a functional (a function of a function) of the electron density. In DFT, the Schrödinger equation is expressed as:

$$\left[-\tfrac{1}{2}\nabla_\mu^2 + \left[\sum_{i=1}^{N} \sum_{s_v} \int \psi_i^*(v)\psi_i(v)/\left|r_\mu - r_v\right| dr_v \right] - \sum_n Z_n/\left|r_\mu - R_n\right| + V_{xc}[n](r) \right] \psi_j(r)$$
$$= \varepsilon_j \psi_j(r) \tag{3.31}$$

The first three terms on the left side of Eq. 3.31 are the same as in the Hartree–Fock equation (Eq. 3.10). The fourth term is the exchange-correlation potential lumping together the many-body effects. This includes exchange (the exchange hole—electrons with parallel spin avoid each other) and dynamic correlation (Coulomb repulsion between electrons with antiparallel spin), the latter of which is neglected in the Hartree–Fock theory. In essence, the exchange term in the Hartree–Fock expression has been replaced with the exchange-correlation potential in the DFT formalism. However, the exact form of this potential is unknown.

Equation 3.31 is known as the Kohn–Sham equation, and the effective one-electron Hamiltonian associated with it is the Kohn–Sham Hamiltonian. Kohn and Sham were the first to evaluate Eq. 3.31 approximately. For the exchange-correlation potential, they started with the exchange-correlation energy of a homogeneous electron gas, evaluated from the electron density at the point r under consideration. In essence, the Hamiltonian then depends on the local value of the density only, even in the presence of strong inhomogeneity. This is called the *local density approximation* (LDA), and it is the approximation adopted in most DFT electronic-structure calculations. Thus, even though the DFT formalism is, in principle, exact, the many-body problem is still only solved approximately in the LDA scheme. The LDA works best for metals. Band gaps tend to be underestimated.

REFERENCES

Bloch, F. 1928, Z. *Phys.*, *52*, 555.

Born, M.; Oppenheimer, J. R. 1927, *Ann. Phys.*, *84*, 457.

Born, M.; von Kármán, Th. 1912, Z. *Phys.*, *13*, 297.

Bouckaert, L. P.; Smoluchowski, R.; Wigner, E. 1936, *Phys. Rev.*, *50*, 58.

Brillouin, L. 1930, *Compt. Rend.*, 191, *198*, 292.

Burdett, J. K. 1996, *J. Phys. Chem.*, *100*, 13263.

Canadell, E. 1998, *Chem. Mater.*, *10*, 2770.

Drude, P. K. L. 1900a, *Ann. Phys.*, *1*, 556.

Drude, P. K. L. 1900b, *Ann. Phys.*, *3*, 369.

Drude, P. K. L. 1902, *Ann. Phys.*, *7*, 687.

Elliot, S. R. 1998, *The Physics and Chemistry of Solids*, John Wiley & Sons, Chichester, UK.

Fock, V. 1930, *Z. Phys.*, *61*, 126.

Harrison, W. A. 1989, *Electronic Structure and Properties of Solids: The Physics of the Chemical Bond*, Dover Publications, New York.

Hartree, D. R. 1928, *Proc. Cambridge Philos. Soc.*, *24*, 89.

Hohenberg, P.; Kohn, W. 1964, *Phys. Rev. B*, *136*, 864.

Hume-Rothery, W.; Abbott, G. W.; Channel-Evans, K. M. 1934, *Philos. Trans. R. Soc.*, *A233*, 1.

Kakehashi, Y.; Tanaka, H.; Yu, M. 1993, *Phys. Rev. B*, *47*, 7736.

Kohn, W.; Sham, L. J. 1965, *Phys. Rev. A*, *140*, 1133.

Lorentz, H. A. *Amsterdam Proc.* 1904–1905.

Margenau, H.; Murphy, G. M. 1956, *The Mathematics of Physics and Chemistry*, 2 ed., D. Van Nostrand Company, Inc. Princeton, NJ.

Mattheiss, L. F. 1969, *Phys. Rev.*, *181*, 987.

Møller, C.; Plesset, M. S. 1934, *Phys. Rev.*, *46*, 618.

Mott, N. F.; Jones, H. 1936, *The Theory of the Properties of Metals and Alloys*, Oxford University Press, Oxford.

Peierls, R. 1930, *Ann. Phys.*, *4*, 121.

Pettifor, D. G.; Ward, M. A. 1984, *Solid. State. Commun.*, *49*, 291.

Pettifor, D. G. 2000. In: Turchi, P. E. A., Shull, R. D., editors, *The Science of Alloys for the 21st Century: A Hume-Rothery Symposium Celebration*, The Minerals, Metals, and Materials Society, Warrendale, PA.

Raynor, G. V. 1947, *An Introduction to the Electron Theory of Metals*, Institute of Metals, London.

Seitz, F. 1936, *Ann. Of Math.*, *37*, 17.

Seitz, F. 1940, *The Modern Theory of Solids*, McGraw-Hill Book Company, New York.

Sonnerfeld, A. 1928, *Z. Phys.*, *47*, 1.

Stroud, D.; Ashcroft, N. W. 1971, *J. Phys. F.: Metal Phys.*, *1*, 113.

Thijssen, J. M. 1999, *Computational Physics*, Cambridge University Press, Cambridge, UK.

Wilson, E. A. 1907, *Vector Analysis: A Textbook for the Use of Students of Mathematics and Physics Founded Upon the Lectures of J. Willard Gibbs*, Charles Scribner's Sons, New York.

CHAPTER 4

Electronic Structure, II: The Tight-Binding Approximation

Before the advent of density-functional theory and advances in computer hardware, the Hartree–Fock theory was used to obtain approximate solutions to the many-body Schrödinger equation. Most band-structure calculations on solids made today are based on density-functional theory, using the local density approximation. However, because of the need to treat large systems and, more importantly for our purposes, because of its similarity to the very familiar LCAO-MO theory, the tight-binding formalism of Bloch's original LCAO method will serve us well in this chapter. Our treatment will be purely qualitative. The reason for this is twofold. First, a quantitative treatment would be too lengthy, taking us well outside the scope of this book. Second, what is of real value to the nonspecialist (i.e., for those of us who are not computational chemists or materials scientists) is the ability to make reasonable estimates without having to carry out time-consuming calculations that, in the end, are just approximations themselves!

Sections 2.3.4 and 3.5 describe how a qualitative energy-level diagram for the smallest repeating chemical-point group, or lattice point (known to crystallographers as the *basis*, or *asymmetric unit*), can be used to approximate the *relative* placement of the energy bands in a solid at the center of the Brillouin zone. This is so because the LCAO-MO theory is equivalent to the LCAO band scheme, minus consideration of the lattice periodicity. In the present chapter, we investigate how the orbital interactions vary for different values of the wave vector over the Brillouin zone.

4.1 THE GENERAL LCAO METHOD

Most chemists are well acquainted with LCAO-MO theory. The numbers of atomic orbitals, even in large molecules, however, are miniscule compared to a nonmolecular solid, where the entire crystal can be considered one giant molecule. In a crystal there are on the order of 10^{23} atomic orbitals, which is, for all practical

Principles of Inorganic Materials Design   By John N. Lalena and David A. Cleary
ISBN 0-471-43418-3   Copyright © 2005 John Wiley & Sons, Inc.

purposes, an infinite number. The principal difference between applying the LCAO approach to solids, versus molecules, is the number of orbitals involved. Fortunately, periodic boundary conditions allow us to study solids by evaluating the bonding between atoms associated with a single lattice point. Thus, the lattice point is to the solid-state scientist what the molecule is to the chemist.

As a prelude to our development of the LCAO treatment of solids, it will be beneficial to briefly review the LCAO-MO method. The cyclic π systems from organic chemistry are familiar, relatively simple and, more importantly, resemble Bloch functions of periodic solids. Thus, we will use them as our introductory examples.

We begin with the independent-electron approximation, which was discussed in the previous chapter. The molecular wave functions, ψ, are solutions of the Hartree–Fock equation, where the Fock operator operates on ψ, but the exact form of the operator is determined by the wave function itself. This kind of problem is solved by an iterative procedure, where convergence is taken to occur at the step in which the wave function and energy do not differ appreciably from the prior step. We denote the effective independent-electron Hamiltonian (the Fock operator) here simply as $\mathcal{H}$. The wave functions are expressed as linear combinations of atomic functions, χ:

$$\psi = \sum_{\mu=1}^{N} c_\mu \chi_\mu \tag{4.1}$$

We also define:

$$H_{\mu\mu} = \int \chi_\mu^* \mathcal{H} \chi_\mu \, d\tau \tag{4.2}$$

$$H_{\mu\nu} = H_{\nu\mu} = \int \chi_\mu^* \mathcal{H} \chi_\nu \, d\tau \tag{4.3}$$

$$S_{\mu\nu} = S_{\nu\mu} = \int \chi_\mu^* \chi_\nu \, d\tau \tag{4.4}$$

The latter two integrals can be represented as a square matrix, for which each matrix element corresponds to a particular combination for the values of μ and ν. We note that because of the Hermitian properties of $\mathcal{H}$, $H_{\mu\nu} = H_{\nu\mu}$ and $S_{\mu\nu} = S_{\nu\mu}$. Equation 4.2 represents the energy of an isolated atomic orbital. It is called the *on-site* integral. In Hückel theory, it is called the *Coulomb integral* and given the symbol α. Equation 4.3 gives the energy of interaction between neighbors, and it is known as the *hopping integral* or *transfer integral* (termed the *exchange* or *resonance integral* in Hückel theory). First nearest-neighbor (adjacent) interactions are conventionally denoted as β in Hückel theory and b in solids. Equation 4.4 is the *overlap integral*, which is a measure of the extent of overlap of the orbitals centered on two adjacent atoms. The transfer integral and overlap integral are proportionally related.

The Ritz variational theorem tells us that:

$$E = \frac{\int \psi^* \mathcal{H} \psi \, d\tau}{\int \psi^* \psi \, d\tau} \tag{4.5}$$

Insertion of Eq. 4.1 into Eq. 4.5 gives

$$E = \sum_\mu \sum_\nu c_\mu^* c_\nu H_{\mu\nu} / \sum_\mu \sum_\nu c_\mu^* c_\nu S_{\mu\nu} \tag{4.6}$$

This expression shall be a minimum if $\partial E / \partial c_\mu$ and $\partial E / \partial c_\nu$ are zero, which leads to a system of equations of the form:

$$\sum_{\mu=1}^{N} c_\mu (H_{\mu\nu} - ES_{\mu\nu}) = 0 \tag{4.7}$$

To illustrate a simple case, let us express Eq. 4.7 explicitly in terms of a two-term LCAO molecular orbital (i.e., for the molecule ethylene), we get:

$$(H_{\mu\mu} - ES_{\mu\mu})c_1 + (H_{\mu\nu} - ES_{\mu\nu})c_2 = 0 \tag{4.8}$$
$$(H_{\nu\mu} - ES_{\nu\mu})c_1 + (H_{\nu\nu} - ES_{\nu\nu})c_2 = 0 \tag{4.9}$$

These two linear algebraic equations in c_1 and c_2 have a nontrivial solution if and only if the determinant of the coefficients vanishes, that is:

$$\begin{vmatrix} H_{\mu\mu} - ES_{\mu\mu} & H_{\mu\nu} - ES_{\mu\nu} \\ H_{\nu\mu} - ES_{\nu\mu} & H_{\nu\nu} - ES_{\nu\nu} \end{vmatrix} = 0 \tag{4.10}$$

Equation 4.10 results in a quadratic equation in the energy, which has two roots corresponding to the energies of the two π-molecular orbitals of ethylene. We know that $H_{\mu\mu} = H_{\nu\nu}$ and $H_{\mu\nu} = H_{\nu\mu}$, and that $S_{\mu\mu} = S_{\nu\nu} = 1$, if χ_μ and χ_ν are normalized. Thus, the two roots of Eq. 4.10 are found to be:

$$E_1 = (H_{\mu\mu} + H_{\mu\nu})/(1 + S) \tag{4.11}$$
$$E_2 = (H_{\mu\mu} - H_{\mu\nu})/(1 - S) \tag{4.12}$$

By substitution of these energies into Eqs. 4.8 and 4.9, one can obtain the orbital coefficients, giving explicit expressions for the MOs.

In general, for a linear combination of N functions (in which N is the number of atomic orbitals in the basis set), we obtain an $N \times N$ secular determinant:

$$\begin{vmatrix} H_{11} - ES_{11} & H_{12} - ES_{12} & H_{13} - ES_{13} & H_{14} - ES_{14} & \cdots & H_{1N} - ES_{1N} \\ H_{21} - ES_{21} & H_{22} - ES_{22} & H_{23} - ES_{23} & H_{24} - ES_{24} & \cdots & H_{2N} - ES_{2N} \\ H_{31} - ES_{31} & H_{32} - ES_{32} & H_{33} - ES_{33} & H_{34} - ES_{34} & \cdots & H_{3N} - ES_{3N} \\ H_{41} - ES_{41} & H_{42} - ES_{42} & H_{43} - ES_{43} & H_{44} - ES_{44} & \cdots & H_{4N} - ES_{4N} \\ \cdots & \cdots & \cdots & \cdots & \cdots & \cdots \\ H_{N1} - ES_{N1} & H_{N2} - ES_{N2} & H_{N3} - ES_{N3} & H_{N4} - ES_{N4} & \cdots & H_{NN} - ES_{NN} \end{vmatrix} = 0 \tag{4.13}$$

Equation 4.13 is often written as:

$$|H_{\mu\nu} - ES_{\mu\nu}| = 0 \qquad (4.14)$$

At this point, the Hückel approximations are often imposed to simplify Equation 4.14. These were introduced by the German physicist Erich Armand Arthur Joseph Hückel (1896–1980). Even though atomic orbitals on neighboring atoms are nonorthogonal (they have nonzero overlap), we can make the approximation that:

$$
\begin{aligned}
S_{\mu\nu} &= 0 \quad && \text{if } \mu \neq \nu \\
&= 1 \quad && \text{if } \mu = \nu
\end{aligned}
\qquad (4.15)
$$

In other words, the overlap integrals, $S_{\mu\nu}$, including those between atomic orbitals on adjacent atoms, are neglected. With this, Eq. 4.13 becomes:

$$
\begin{vmatrix}
H_{11} - E & H_{12} & H_{13} & H_{14} & \cdots & H_{1N} \\
H_{21} & H_{22} - E & H_{23} & H_{24} & \cdots & H_{2N} \\
H_{31} & H_{32} & H_{33} - E & H_{34} & \cdots & H_{3N} \\
H_{41} & H_{42} & H_{43} & H_{44} - E & \cdots & H_{4N} \\
\cdots & \cdots & \cdots & \cdots & \cdots & \cdots \\
H_{N1} & H_{N2} & H_{N3} & H_{N4} & \cdots & H_{NN} - E
\end{vmatrix} = 0 \qquad (4.16)
$$

Neglecting the overlap integral is a severe approximation. However, this approach is still useful because it allows us to obtain a general picture of the relative MO energy levels utilizing primarily symmetry arguments.

The secular determinant can be further simplified by considering only interactions between first nearest neighbors. In this case, all the other $H_{\mu\nu}$ matrix elements become equal to zero. Using the familiar Hückel notation, Eq. 4.16 then looks like:

$$
\begin{vmatrix}
\alpha - E & \beta & 0 & 0 & \cdots & 0 \\
\beta & \alpha - E & \beta & 0 & \cdots & 0 \\
0 & \beta & \alpha - E & \beta & \cdots & 0 \\
0 & 0 & \beta & \alpha - \beta & \cdots & 0 \\
\cdots & \cdots & \cdots & \cdots & \cdots & \cdots \\
0 & 0 & 0 & 0 & \cdots & \alpha - E
\end{vmatrix} = 0 \qquad (4.17)
$$

Even with these simplifications, we still must solve an $N \times N$ secular equation with nonzero off-diagonal matrix elements, which becomes a formidable task for large molecules. An $N \times N$ determinant will give an equation of the Nth degree in the energy, which has N roots.

At this point, we show how the calculations can be simplified considerably by adapting the atomic orbitals to the symmetry of the molecule. This is most easily illustrated with the molecule benzene, a conjugated cyclic π system. Suppose that, instead of using an atomic orbital basis set, we write the secular equation for benzene as an $N \times N$ array of SALCs of the atomic orbitals. Each SALC corresponds to an irreducible representation of the point group for the molecule. Only orbitals of the same irreducible representation can interact; SALCs that have different irreducible representations are orthogonal. Thus, if the SALCs of a given representation are grouped together in the secular determinant, the only nonzero matrix elements will lie in blocks along the principal diagonal, thereby factoring the equation into smaller determinants, each of which is solved separately. The secular equation is said to be block-diagonalized. Since the entire equation is to have the value of zero, each of the smaller block factor determinants must also equal zero. It should be noted that the orbital coefficients are obtained by this process, rather than from solution of the secular equation, as before.

In general, the SALCs themselves are generated by the use of what are known as projection operators. A thorough description of this procedure can be found elsewhere (Cotton, 1990). Fortunately, it will not be necessary to concern ourselves with the details of this process. This is because we can make use of the extremely useful simplification that, for cyclic π systems, C_nH_n, there will always be n π MOs, one belonging to each irreducible representation of the C_n pure rotation group. In other words, benzene belongs to the D_{6h} point group, but the essential symmetry elements needed for constructing our SALCs are contained in the C_6 pure-rotation subgroup. Furthermore, the coefficients of the MOs *are* the characters of the irreducible representations of C_6. Thus, from inspection of the character table for the C_6 point group, the SALCs for benzene can immediately be written as (Cotton, 1990):

$$\psi(A) = 1/6^{0.5}(\chi_1 + \chi_2 + \chi_3 + \chi_4 + \chi_5 + \chi_6)$$
$$\psi(B) = 1/6^{0.5}(\chi_1 - \chi_2 + \chi_3 - \chi_4 + \chi_5 - \chi_6)$$
$$\psi(E_1) = 1/6^{0.5}(\chi_1 + \varepsilon\chi_2 - \varepsilon^*\chi_3 - \chi_4 - \varepsilon\chi_5 + \varepsilon^*\chi_6)$$
$$\psi(E_1') = 1/6^{0.5}(\chi_1 + \varepsilon^*\chi_2 - \varepsilon\chi_3 - \chi_4 - \varepsilon^*\chi_5 + \varepsilon\chi_6) \quad (4.18)$$
$$\psi(E_2) = 1/6^{0.5}(\chi_1 - \varepsilon^*\chi_2 - \varepsilon\chi_3 + \chi_4 - \varepsilon^*\chi_5 - \varepsilon\chi_6)$$
$$\psi(E_2') = 1/6^{0.5}(\chi_1 - \varepsilon\chi_2 - \varepsilon^*\chi_3 + \chi_4 - \varepsilon\chi_5 - \varepsilon^*\chi_6)$$

where ε is equal to $\exp(2\pi i/6)$; A, B, E_1, and E_2 are the Mulliken symmetry labels for the irreducible representations; and the $1/6^{0.5}$ factor is a normalization constant such that the square of the coefficients on each atomic orbital in a given MO add up to one. Each wave function in Eq. 4.18 can be written (Worked Example 4.1) as (Albright et al., 1985):

$$\psi_j = \sum_{\mu=1}^{N} c_{j\mu}\chi_\mu = N^{-1/2} \sum_{\mu=1}^{N} [\exp{(2\pi i j(\mu - 1)/N)}]\chi_\mu \quad (4.19)$$

where j runs from $0, \pm 1, \pm 2, \ldots, \pm N/2$ and. i is the square root of -1. The term in brackets is the orbital coefficient for the μth atomic orbital. It is possible to write the wave functions in Eq. 4.19 as linear combinations that have real coefficients instead of imaginary ones. However, we do not do this, because leaving them in this form will reveal their similarity to Bloch functions later. Equation 4.20 below gives the resultant block-diagonalized secular equation, where the subscripts of the matrix elements now refer to the SALCs.

$$
\begin{array}{ccccccc}
 & \psi(A) & \psi(B) & \psi(E_1) & \psi(E_1') & \psi(E_2) & \psi(E_2') \\
\psi(A) & H_{11} - E \\
\psi(B) & & H_{22} - E \\
\psi(E_1) & & & H_{33} - E & & & & = 0 \\
\psi(E_1') & & & & H_{44} - E \\
\psi(E_2) & & & & & H_{55} - E \\
\psi(E_2') & & & & & & H_{66} - E
\end{array}
$$

(4.20)

The energy of each MO is obtained by solving its respective block factor in Eq. 4.20. Thus, for each of the six MOs of benzene, we have:

$$
E_j = \int \psi_j^* \mathcal{H} \psi_j \, d\tau \tag{4.21}
$$

In order to calculate the energy, we use the simple Hückel theory for π systems. That is, we consider only the π interactions between adjacent p orbitals and neglect the overlap integrals. Equation 4.21 then becomes equal to:

$$
E_j = \int \chi_\mu^* \mathcal{H} \chi_\mu \, d\tau + \sum 1/N\{[\exp(-2\pi ij/N)] + [\exp(2\pi ij)/N]\} \int \chi_\mu^* \mathcal{H} \chi_\nu \, d\tau \tag{4.22}
$$

The first integral is simply α, and the second integral is β. The sum of the two terms in brackets is equal to $2\cos(2\pi j/N)$, and since there are N atoms in the molecule, the energy is finally found to be:

$$
E_j = \alpha + 2\beta \cos[2\pi j/N] \tag{4.23}
$$

where j runs from $0, \pm 1, \pm 2, \ldots, \pm N/2$.

Example 4.1   Show that Eq. 4.19 does indeed give the wave functions of Eq. 4.18

Solution   Since j runs from $0, \pm 1, \pm 2, \ldots, \pm N/2$, and the wave functions of E_1 and E_2 symmetry are doubly degenerate, we must evaluate the expression in Eq. 4.19 for $j = 0, \pm 1, \pm 2$, and $+3$, and $\mu = 1 - 6$. Replacing j with 0, for example, results in $c_j = 1$ for every value of j, leading to $\psi_0 = 1/6^{0.5}(\chi_1 + \chi_2 + \chi_3 + \chi_4 + \chi_5 + \chi_6)$, which is the expression for $\psi(A)$. When $j = +1$, we get, for $\mu = 1 - 6$,

$$
\begin{aligned}
\psi_1 = {}& 1/6^{0.5}(\exp(0)\chi_1 + \exp(2\pi i/6)\chi_2 + \exp(4\pi i/6)\chi_3 + \exp(6\pi i/6)\chi_4 \\
& + \exp(8\pi i/6)\chi_5 + \exp(10\pi i/6)\chi_6)
\end{aligned}
$$

Recognizing the following relations:

$$e^{i\theta} = \cos\theta + i\sin\theta \qquad e^{-i\theta} = \cos\theta - i\sin\theta,$$

the expression for ψ_1 is found to be equivalent to:

$$\psi_1 = 1/6^{0.5}(\chi_1 + \exp(2\pi i 6)\chi_2 + \exp(4\pi i/6)\chi_3 - \chi_4 + \exp(8\pi i/6)\chi_5 + \exp(10\pi i/6)\chi_6)$$

We also see that:

$$\exp(4\pi i/6)\chi_3 = -\exp(-2\pi i/6)\chi_3 = -\varepsilon^*\chi_3$$
$$\exp(8\pi i/6)\chi_5 = -\exp(2\pi i/6)\chi_5 = -\varepsilon\chi_5$$
$$\exp(10\pi i/6)\chi_6 = \exp(-2\pi i/6)\chi_6 = \varepsilon^*\chi_6$$

Making the substitutions, gives:

$$\psi_1 = 1/6^{0.5}(\chi_1 + \varepsilon\chi_2 - \varepsilon^*\chi_3 - \chi_4 - \varepsilon\chi_5 + \varepsilon^*\chi_6)$$

which is $\psi(E_1)$.

The reader should show that proceeding in an analogous fashion will give the remaining expressions in Eq. 4.18.

4.2 EXTENSION OF THE LCAO TREATMENT TO CRYSTALLINE SOLIDS

Bloch orbitals are comprised of an enormous number ($\sim 10^{23}$) of atomic orbitals. Such a gigantic basis set is handled by making use of the crystalline periodicity. Because the crystal structure is periodic, so too is the electron density. The presence of structural periodicity imposes translational periodicity on the wave functions. Consider the 1D chain of N atoms, with spacing a, in Figure 3.4 to be of finite length L, where $L = Na$. For running wave solutions to Schrödinger's equation, describing the motion of an electron in the array, periodic boundary conditions are appropriate. An integral number of wavelengths must fit into L:

$$\psi(r) = \psi(r + L) \tag{4.24}$$

Equation 4.24 can be achieved for a chain of finite length by joining the two ends of the chain (albeit, the analogous situation in three dimensions is difficult to visualize). In other words, periodic boundary conditions require us to treat the chain of atoms as an imperceptibly bent ring. In reality, the atoms at the ends (surfaces in 3D) experience different forces from those of the bulk. We are not concerned, however, with surface effects and the advantages of using periodic boundary conditions far outweigh any inaccuracies in our picture of the surface states. Certainly, as $N \to \infty$, the atoms deep in the bulk will be unaffected by the surface conditions, anyway.

The most general expression for a periodic function is the plane wave, $e^{i\theta}$, in which θ is a parameter we let be equal to the vector dot product $\mathbf{k} \cdot \mathbf{R}$. Hence, the wave function, $\psi(\mathbf{r})$, of Eq. 4.24 is of the form:

$$\psi(\mathbf{r}) = e^{i\mathbf{k}\cdot\mathbf{r}} u(\mathbf{r}) \tag{4.25}$$

where $u(\mathbf{r})$ has the periodicity of the lattice, $u(\mathbf{r} + \mathbf{R}) = u(\mathbf{r})$, and $e^{i\mathbf{k}\cdot\mathbf{r}}$ is simply a phase factor that depends on the separation, $\mathbf{R}$, between points. Equation 4.25 states that $\psi(\mathbf{r})$, which is an eigenstate of the one-electron Hamiltonian, can be written in the form of a plane wave times a function periodic in the Bravais lattice of the solid. If Eq. 4.25 holds, then so does the following relation:

$$\psi(\mathbf{r} + \mathbf{R}) = e^{i\mathbf{k}\cdot\mathbf{R}} \psi(\mathbf{r}) \tag{4.26}$$

Equations 4.25 and 4.26 are equivalent expressions. If the physical location in real space is shifted by $\mathbf{R}$, only the phase of the wave function will change. Because we know $\psi(\mathbf{r})$ over any unit cell, we can calculate it for any other unit cell using $\psi(\mathbf{r} + \mathbf{R}) = e^{ikR}\psi(\mathbf{r})$. Thus, e^{ikR} is an eigenfunction of the translation operator. Bloch's theorem states that the eigenfunctions of the Hamiltonian have the same form—the Hamiltonian commutes with the translation operator. Therefore, we only have to solve the Hamiltonian for one unit cell.

For each type of atomic orbital in the basis set, which is the chemical-point group, or lattice point, one defines a Bloch sum (also known as Bloch orbital or Bloch function). A Bloch sum is simply a linear combination of all the atomic orbitals of that type, under the action of the infinite translation group. These Bloch sums are of the exact same form in Eq. 4.19, but with χ_μ replaced with the atomic orbital located on the atom in the nth unit cell $\chi(\mathbf{r} - \mathbf{R}_n)$:

$$\phi(\mathbf{k}) = N^{-1/2} \sum_{n=1}^{N} \exp\left(i\mathbf{k} \cdot \mathbf{R}_n\right)\chi(\mathbf{r} - \mathbf{R}_n) \tag{4.27}$$

In this equation, N is equal to the number of unit cells in the crystal. Note how the function in Eq. 4.27 is the same as that of Eq. 4.19 for cyclic π molecules, if we define a new index: $k = 2\pi j/Na$. Bloch orbitals are simply symmetry-adapted linear combinations of atomic orbitals. However, whereas the exponential term in Eq. 4.19 is the character of the jth irreducible representation of the cyclic group to which the molecule belongs, in Eq. 4.27 the exponential term is related to the character of the kth irreducible representation of the cyclic group of infinite order (Albright et al., 1985). This, in turn, may be replaced with the infinite linear translation group because of our periodic boundary conditions. It turns out that SALCs for any system with translational symmetry are constructed in this same manner. Thus, as with cyclic π systems, we should never have to use the projection operators referred to earlier to generate a Bloch orbital.

We now must consider how Bloch orbitals combine to form crystal orbitals, for example, like the σ- or π-combinations between the Bloch sums of metal d orbitals and oxygen p orbitals in a transition metal oxide. Bloch sums are used as the basis

ERRATA
Principles of Inorganic Materials Design
By J.N. Lalena and David Cleary
©2004 John Wiley & Sons, Inc. ISBN 0-471-43418-3

Chapter 1
p. 24:
As in the case for pure *metals*, the positive temperature gradient criterion for planar interface stability still holds.

p. 34, Figure 1.15(b) Legend:
Below T_M, transport along the c-axis is semiconducting, while transport along the ab axis is metallic across the entire temperature range.

Chapter 2
p.48:
Lattice points must belong to one of the 32 crystallographic point groups derived by Johann Hessel (1796-1872) in 1830.

p. 50:
Figure 2.4 Legend:
14 3D Bravais lattices arranged into the *seven* crystal systems.

Chapter 4
p. 165, 173, 174:
$\lambda = 2\pi/0 = \infty$ (infinity)

Chapter 6
p. 212:
In the limit $a \to \infty$ *(infinity)*, the excitation energy required for electron transfer from an atomic orbital on one of the atoms to an atomic orbital on the other atom, already containing an electron, is defined by Eq. 6.4.

Chapter 7
p. 249:
In the case of 180° M-X-M angles (*see Figure 7.5*), the anion pσ orbitals are orthogonal to the cation t_{2g} orbitals, but overlap strongly with the e_g orbitals.

Chapter 9
p. 318:

"These belong to the set with indices {hkl} = {111} and include the (111), (one-bar one one), (one one-bar one), and (one one one-bar) planes."

p. 318, in the next sentence:
"...(with [uvw] = [one-bar one zero], [one zero one-bar], and [zero one-bar one])..."

p. 319:
all the [1120]'s should be changed to [one-bar one-bar two zero]
the [2110]s should be changed to [two one-bar one-bar zero]
the [1210]s should be changed to [one-bar two one-bar zero]

p. 322:
Similar considerations show the smallest possible Burgers vectors in the hcp and bcc lattices are 1/3a[two *one-bar one-bar* zero] and 1/2a[one one one], respectively.

p. 326, Example 9.5:
1/2a[110] should be changed to 1/2a[one-bar one zero]
1/2a[101] should be changed to 1/2a[one-bar zero one]
1/2a[011] should be changed to [zero one-bar one]

p. 326, Example 9.6:
1/2a[011] should be changed to 1/2a[zero one one-bar]

for such crystal orbitals:

$$\psi_m(\mathbf{k}) = \sum_{\mu} c_{m\mu}(\mathbf{k})\phi_{\mu}(\mathbf{k}) \tag{4.28}$$

where we note that a linear combination of Bloch functions is also a Bloch function. The eigenvalue problem that we are trying to solve can be represented as:

$$\mathcal{H}\psi_m(\mathbf{k}) = E\psi_m(\mathbf{k}) \tag{4.29}$$

We saw earlier how using SALCs to construct molecular orbitals resulted in a block-diagonalized secular equation. Exactly the same thing happens with solids. We end up with an $N \times N$ determinant (where N is the number of unit cells, $\sim 10^{23}$) diagonalized with $n \times n$ block factors (where n is the number of atomic orbitals in the basis set), each having a particular value of $\mathbf{k}$. For example, a substance containing valence s, p, and d atomic orbitals, having a Bravais lattice with a one-atom basis (one atom per lattice point), gives a 9×9 block factor, or an 18×18 with a two-atom basis. Within each block factor, the matrix elements can be written as (Canadell and Whangbo, 1991):

$$
\begin{aligned}
H_{\mu\nu}(\mathbf{k}) &= \int \phi_{\mu}^*(\mathbf{k})\mathcal{H}\phi_{\nu}(\mathbf{k})\,d\tau \\
&= \int \chi_{\mu}^*(\mathbf{r})\mathcal{H}\chi_{\nu}(\mathbf{r})\,d\tau + N^{-1}\sum_m\sum_n \exp[i\mathbf{k}\cdot\mathbf{r}_m] \\
&\qquad \cdot \exp[i\mathbf{k}\cdot\mathbf{r}_n]\int \chi_{\mu}^*(\mathbf{r}-\mathbf{r}_m)\mathcal{H}\chi_{\nu}(\mathbf{r}-\mathbf{r}_n)\,d\tau
\end{aligned} \tag{4.30}
$$

$$
\begin{aligned}
S_{\mu\nu}(\mathbf{k}) &= \int \phi_{\mu}^*(\mathbf{k})\phi_{\nu}(\mathbf{k})\,d\tau \\
&= \int \chi_{\mu}^*(\mathbf{r})\chi_{\nu}(\mathbf{r})\,d\tau + N^{-1}\sum_m\sum_n \exp[i\mathbf{k}\cdot\mathbf{r}_m]\exp[i\mathbf{k}\cdot\mathbf{r}_n] \\
&\qquad \cdot \int \chi_{\mu}^*(\mathbf{r}-\mathbf{r}_m)\chi_{\nu}(\mathbf{r}-\mathbf{r}_n)\,d\tau
\end{aligned} \tag{4.31}
$$

where the indices m and n denote the mth and nth unit cells, and ϕ_{μ} and ϕ_{ν} represent two different Bloch sums (e.g., one formed from, say, all the oxygen p orbitals and one formed from all the transition-metal d orbitals in a metal oxide). If we have only one atom at each lattice point, the primitive lattice translation vectors, $\mathbf{R}_n = (\mathbf{r}_n - \mathbf{r}_m)$, are the basis vectors, which give the displacement from the atom on which the orbital χ_{μ} is centered to the atom on which χ_{ν} is centered. We also recognize that, since there is translational invariance in a Bravais lattice, the sum over m atoms is done N times. Hence, the factor N^{-1} is canceled by the sum over m. Equations 4.30 and 4.31 can now be written as:

$$H_{\mu\nu}(\mathbf{k}) = \int \chi_{\mu}^*(\mathbf{r})\mathcal{H}\chi_{\nu}(\mathbf{r})\,d\tau + \sum_n \exp[i\mathbf{k}\cdot\mathbf{R}_n]\int \chi_{\mu}^*(\mathbf{r})\mathcal{H}\chi_{\nu}(\mathbf{r}-\mathbf{R}_n)\,d\tau \tag{4.32}$$

$$S_{\mu\nu}(\mathbf{k}) = \int \chi_{\mu}^*(\mathbf{r})\chi_{\nu}(\mathbf{r})\,d\tau + \sum_n \exp[i\mathbf{k}\cdot\mathbf{R}_n]\int \chi_{\mu}^*(\mathbf{r})\chi_{\nu}(\mathbf{r}-\mathbf{R}_n)\,d\tau \tag{4.33}$$

The variational principle can be used to estimate the energy. If we are considering only first nearest-neighbor interactions and an orthonormal (Eq. 4.15) set of atomic orbitals, substitution of Eqs. 4.32 and 4.33 into Eq. 4.5 yields:

$$E(\mathbf{k}) = E^0(\mathbf{k}) + \sum_n \exp[i\mathbf{k} \cdot \mathbf{R}_n] \int \chi_\mu^*(\mathbf{r}) \mathcal{H} \chi_\nu(\mathbf{r} - \mathbf{R}_n) \, d\tau \qquad (4.34)$$

where $E_0(\mathbf{k})$ is $\int \chi_\mu^*(\mathbf{r}) \mathcal{H} \chi_\nu(\mathbf{r}) \, d\tau$, the on-site or Coulomb integral, α.

The vector notation we have used to this point is concise, but it will be instructive to resolve the vectors into their components. Each atom is located at a position $pa\mathbf{i} + qb\mathbf{j} + rc\mathbf{k}$, where a, b, c are lattice parameters and $\mathbf{i}, \mathbf{j}, \mathbf{k}$ are unit vectors along the x, y, z axes. A Bloch orbital may thus be written as:

$$\begin{aligned}
\phi(k_x, k_y, k_z) &= N^{-1/2} \sum_p \sum_q \sum_r [\exp(ik_x pa + ik_y qb + ik_z rc)] \chi_\mu(\mathbf{r} - p\mathbf{a} - q\mathbf{b} - r\mathbf{c}) \\
&= N^{-1/2} \sum_p \sum_q \sum_r [\exp(ik_x pa)][\exp(ik_y qb)][\exp(ik_z rc)] \\
&\quad \cdot \chi_\mu(\mathbf{r} - p\mathbf{a} - q\mathbf{b} - r\mathbf{c})
\end{aligned}$$

$$(4.35)$$

At the risk of being superfluous, let us go ahead and write out the corresponding expression for the energy. Upon resolving the vectors in the matrix elements (Eqs. 4.32 and 4.33), Eq. 4.35 becomes:

$$\begin{aligned}
E(k_x, k_y, k_z) &= \int \chi_\mu^*(\mathbf{r}) \mathcal{H} \chi_\nu(\mathbf{r}) \, d\tau \\
&\quad + \sum_p \sum_q \sum_r [\exp(ik_x pa)][\exp(ik_y qb)][\exp(ik_z rc)] \\
&\quad \cdot \int \chi_\mu^*(\mathbf{r}) \mathcal{H} \chi_\nu(\mathbf{r} - p\mathbf{a} - q\mathbf{b} - r\mathbf{c}) \, d\tau \\
&= \int \chi_\mu^*(\mathbf{r}) \mathcal{H} \chi_\nu(\mathbf{r}) \, d\tau \\
&\quad + \left\{ \sum_p \exp(ik_x pa) \sum_q \exp(ik_y qa) \sum_r \exp(ik_z ra) \right\} \\
&\quad \cdot \int \chi_\mu^*(\mathbf{r}) \mathcal{H} \chi_\nu(\mathbf{r} - p\mathbf{a} - q\mathbf{b} - r\mathbf{c}) \, d\tau \qquad (4.36)
\end{aligned}$$

The phase factor sum tells us that the amplitude of the plane wave at the lattice point in question is the sum of contributions from all the atoms. Indeed, in the LCAO method we are really just expressing the electronic wave function in a solid as a superposition of all the atomic wave functions.

4.3   ORBITAL INTERACTIONS IN MONATOMIC SOLIDS

4.3.1   σ-Bonding Interactions

Let us begin by considering a primitive Bravais lattice with one atomic orbital of spherical symmetry (one s atomic orbital) per lattice point. For example, in an sc lattice each atom has six first nearest neighbors. Relative to the atom in question, the six neighbors are at coordinates $(a, 0, 0)$, $(-a, 0, 0)$, $(0, a, 0)$, $(0, -a, 0)$, $(0, 0, a)$, and $(0, 0, -a)$. The phase-factor sums for each of the planes defined by these six points are:

$$(\pm a, 0, 0) \quad \{\exp[ik_x a] + \exp[ik_x(-a)]\}\exp(0)\exp(0) \quad 2\cos(k_x a)$$
$$(0, \pm a, 0) \quad \exp(0)\{(\exp[ik_y a] + \exp[ik_y(-a)]\}\exp(0) \quad 2\cos(k_y a)$$
$$(0, 0, \pm a) \quad \exp(0)\exp(0)\{\exp[ik_z a] + \exp[ik_z(-a)]\} \quad 2\cos(k_z a)$$

Thus, the energy of a Bloch sum of s atomic orbitals in the simple cubic lattice (with one atom per lattice point), is:

$$E = E_{s,s}(000) + \{[2\cos(k_x a)] + [2\cos(k_y a)] + [2\cos(k_z a)]E_{s,s}(100)$$
$$= E_{s,s}(000) + 2E_{s,s}(100)[\cos(k_x a) + \cos(k_y a) + \cos(k_z a)] \quad (4.37)$$

The first term on the right, $E_{s,s}(000)$, represents the first integral on the right side of Eq. 4.36, the energy of an isolated atomic orbital. In Hückel theory, $E_{s,s}(000)$ is given the symbol α. The term $E_{s,s}(100)$ represents the second integral on the right side of Eq. 4.36, the interaction between an atomic orbital with its first-nearest neighbors. In the simple cubic lattice, these are located at the $(1, 0, 0)$, $(-1, 0, 0)$, $(0, 1, 0)$, $(0, -1, 0)$, $(0, 0, 1)$, and $(0, 0, -1)$. In Hückel theory, the integral representing first nearest-neighbor interactions is given the symbol β, but it is often represented simply as b in solids.

With lower dimensional systems the results are, of course, analogous. For example, with a 2D square lattice of s atomic orbitals, if we consider only interactions between first nearest neighbors, $E(k_x, k_y)$ is given as $E_{s,s}(00) + 2E_{s,s}(00)[\cos(k_x a) + \cos(k_y a)]$ or, in Hückel theory, as $\alpha + 2\beta[\cos(k_x a) + \cos(k_y a)]$. For the 1D chain illustrated back in Figure 3.4, $E(k_x) = E_{x,x}(0) + 2E_{x,x}(1)\cos(k_x a) \equiv \alpha + 2\beta\cos(k_x a)$. Using this simple expression, we can construct a *qualitative* band-structure diagram corresponding to the 1D array of Figure 3.4. As k changes from 0 to π/a, the energy of the crystal orbital changes from $[E_{x,x}(0) + 2E_{x,x}(1)]$ to $[E_{x,x}(0) - 2E_{x,x}(1)]$. Since $E_{x,x}(0)$ is the energy of an electron in a nonbonding atomic orbital, we can use it to set our zero of energy. Therefore, the two energies $E_{x,x}(0) \pm 2E_{x,x}(1)$ must correspond to the maximum bonding and maximum antibonding orbitals. Between these two energies, there is a quasi-continuum of levels, which gives rise to the curve shown in Figure 4.1.

The tight-binding bandwidth, W, or *band dispersion*, is given by:

$$W = 2z\beta \quad (4.38)$$

where z is the coordination number (the number of first nearest neighbors) and $\beta = -E_{s,s}(1)$. In the case of the 1D chain, $z = 2$, so W equals 4β. For the

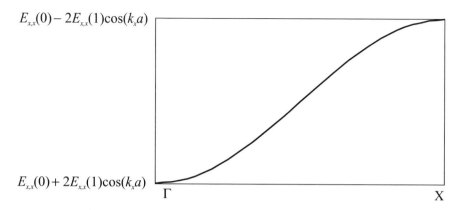

$E_{x,x}(0) - 2E_{x,x}(1)\cos(k_x a)$

$E_{x,x}(0) + 2E_{x,x}(1)\cos(k_x a)$

Γ — X

Figure 4.1   The band-dispersion diagram for a 1D Bloch sum of σ-bonded p atomic orbitals.

2D square lattice, each atom has four nearest neighbors and W equals 8β. In the simple cubic lattice, each atom has six nearest neighbors, so the bandwidth is equal to 12β.

The band dispersion depends on the atomic arrangement in the unit cell. We have discussed the sc system. Let us now evaluate some other structure types. For the time being, we shall continue to restrict consideration to a one-atom basis of s atomic orbitals. The bcc lattice contains eight first nearest neighbors located, relative to the atom in question, at the coordinates $(\pm 1/2a, \pm 1/2a, \pm 1/2a)$. Equation 4.36 gives:

$$
\begin{aligned}
E(k_x, k_y, k_z) &= \int \chi_\mu^*(r)\mathcal{H}\chi_\mu(r)\, d\tau + [\exp(ik_x a/2) + \exp(-ik_x a/2)][\exp(ik_y a/2) \\
&\quad + \exp(-ik_y a/2)][\exp(ik_z a/2) + \exp(-ik_z a/2)] \\
&\quad \times \int \chi_\mu^*(r)\mathcal{H}\chi_\nu(r - pa - qb - rc)\, d\tau \\
&= \int \chi_\mu^*(r)\mathcal{H}\chi_\mu(r)\, d\tau + [2\,\cos(k_x a/2)][2\,\cos(k_y a/2)] \\
&\quad \times [2\,\cos(k_z a/2)] \int \chi_\mu^*(r)\mathcal{H}\chi_\nu(r - pa - qb - rc)\, d\tau \\
&= E_{s,s}(000) + 8E_{s,s}(111)[\cos(k_x a/2)\cos(k_y a/2)\cos(k_z a/2)] \quad (4.39)
\end{aligned}
$$

For the fcc lattice, with twelve first nearest neighbors located at coordinates $(\pm 1/2a, \pm 1/2a, 0)$, $(0, \pm 1/2a, \pm 1/2a)$, and $(\pm 1/2a, 0, \pm 1/2a)$, the energy is given by (see Worked Example 4.2):

$$
\begin{aligned}
E(k_x, k_y, k_z) &= E_{s,s}(000) + 4E_{s,s}(110)[\cos(k_x a/2)\cos(k_y a/2) \\
&\quad + \cos(k_y a/2)\cos(k_z a/2) + \cos(k_x a/2)\cos(k_z a/2)]
\end{aligned}
\quad (4.40)
$$

Example 4.2   Verify that the sum of the phase factors given in Equation 4.40 is correct.

Solution

$(\pm a/2, \pm a/2, 0)$ $\{\exp[ik_x a/2] + \exp[ik_x(-a/2)]$ $2\cos(k_x a/2)2\cos(k_y a/2)$
$(\exp[ik_y a/2] + \exp[ik_y(-a)/2]\}$
$\exp(0)$

$(\pm a/2, 0, \pm a/2)$ $\{\exp[ik_x a/2] + \exp[ik_x(-a)/2]\}$ $2\cos(k_x a/2)2\cos(k_z a/2)$
$\exp(0)$
$(\exp[ik_z a/2] + \exp[ik_z(-a)/2]\}$

$(0, \pm a/2, \pm a/2)$ $\exp(0)$ $2\cos(k_y a/2)2\cos(k_z a/2)$
$\{\exp[ik_y a/2] + \exp[ik_y(-a)/2]\}$
$(\exp[ik_z a/2] + \exp[ik_z(-a)/2]\}$

$$\sum_n \exp[i\boldsymbol{k} \cdot \boldsymbol{R_n}] = 4\cos(k_x a/2)(k_y a/2) + 4\cos(k_x a/2)(k_z a/2) + 4\cos(k_y a/2)(k_z a/2)$$

$$= 4[\cos(k_x a/2)\cos(k_y a/2) + \cos(k_y a/2)\cos(k_z a/2) + \cos(k_x a/2)\cos(k_z a/2)]$$

It is possible to consider interactions between atoms separated by any distance, of course. For example, returning to the simple cubic lattice, had we chosen to consider second nearest-neighbor interactions as well, we would have:

$$E(\boldsymbol{k}) = E_{s,s}(000) + 2E_{s,s}(100)[\cos(k_x a) + \cos(k_y a) + \cos(k_z a)]$$
$$+ 4E_{s,s}(110)[\cos(k_x a)\cos(k_y a) + \cos(k_x a)\cos(k_z a) + \cos(k_y a)\cos(k_z a)]$$
$$(4.41)$$

where $E_{s,s}$ (110) is the contribution to the energy that arises from the interaction of an atom with its twelve second nearest neighbors, at positions $(\pm a, \pm a, 0)$, $(0, \pm a, \pm a)$, and $(\pm a, 0, \pm a)$.

The inclusion of second (and often third terms) is particularly important for certain structure types. For example, in the bcc (and CsCl) lattice, the second nearest neighbors are only 14% more distant than the first nearest (Harrison, 1989). Accounting for the six second nearest-neighbor interactions in the fcc and bcc lattices, the energies of a Bloch sum of s atomic states are given by Eqs. 4.42 and 4.43, respectively:

$$E(k_x, k_y, k_z) = E_{s,s}(000)$$
$$+ 4E_{s,s}(110)[\cos(k_x a/2)\cos(k_y a/2) + \cos(k_x a/2)\cos(k_z a/2)$$
$$+ \cos(k_y a/2)\cos(k_z a/2)]$$
$$+ 2E_{s,s}(200)[\cos(k_x a) + \cos(k_y b) + \cos(k_z c)]$$
$$(4.42)$$

$$E(k_x, k_y, k_z) = E_{s,s}(000)$$
$$+ 8E_{s,s}(111)[\cos(k_x a/2)\cos(k_y b/2))\cos(k_z c/2))$$
$$+ 2E_{s,s}(200)[\cos(k_x a) + \cos(k_y a) + \cos(k_z a)]$$
$$(4.43)$$

Second nearest-neighbor interactions are also important for solids with non-primitive lattices. Up to now, we have only evaluated structures where there is only one atom associated with each lattice point. Let us now consider diamond, which is equivalent to zinc blende, but with all the atoms of the same type. The diamond lattice is not a Bravais lattice, since the environment of each carbon differs from that of its nearest neighbors. Rather, diamond is an fcc lattice with a two-atom basis: four lattice points and eight atoms per unit cell. The structure consists of two interlocking fcc Bravais sublattices, displaced by a quarter of the body diagonal. The displacement vectors from a given site to its four nearest neighbors, belonging to the other sublattice, are: (1/4a, 1/4a, 1/4a), (1/4a, −1/4a, −1/4a), (−1/4a, 1/4a, −1/4a), and (−1/4a, −1/4a, 1/4a), where a is the lattice parameter. These four points are the corners of a tetrahedron whose center is taken as the origin, (0, 0, 0).

Since there are two atoms per primitive cell, or lattice point, and we are still considering just s atomic orbitals, two separate Bloch orbitals are required. These combine to form two crystal orbitals with energies given by the sum and difference in energy between the nearest-neighbor and second nearest-neighbor interactions. The nearest-neighbor interactions are between atoms on the two different sublattices, that is, between an atom on one sublattice and its four neighbors on the other sublattice, which form a tetrahedron around it.

$$4E_{s,s}(111)(\cos(k_x a/4)\cos(k_y a/4)\cos(k_z a/4) - i\,\sin(k_x a/4)\sin(k_z a/4)\sin(k_z a/4)]$$

$$(4.44)$$

The twelve second nearest-neighbor interactions are between atoms all on the same sublattice:

$$E_{s,s}(000) + 4E_{s,s}(110)[\cos(k_x a/2)\cos(k_y a/2) + \cos(k_y a/2)\cos(k_z a/2)$$
$$+ \cos(k_x a/2)\cos(k_z a/2)]$$

$$(4.45)$$

4.3.2 π-Bonding Interactions

In the foregoing examples, it was not necessary to include π or δ interactions. This is not generally the case for atomic orbitals with a nonzero angular momentum quantum number. Consider the 2D square lattice of p atomic orbitals shown in Figure 4.2. The p orbitals bond in a σ-fashion in one direction and in a π-fashion in the perpendicular direction. As we might expect, these two interactions are not degenerate for every value of $\mathbf{k}$.

At Γ, $\mathbf{k} = (0, 0)$, the σ-interactions are antibonding (overlapping lobes on neighboring sites have opposite sign), and the π-interactions are bonding (overlapping lobes have the same sign) for both the p_x and p_y orbitals. At M, $\mathbf{k} = (\pi/a, \pi/a)$, the reverse is true. At X, $\mathbf{k} = (\pi/a, 0)$; however, both σ- and π-interactions are bonding for the p_x, but both are antibonding in the p_y orbitals.

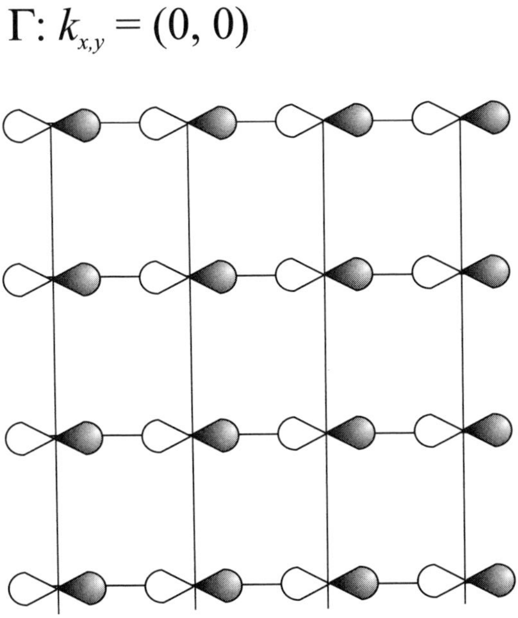

$$\Gamma: k_{x,y} = (0, 0)$$

Figure 4.2   A 2D square array of p_x atomic orbitals. The bonding is by σ-interactions in the horizontal direction and by π-interactions in the vertical direction.

For the simple cubic lattice, the energy of a Bloch sum of p_x atomic states, including first and second nearest-neighbor interactions, is:

$$
\begin{aligned}
E(\mathbf{k}) = {} & E_{x,x}(000) + 2E_{x,x}(100)[\cos(k_x a)] + 2E_{y,y}(100)[\cos(k_y a) + \cos(k_z a)] \\
& + 4E_{x,x}(110)[\cos(k_x a)\cos(k_y a) + \cos(k_x a)\cos(k_z a)] \\
& + 4E_{x,x}(011)[\cos(k_y a)\cos(k_z a)]
\end{aligned}
\tag{4.46}
$$

The energy of the p_y (or p_z) band is obtained by cyclic permutation. The behavior of Eq. 4.46, for 2D square lattices of p_x and p_y atomic orbitals, using the first nearest-neighbor approximation, is shown later in Figure 4.7.

A very noteworthy example involving π-interactions is the single graphite sheet (graphene), with the honeycomb structure. This has been of renewed interest since the discovery of nanographites (graphene ribbons of finite width) and carbon nanotubes (slices of graphene rolled into cylinders). Each carbon atom is sp^2-hybridized and bonded to its three first nearest neighbors in a sheet via sp^2 σ-bonds. The fourth electron of each carbon atom is in a π (p_z) orbital perpendicular to the sheet. The electronic properties are well described from just consideration of the π-interactions.

Like the diamond structure discussed earlier, the honeycomb structure is not itself a Bravais lattice. If we translate the lattice by one nearest-neighbor distance, the lattice does not go into itself. There are two nonequivalent, or distinct types of sites per unit cell, atoms a and b, separated by a distance a_0, as later shown in Figure 4.6. However, a Bravais lattice can be created by taking this pair of distinct atoms to serve as the basis. Doing so, we can see that the vectors of the 2D hexagonal lattice, a_1 and a_2, are primitive translation vectors. A given site on one sublattice with coordinates $(0, 0)$, has three nearest neighbors on the other sublattice. They are located at $(0, a_2)$, $(a_1, 0)$, and $(-a_1, 0)$.

As in the case of diamond, two different Bloch sums of p_z atomic orbitals are needed, one for each distinct atomic site:

$$\phi_\mu(k) = N^{-1/2} \sum_{n=1}^{N} \exp(ik \cdot r_\mu)\chi(r - r_\mu) \tag{4.47}$$

$$\phi_\nu(k) = N^{-1/2} \sum_{n=1}^{N} \exp(ik \cdot r_\nu)\chi(r - r_\nu) \tag{4.48}$$

The secular equation, with the usual approximations, is:

$$\begin{vmatrix} H_{\mu\mu} - E & H_{\mu\nu} \\ H_{\nu\mu} & H_{\nu\nu} - E \end{vmatrix} = 0 \tag{4.49}$$

Noting the Hermitian properties of the Hamiltonian ($H_{\nu\mu} = H_{\mu\nu}^*$), and setting our zero of energy at zero ($H_{aa} = H_{bb} = 0$), Eq. 4.49 becomes:

$$\begin{vmatrix} E & H_{\mu\nu} \\ H_{\mu\nu}^* & E \end{vmatrix} = 0 \tag{4.50}$$

Thus, for the energies, we have the very simple expression:

$$E_\pm = \pm[H_{\mu\nu}^* H_{\mu\nu}]^{1/2} \tag{4.51}$$

As expected, we will have a phase-factor sum and an integral representing the interactions between an atom with its three nearest neighbors, which are of the other type. Thus, Eq. 4.51 can be written as:

$$E(k) = \pm[(1 + \exp[-ik \cdot a_1] + \exp[-ik \cdot a_2])(1 + \exp[ik \cdot a_1]$$
$$+ \exp[ik \cdot a_2])]^{1/2}E(10) \tag{4.52}$$

It can readily be shown that Eq. 4.52 for the band dispersion simplifies to:

$$E(k) = \pm E(10)[3 + 2\cos(k \cdot a_1) + 2\cos(k \cdot a_2) + 2\cos[k \cdot (a_2 - a_1)]^{1/2} \tag{4.53}$$

The primitive lattice vectors in the (x, y) coordinate system are $a_1 = (\surd 3 a_0/2, 1/2)$ and $a_2 = (\surd 3 a_0/2, -1/2)$, where a_0 is the basis vector. Thus, Equation 4.53

can also be written as:

$$E(k_x, k_y) = \pm E(\sqrt{3}/2,\ 1/2)[1 + 4\cos^2(k_x a/2) + 4\cos(k_x a/2)\cos(3^{1/2}k_y a/2)]^{1/2}$$
$$(4.54)$$

In two dimensions, these functions are surfaces. Figure 4.3 shows plots of the π^* band (positive root) and π band (negative root) made with the freeware program *gnuplot for windows*®.

In graphene, the top of the valence band just touches the bottom of the conduction band at the corner point of the hexagonal Wigner–Seitz cell (the K-point in

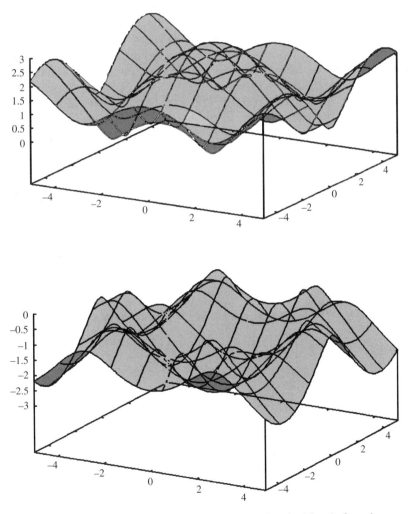

Figure 4.3 Surface plots of the 2D π band (bottom) and π^* band of graphene.

the 2D Brillouin zone) where the energy goes to zero; the π and π^* bands are degenerate at this point. The Fermi level, E_F, which is analogous to the highest occupied molecular orbital, passes right through this intersection. There is thus a zero density of states at E_F, but no band gap, which is the defining characteristic of a semimetal. Semimetals differ from semiconductors in that their resistivities have a metallic-like temperature dependency.

Single-walled carbon nanotubes (SWNTs) are single slices of graphene containing several hundred or more atoms rolled into seamless cylinders, usually with a removable polyhedral cap on each end. Multiwalled carbon nanotubes (MWNTs) are several slices rolled into concentric cylinders. The direction along which a graphene sheet is rolled up is related to the 2D hexagonal-lattice translation vectors a_1 and a_2 via: $C = na_1 + ma_2$, where C is called the chiral or wrapping vector. Thus, C defines the relative location of the two lattice points in the planar graphene sheet that are connected to form the tube in terms of the number of hexagons along the directions of the two translation vectors, as illustrated in Figure 4.4.

Each pair of (n, m) indices corresponds to a specific chiral angle (helicity) and diameter, which give the bonding pattern along the circumference. The chiral angle is determined by the relation $\theta = \tan^{-1}[\sqrt{3}n/(2m + n)]$ and the diameter by the relation $d = (\sqrt{3}\pi)a_{c-c}(m^2 + mn + n^2)^{1/2}$, where a_{c-c} is the distance between carbon atoms. The indices $(n, 0)$ or $(0, m)$, $\theta = 0°$, correspond to the "zigzag" tube (so-called because of the ∧∨∧∨ shape around the circumference, perpendicular to the tube axis). The indices (n, m) with $n = m$ ($\theta = 30°$) corresponds to the "armchair" tube (with a ⌊⌒⌋⌒ shape around the circumference, perpendicular to the tube axis). If one of the two indices n or m is zero, the tube is *nonchiral*; it is superimposable on its mirror image. A general *chiral* nanotube (nonsuperimposable

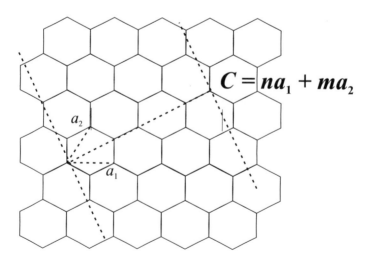

Figure 4.4 The chiral vector, C, defines the location of the two lattice points that are connected to form a nanotube.

on its mirror image) occurs for all other arbitrary angles. Common nanotubes are the armchair (5,5) and the zigzag (9,0).

Unlike graphene, which is a 2D semimetal, carbon nanotubes are either metallic or semiconducting along the tubular axis. In graphene, which is regarded as an infinite sheet, artificial periodic boundary conditions are imposed on a macroscopic scale. This is also true for the nanotube axis. However, the periodic boundary condition is imposed for a *finite* period along the circumference. One revolution around the very small circumference introduces a phase shift of 2π. Electrons are thus confined to a discrete set of energy levels along the tubular axis. Only wave vectors satisfying the relation $C \cdot k = 2\pi q$ (q is an integer) are allowed in the corresponding reciprocal space direction (see Eq. 4.27 and accompanying discussion; let $C = N$ and $q = n$). This produces a set of 1D subbands and what are known as *van Hove singularities* in the density of states (if the tube is very long) for a nanotube. Transport can only propagate along parallel channels down the tubular axis, making a nanotube a 1D *quantum wire*. It is not semimetallic like graphene because the degenerate point is slightly shifted away from the K-point in the Brillouin zone due to the curvature of the tube surface and the ensuing hybridization between the σ^* and π^* bands, which modify the dispersion of the conduction band (Saito et al., 1992; Hamada et al., 1992, Blasé et al., 1994).

These factors, in turn, are dependent on the diameter and helicity. It has been found that metallicity occurs whenever $(2n + m)$ or $(2 + 2m)$ is an integer multiple of three. Hence, the armchair nanotube is metallic. Metallicity can only be exactly reached in the armchair nanotube. The zigzag nanotubes can be semimetallic or semiconducting with a narrow band gap that is approximately inversely proportional to the tube radius, typically between 0.5 and 1.0 eV. As the diameter of the nanotube increases, the band gap tends to zero, as in graphene. We should point out that, theoretically, in sufficiently short nanotubes electrons are predicted to be confined to a discrete set of energy levels along all three orthogonal directions. Such nanotubes could be classified as zero-dimensional *quantum dots*.

Nanometer-width ribbons of graphene, the so-called nanographites, may also have armchair or zigzag edges. The edge shape and system size are critical to the electronic properties. Zigzag edges exhibit a localized nonbonding π electronic state with flat bands that result in a very sharp peak near E_F in the density of states. This is not found in armchair nanographites.

4.4 TIGHT-BINDING ASSUMPTIONS

The original LCAO method by Bloch (Bloch, 1928) is difficult to carry out with full rigor because of the large number of complicated integrals that must be computed. In their landmark paper of 1954, John Clarke Slater (1900–1976) and George Fred Koster (b. 1926) of M. I. T. proposed a scheme for interpolating the band structure over the entire Brillouin zone by fitting to first-principles calculations carried out at high symmetry points (Slater and Koster, 1954). This became known as the *tight-binding method* (sometimes called the SK scheme), and it has been applied to monatomic metals, semiconductors, and compounds.

Actually, what we have been describing all along is the tight-binding method. Some important simplifications are made in the tight-binding scheme, which we now state explicitly. First, in the original SK method, it is assumed that the atomic orbitals and the Bloch sums formed from them are orthogonal. This is not generally true, but the atomic orbitals can be transformed into orthogonal orbitals. One such way this can be accomplished was due to Swedish physicist Per-Olov Löwdin (1916–2000) (Löwdin, 1950). Because of this simplification, the original SK method is sometimes referred to as the linear combination of orthogonalized atomic orbitals (LCOAO) method. Mattheiss later made a modification that does away with the Löwdin orthogonalization, transforming the method into a generalized eigenvalue problem, involving overlap integrals in addition to the on-site and exchange integrals (Mattheiss, 1972).

Second, Slater and Koster treated the potential energy term in the Hamiltonian like that of a diatomic molecule, that is, as being the sum of spherical potentials located on the two atoms at which the atomic orbitals are located. In reality, the potential energy is a sum of potentials (they are not necessarily spherically symmetric) located at *all* the atoms in the crystal. Hence, there are three-center integrals present. The two-center approximation, however, allows one to formulate the problem using a smaller number of parameters. Matrix elements that are two-center integrals use orbitals that are space quantized with respect to the axis between them, so they have a form that is dependent only on the internuclear separation and the symmetry properties of the atomic orbitals (see below). The SK parameters are somewhat independent of the crystal structure and, thus, transferable from one structure type to another, which is a major advantage of the SK method. Two-center SK parameterizations have also been formulated within the density functional theory (Cohen et al., 1994; Mehl and Papaconstantopoulos, 1996), which has become the predominant method of calculating band structures.

Slater and Koster introduced notation that clearly distinguishes σ, π, and δ interactions. For example, referring back to Figure 4.2, we see that in the y direction, the p_x orbitals bond in a π fashion. The third term on the right side of Eq. 4.46 must correspond to the energy of π interactions to first nearest neighbors. Hence, the integral $E_{y,y}(100)$ is replaced with the two-center integral symbolized as $(pp\pi)_1$. Making similar substitutions throughout Eq. 4.46 allows us to rewrite that equation as:

$$
\begin{aligned}
E(\mathbf{k}) = {} & E_{p,p}(0) + 2(pp\sigma)_1 \cos(k_x a) + 2(pp\pi)_1 [\cos(k_y a) + \cos(k_z a)] \\
& + 2(pp\sigma)_2 [\cos(k_x a)\cos(k_y a) + \cos(k_x a)\cos(k_z a)] \\
& + 2(pp\pi)_2 [\cos(k_x a)\cos(k_y a) + \cos(k_x a)\cos(k_z a) + 2\cos(k_y a)\cos(k_z a)]
\end{aligned}
$$

$$(4.55)$$

Of course, in the tight-binding scheme, all such integrals are evaluated only at the high symmetry points in the Brillouin zone. Fitted parameters are then used to interpolate the band structure between these points.

John Slater (Courtesy AIP Emilio Segrè Visual Archives. © Massachusetts Institute of Technology. Reproduced with permission.)

JOHN CLARKE SLATER

(1900–1976) received his Ph.D. in physics from Harvard University in 1923 under P. W. Bridgeman. He then studied at Cambridge and Copenhagen, returning to Harvard in 1925. In 1929, he introduced the Slater determinant in a paper on the theory of complex spectra. Slater was appointed head of the physics department at the Massachusetts Institute of Technology by Karl Taylor Compton. He remained at M. I. T. from 1930 to 1966, during which time he started the school of solid-state physics. Among Slater's Ph.D. students were Nobel laureates Richard Feynman and William Shockley. In World War II, Slater was drawn into the war effort, having been involved in the development of the electromagnetic theory of microwaves, which eventually led to the development of radar. In 1953, Slater and Koster published their now famous simplified LCAO interpolation scheme for determining band structures. After retirement from M.I.T., Slater was Research Professor in Physics and Chemistry at the University of Florida. Slater authored fourteen books on a variety of topics from chemical physics to microwaves to quantum theory. Slater is also widely recognized for calculating algorithms, known as Slater-type orbitals (STOs), which describe atomic orbitals. He was elected to the U.S. National Academy of Sciences in 1932.

(*Primary source*: P. M. Morse, *Biographical Memoirs of the U.S. National Academy of Sciences*, Vol. 53, **1982**, pp. 297-322.)

4.5 QUALITATIVE LCAO BAND STRUCTURES

Unlike the prior two examples—diamond and graphite—compounds have more than one atom *type* per lattice point. Accordingly, it is usually necessary to consider interactions between atomic orbitals with different angular-momentum quantum numbers (e.g., p–$d\pi$, s–$p\sigma$ bonding). Slater and Koster gave two-center integrals for s–p, s–d, and p–d interactions expressed in terms of the direction cosines (l, m, n) of the interatomic vector.

Let us examine a relatively simple structure that has been of enormous interest for decades, the transition metal oxide perovskite, ABO_3. Figure 4.5a shows the real-space unit cell, which contains five atoms. Figure 4.5b shows the Brillouin zone for the simple cubic lattice. It is important to note that the unit cell is not bcc, since the body-centered atom is different from the atoms at the corner positions. Perovskite can also be considered a 3D network of vertex-sharing octahedra. Many possible types of interactions between the various atomic orbitals can be deduced. Table 4.1 lists some of the first and second nearest-neighbor two-center matrix elements that would have to be considered in a rigorous analysis.

Nowhere in this chapter have we actually made a band-structure calculation. Although a sophisticated mathematical formulation has been presented for some simple cases, we have yet to provide any numerical values for the energies. Since nothing to this point has been quantitative, rather than continue to develop the mathematical treatment, we switch to a conceptual treatment of the tight-binding method for compounds, in which symmetry and overlap are considered *qualitatively*. This is a great simplification, which allows one to obtain *approximate* band-structure diagrams without becoming bogged down in tedious mathematics.

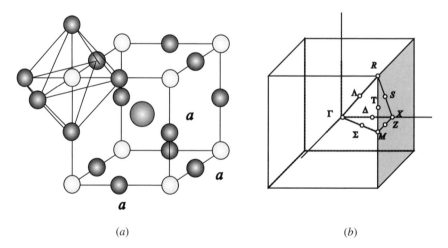

(a)   (b)

Figure 4.5   (a) The perovskite structure corresponds to a simple cubic real-space Bravais lattice and simple cubic reciprocal lattice. (b) The first Brillouin zone for the simple cubic reciprocal lattice.

TABLE 4.1   Interatomic Matrix Elements for the Transition-Metal Perovskite Oxides

Atomic Orbitals Involved	Integral	Two-Center Approx
$A\ s, O\ s$	$E_{s,s}(1/2, 1/2, 0)$	$(ss\sigma)_1$
$A\ x, A\ x$	$E_{x,x}(1, 0, 0)$	$(pp\sigma)_2$
$A\ x, A\ x$	$E_{x,x}(0, 1, 0)$	$(pp\pi)_2$
$A\ y, A\ y$	$E_{y,y}(1, 0, 0)$	$(pp\pi)_2$
$A\ y, A\ y$	$E_{y,y}(0, 1, 0)$	$(pp\sigma)_2$
$A\ y, A\ y$	$E_{y,y}(0, 0, 1)$	$(pp\pi)_2$
$A\ x, O\ x$	$E_{x,x}(1/2, 1/2, 0)$	$1/2[(pp\sigma)_1 - (pp\pi)_1)]$
$A\ y, O\ x$	$E_{y,x}(1/2, 1/2, 0)$	$1/2[(pp\sigma)_1 - (pp\pi)_1)]$
$A\ z, O\ z$	$E_{z,z}(1/2, 1/2, 0)$	$(pp\pi)_1$
$B\ xy, B\ xy$	$E_{xy,xy}(1, 0, 0)$	$(dd\pi)_2$
$B\ xy, B\ xy$	$E_{xy,xy}(0, 0, 1)$	$(dd\delta)_2$
$B\ z^2, B\ z^2$	$E_{z2,z2}(0, 0, 1)$	$(dd\sigma)_2$
$B\ x^2 - y^2, B\ x^2 - y^2$	$E_{z2-y2,z2-y2}(0, 0, 1)$	$(dd\delta)_2$
$B\ z^2, O\ s$	$E_{z2,s}(0, 0, 1/2)$	$(sd\sigma)_1$
$B\ z^2, O\ z$	$E_{z2,z}(0, 0, 1/2)$	$(pd\sigma)_1$
$B\ xy, O\ x$	$E_{xy,x}(0, 1/2\ 0)$	$(pd\pi)_1$

In principle, the scheme is applicable to any periodic solid. However, in practice, the types of structures that are most conducive to the methodology tend to have high symmetry, a small number of atoms per unit cell, and/or a covalent framework with linear cation–anion–cation linkages (180° M–X–M bond angles). These are the only types considered here. The utility of the conceptual approach to low-dimensional systems (including layered perovskite-like oxides) has been exhaustively treated in several papers by Whangbo, Canadell, and co-workers (Rousseau et al., 1996; Whangbo and Canadell, 1989; Canadell and Whangbo, 1991).

For transport properties, one is primarily interested in the nature of the bands near the Fermi energy, in which case only the dispersion at the bottom of the conduction band needs evaluation, at least for systems with a low electron count. How does one know where the Fermi level is? Well, the *relative* band energies at the center and corner of the Brillouin zone in a band-structure diagram usually correspond reasonably well to the order of the localized states in the energy-level diagram for the geometric arrangement at a discrete lattice point (i.e., molecular orbital diagram). The Fermi level in the solid can be approximated by filling the orbitals in such a diagram with all the available electrons, two (each of opposite spin) per orbital, and noting where the HOMO falls. This is precisely what was done in Chapter 3, where it was shown how the energy-level diagram for a discrete MX_6 octahedral unit approximated the relative band energies of a 3D array of vertex-sharing octahedra, that is, perovskite. It must be stressed that lattice periodicity is not part of a "molecular orbital" picture—the band energies in a solid *will* vary over the Brillouin zone. A generic "molecular orbital" energy-level diagram is in no way equivalent to a band-structure diagram! We are simply trying to use the knowledge acquired from the molecular orbital diagram to determine *which* of the bands to assess.

Returning to our perovskite example, it was revealed in the discussion accompanying Figures 3.4 and 3.5 that, for the transition metal oxides with this structure, the Fermi level lies within the t_{2g}- or e_g-block bands. So let us look at just these bands. The five d orbitals from the transition metal atom and three p orbitals on each of the three oxygen atoms give us the 14 tight-binding basis functions listed in Table 4.2. Hence, there is still a 14×14 secular equation to solve at various k points. This equation contains diagonal transition-metal d–d and oxygen p–p Hamiltonian matrix elements. We are interested in the dispersion of the d bands resulting from σ- and π-interactions between the d and oxygen p atomic orbitals. These p–d interactions are off-diagonal matrix elements. The mathematical procedure is beyond the scope of the present discussion, but it is possible by means of a unitary transformation to reduce this matrix at certain high-symmetry points in the Brillouin zone to a group of 1×1 and 2×2 submatrices.

Taking the B atom as the origin of a Cartesian coordinate system, along any axis there is a dimer unit consisting of a B atom and an O atom at each lattice point.

TABLE 4.2   Transition Metal d and Oxygen p Basis Functions for Perovskite

	No.	Position	Function
Metal d	1	(0, 0, 0)	xy
	2	(0, 0, 0)	yz
	3	(0, 0, 0)	xz
	4	(0, 0, 0)	z^2
	5	(0, 0, 0)	$x^2 - y^2$
Oxygen p	6	$1/2a(1, 0, 0)$	x
	7	$1/2a(1, 0, 0)$	y
	8	$1/2a(1, 0, 0)$	z
	9	$1/2a(0, 1, 0)$	x
	10	$1/2a(0, 1, 0)$	y
	11	$1/2a(0, 1, 0)$	z
	12	$1/2a(0, 0, 1)$	x
	13	$1/2a(0, 0, 1)$	y
	14	$1/2a(0, 0, 1)$	z

Metal d and Oxygen p Hamiltinian Matrix Elements

$H_{7,1} = 2i(pd\pi)_1 \sin(k_x a/2)$ $H_{9,1} = 2i(pd\pi)_1 \sin(k_y a/2)$
$H_{11,2} = 2i(pd\pi)_1 \sin(k_y a/2)$ $H_{13,2} = 2i(pd\pi)_1 \sin(k_z a/2)$
$H_{8,3} = 2i(pd\pi)_1 \sin(k_x a/2)$ $H_{12,3} = 2i(pd\pi)_1 \sin(k_z a/2)$
$H_{6,4} = -i(pd\sigma)_1 \sin(k_x a/2)$ $H_{10,4} = -i(pd\sigma)_1 \sin(k_y a/2)$
$H_{14,4} = 2i(pd\sigma)_1 \sin(k_z a/2)$ $H_{6,5} = -\sqrt{3}i(pd\sigma)_1 \sin(k_x a/2)$
$H_{10,5} = -\sqrt{3}i(pd\sigma)_1 \sin(k_y a/2)$

Consequently, each dimer unit has a bonding and antibonding π-level, given in Table 4.2, as:

$$E(k) = E(000) \pm 2i(pd\pi)_1 \sin(ka/2) \qquad (4.56)$$

The bonding combinations, $E(000) + 2i(pd\pi)_1\sin(ka/2)$ correspond to lower energy states located at the top of the valence band (the fully occupied group of states below the Fermi level). We can think of the dispersion of the oxygen p bands as being driven by the *p–d bonding* interactions (see Figure 3.5). The crystal orbitals composed of these bonding combinations have a lower energy and longer wavelength than those of the antibonding combinations. The antibonding combinations, $E(000) - 2i(pd\pi)_1\sin(ka/2)$, are higher energy (shorter wavelength) states, at the bottom of the conduction band (the vacant or partially occupied bands above the Fermi level). In perovskite transition metal oxides, the *antibonding p–d* interactions cause the dispersion of the metal d bands, and this is what we are interested in evaluating.

Qualitative information on the d bandwidth in perovskite can be acquired without carrying out these mathematical operations. This is accomplished simply by evaluating the p–d orbital interactions for some of the special points in the Brillouin zone. For now, let us just consider the π-interactions between the metal t_{2g} and oxygen p orbitals. From Figure 3.5, it is expected that the Fermi level will lie in one of the t_{2g}-block bands for $A^{II}B^{IV}O_3$ oxides, if B is an early transition metal with six or fewer d electrons.

First, Bloch sums of atomic orbitals are constructed for each of the atoms in the basis. For example, at $\Gamma(k_x = k_y = k_z = 0)$, $\lambda = 2\pi/0 = 8$. Hence, in perovskite, the metal d_{xy} atomic orbitals forming the Bloch sum at Γ, viewed down the [001] direction, must have the sign combination shown in Figure 4.6a, where the positive

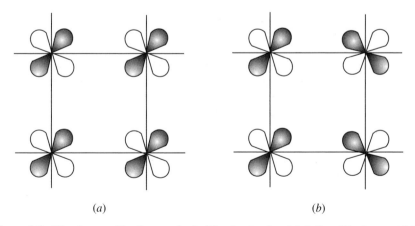

(a) (b)

Figure 4.6 The sign combination required of the d_{xy} atomic orbitals for a Bloch sum at (a) Γ and (b) X viewd down the [001] direction. Positive or negative sign is indicated by the presence or absence of shading.

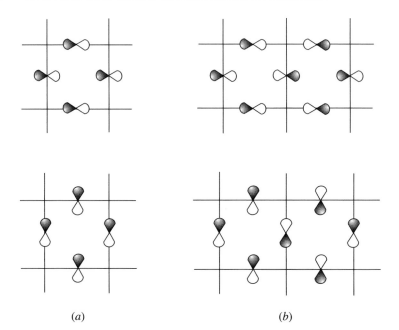

(a)   (b)

Figure 4.7   The sign combinations required of the p_x (top) and p_y (bottom) atomic orbitals for a Bloch sum at (a) Γ and at (b) X.

or negative sign of the electron wave function is indicated by the presence or absence of shading. By comparison, at X for the sc lattice, we have $k_x = \pi/a, k_y = k_z = 0$, and $\lambda = 2a$. Thus, the Bloch sum is formed from d_{xy} atomic orbitals with the sign combination like Figure 4.6b. Similarly, the Bloch sums of the p_y and p_x orbitals at Γ, viewed down the [001] direction, are constructed from atomic orbitals with the signs given in Figure 4.7a. Those at X are like Figure 4.7b. By simply knowing something about the nature of the interactions between given pairs of Bloch sums for a few k-points, we can draw a qualitative band-structure diagram! Let us now look at how Bloch sums interact, in a symmetry-allowed manner.

There are some guiding principles that aid the construction of LCAOs in solids, which are analogous to LCAO molecular orbitals. The first is that the combining atomic orbitals must have the same symmetry about the internuclear axis. Second, the strength of the interactions generally decreases in going from σ to π to δ symmetry. Third, orbitals of very different energies give small interactions. The major principle, however, is from group theory and states that *any electron wave function in a crystal forms a basis for some irreducible representation of the space group*. Essentially, this principle means that the wave function, with a wave vector k, is left invariant under the symmetry elements of the crystal class (e.g., translations, rotations, reflections) or transformed into a new wave function with the same wave vector k.

Γ   X   M

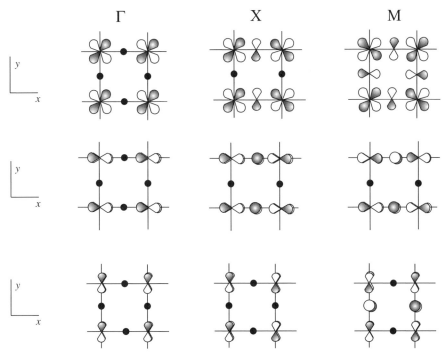

Figure 4.8 The oxygen p–metal d π*-interactions in a transition metal oxide with vertex-sharing octahedra (e.g., perovskite) at the k-points Γ, X, and M, viewed down the [001] direction. Top row: d_{xy}. Middle row: d_{xz}. Bottom row: d_{yz}.

4.5.1 Illustration 1: Transition Metal Oxides with Vertex-Sharing Octahedra

We can use the aforementioned principles to construct qualitative dispersion curves for the conduction band in transition metal oxides with vertex-sharing octahedra (e.g., perovskite, tungsten bronzes) rather easily. Figure 4.8 shows the metal t_{2g}-block and O $2p$ orbital interactions in transition metal oxides with the perovskite structure. At $\Gamma(k_x = k_y = k_z = 0)$, no p–d interactions are symmetry-allowed, including those with the two axial oxygen atoms above and below the plane of the figure, for any of the t_{2g}-block orbitals. The reader may want to expand the diagrams in Figure 4.8 to show the signs of the axial p orbitals at the various k-points. By contrast, at X ($k_x = \pi/a, k_y = k_z = 0$), an atomic d_{xy} orbital can interact with two equatorial oxygen p_y orbitals on either side, as can the d_{xz} and oxygen p_z orbitals. Another high-symmetry point of the first Brillouin zone for the simple cubic lattice is M ($k_x = k_y = \pi/a, k_z = 0$). At this point, the d_{xy} orbitals can interact with all four equatorial p_x and p_y orbitals. The d_{xz} and d_{yz} orbitals each interact with only two of the equatorial p orbitals and no axial orbitals.

The number of interactions per metal atom can be used to plot the band dispersion, as we move between the high-symmetry points in the Brillouin zone. This is

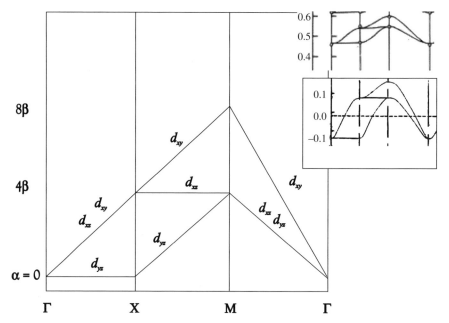

Figure 4.9 Dispersion curves for the d_{xy}, d_{xz}, and d_{yz} bands in a transition metal oxide with octahedra sharing vertices in three dimensions (e.g., perovskite). Top upper insert: calculated band dispersion for SrTiO$_3$. Bottom upper insert: calculated band dispersion for ReO$_3$. The dashed line is the Fermi level.

done in Figure 4.9. At Γ, the d_{xy}, d_{xz}, and d_{yz} bands are all nonbonding—all three are degenerate—with an energy equal to α. In moving from Γ to X, the d_{xy} has two antibonding interactions per metal atom, as does the d_{yz} and the d_{xz} orbitals. This amounts to a destabilization of each of these bands, by an amount we can estimate from Eq. 4.38 as 4β. These two bands are thus degenerate at X. The d_{yz} band is nondispersive, as it still has no symmetry-allowed interactions, just like at Γ.

In moving from X to M, the d_{xz} and d_{yz} bands now become degenerate, with two antibonding interactions (destabilization $= 4\beta$) per metal atom with the oxygen p orbitals. These two bands are lower in energy than the d_{xy} band, which has four antibonding interactions per metal atom and a destabilization of 8β (Eq. 4.38). Figure 4.9 also shows two insets in the upper right corner: the top is the calculated band dispersion in SrTiO$_3$, and the bottom is the calculated band dispersion in ReO$_3$.

Bearing in mind that the qualitative t_{2g} dispersion behavior was obtained by considering only the symmetry and overlap between first nearest-neighbor metal and oxygen atoms, the agreement with the calculated dispersion curves is reasonable. Indeed, the d-block band dispersion in transition metal perovskite oxides is due almost entirely to p–d hybridization effects. In a real calculation, however, second nearest-neighbor interactions as well as metal–metal and oxygen–oxygen contributions, would be included.

Second, although the t_{2g}-block bands are vacant in the insulating phase SrTiO$_3$, they are partially occupied in the metallic phase ReO$_3$ (note the location of E$_F$ in the latter). Nonetheless, the general shape is unchanged, although a wider bandwidth is observed in ReO$_3$. The latter feature is expected for second- and third-row transition metals, however, because the larger radial extent of the $4d$ and $5d$ orbitals should result in stronger overlap. We conclude that the dispersion is independent of the electron count, which is consistent with our earlier claims.

Last, ReO$_3$ has the octahedral framework of SrTiO$_3$ structure, minus the twelve-coordinate atom in the center of the unit cell. However, the orbitals on this atom are of such high energy (Sr electron configuration $= 4s^2 4p^6 5s^2$) that they do not hybridize with the Ti $3d$–O $2p$ bands. In the perovskite structure, this atom simply provides electrons to the system, which can occupy the valence or conduction bands. Hence, there is little change to the band dispersion, directly resulting from the presence of the A cation.

4.5.2   Illustration 2: Reduced Dimensional Systems

Consider now a layered structure that consists of covalent vertex-sharing single-layer octahedral slabs alternating with ionic rock-salt-like slabs, such as the $n = 1$ Ruddlesden–Popper or Dion–Jacobsen phases described in Chapter 2. The unit cells are typically tetragonal, with an elongated c-axis. This constitutes a 2D system, in that electronic transport can only occur within the ab plane of the vertex-sharing octahedral layers occurring along the c-axis of the unit cell. Electronic conduction does not occur in the ionic rock-salt slabs.

How does the reduced dimensionality affect the dispersion relations? Whangbo, Canadell, and co-workers have qualitatively evaluated the band dispersion in such systems in the same way as was done for the 3D perovskite structure in the last section, by counting the number of oxygen p orbital contributions present in the crystal orbitals at certain points in the Brillouin zone.

The Fermi surface of a low-dimensional transport system has a special topology. For example, because wave vectors in the direction perpendicular to the perovskite layer—the c direction in real space—do not cross the Fermi surface, there are no electrons at the Fermi level having momentum in that direction, and the system is nonmetallic along c. Nevertheless, the axial oxygen contributions still must be considered when constructing the band-structure diagram. For a discrete 2D sheet of vertex-sharing octahedra, or for two such sheets separated by, say, nonconducting rock-salt-like slabs, the equatorial and axial oxygen atoms are in two nonequivalent positions, as they have different chemical environments. The equatorial oxygen atoms are in bridging positions—shared by two octahedra—while the axial atoms belong to a single octahedron, as shown in Figure 4.10. It has been shown that the amount of destabilization that results from an axial contribution is one-fourth that due to an equatorial contributions. (Rousseau et al., 1996).

For the d_{xy} orbitals, there are no axial contributions. However, for the d_{xz} and d_{yz} orbitals, there are two axial contributions per metal atom at all three of the $\boldsymbol{k}$-points of Figure 4.8, which are not shown in that figure. Thus, the band dispersion diagram

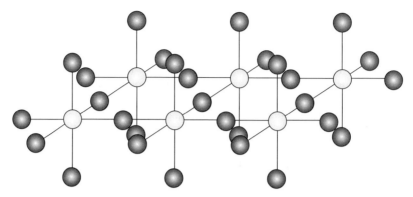

Figure 4.10   A layer of octahedra sharing vertices in only two dimensions, as might be found, for example, in perovskite intergrowths like the Ruddlesden–Popper oxides. The axial and equatorial oxygen atoms are not equivalent in that the former belong to single octhedra, whereas the latter are in bridging positions, shared by two octahedra.

for this case is obtained simply by shifting the d_{xz} and d_{yz} bands upward by $E/2$, or β, on the energy axis, as shown in Figure 4.11. The difference between the dispersion of these three d bands in a 3D system, like perovskite, and a 2D system, like the Ruddlesden–Popper series, is that the d_{xy} band in the latter is of the lowest energy (at Γ). Since the "easy" axis for electronic conduction is the ab plane of the unit cell, the Fermi level is expected to lie in the d_{xy} band for low d electron counts.

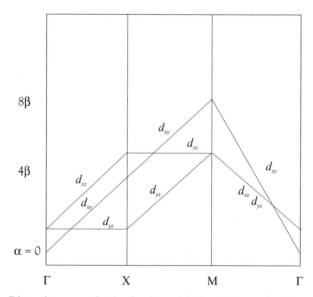

Figure 4.11   Dispersion curves for the d_{xy}, d_{xz}, and d_{yz} bands from a single layer of octahedra sharing vertices in only two dimensions, as in Figure 4.10.

Example 4.3   If the lowest energy point lies at Γ, predict the lowest-lying t_{2g}-block band for the $n = 3$ member of the Ruddlesden–Popper phase, which exhibits out-of-center octahedral distortion in the outer layers. Discuss the implications on the electronic properties of oxides with low d electron counts.

Solution   The $n = 3$ member of the Ruddlesden–Popper phases have *triple-layer* corner-sharing octahedral slabs separated by rock-salt-like layers. It is a low-dimensional transport system with nine t_{2g} bands. These can be denoted xz', xz'', xz''', and so on, where " denotes the d band of the central layer of an M–O–M–O–M linkage.

Drawing the orbitals in the M–O–M–O–M linkage allows one to count the antibonding contributions in each of these bands. Compared to that of the single-layer octahedral slab, each xy band in the triple-layer slab will have three times as many O p orbital contributions. The xz' and yz' bands will have three times as many *minus* two axial contributions. The xz'' and yz'' will have two times as many *plus* one bridging and *minus* two axial contributions. The xz''' and yz''' bands are equal to those of the xz' and yz' *plus* two bridging contributions.

Since we are assuming that the lowest energy states occur at Γ, we only have to estimate the energy at this one k-point. Using Canadell's relation, which states that the energy destabilization of an oxygen p contribution in a bridging position is four times that of an oxygen p orbital contribution in an axial position, it is found that the lowest-lying bands in the *undistorted* triple layer are three degenerate xy bands. However, shortened apical M–O bonds in the outer layers exhibiting out-of-center distortion raise the energies of those antibonding states so that *the xy'' band of the inner octahedral layer is the lowest band state in the oxide of Figure 1.36.*

The implications are that conduction electrons confined to the inner-layer slab, in oxides with low d electron counts, *may* be more spatially screened from electron localizing effects such as chemical or structural disorder in the rock-salt-like slabs, as compared to conduction electrons in single-layer slabs.

4.5.3   Illustration 3: Transition Metal Monoxides with Edge-Sharing Octahedra

Let us now look at the transition metal monoxides with the rock-salt structure. Since the rock-salt structure is a 3D network of edge-sharing MX_6 octahedra, in which the metal may possess an incomplete d shell, we conclude that the Fermi level should reside in the metal t_{2g}- or e_g-block bands.

Considering the similarity of this structure to the transition-metal perovskite oxides, one might expect to see the same dispersion curves for the d bands. However, because of the proximity of the metal atoms to each other, neglecting the metal d–d interactions in the monoxides produces dispersion curves that are grossly in error. The d–d interactions in the Bloch sums are important in affecting the dispersion of the d–p crystal orbitals.

We start by examining the dispersion of one of the t_{2g}-block bands, the d_{xz} band. We can draw the d_{xz} and p_z Bloch orbitals at Γ to include the atomic orbitals in the

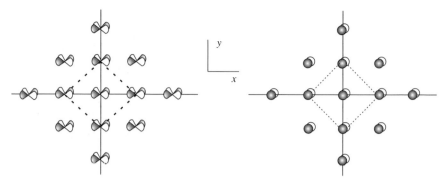

Figure 4.12 The sign combinations required of the atomic orbitals at Γ ($k_x = k_y = k_z = 0$, $\lambda = \infty$) for the d_{xz} Bloch sum (left) and p_z Bloch sum in a single plane viewed down the [001] direction.

face-center positions, as shown in Figure 4.12, where we have chosen the Cartesian coordinate system shown. The rock-salt structure can be considered two interpenetrating fcc lattices, one of cations and one of anions, displaced by one half a unit cell dimension along the $\langle 100 \rangle$ direction. This can be done with the two Bloch orbitals of Figure 4.12, as shown in Figure 4.13.

At Γ, the d_{xz} and p_z Bloch orbital interactions are not symmetry-allowed. Nor is the hybridization between the d_{xz} and p_y Bloch orbitals. The d_{xz} orbital at the origin does, however, have a π^* antibonding interaction with the axial p_x orbitals (not shown) directly above and below it. In other words, the d_{xz} orbital at the origin only interacts with two of the six first nearest-neighbor oxygen atoms.

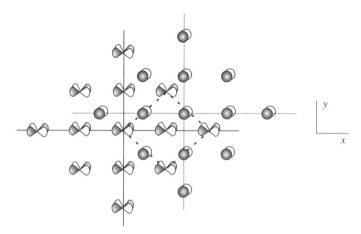

Figure 4.13 A schematic illustrating the superposition of the d_{xz} and p_z Bloch sums at Γ viewed down the [001] direction. The only symmetry-allowed d_{xz}–p_z interactions at Γ are those between the d_{xz} orbitals and the axial oxygen p_z orbitals directly above and below them (not shown). Thus, the dispersion at Γ is driven primarily by d–d interactions.

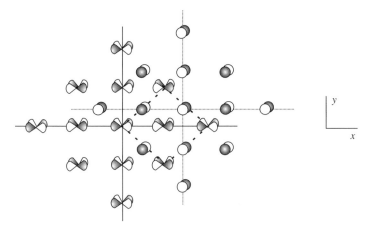

Figure 4.14   A schematic illustration of the superposition of the d_{xz} and p_z Bloch sums at X viewed down the [001] direction. The dispersion at X is driven primarily by π-type and δ-type d–d interactions.

In the rock-salt structure, there are d–d orbital interactions between the twelve second nearest-neighbor metal atoms that must also be included. Figure 4.13 shows four of these d_{xz}–d_{xz} interactions, two of which are π*-antibonding and two that are δ-bonding. The eight remaining d–d interactions are much weaker because of the poorer overlap between the orbitals on these atoms with the orbital at the origin. Hence, at Γ we have a roughly nonbonding d_{xz}–d_{xz} net interaction plus an antibonding p–d contribution.

The interactions at X are shown in Figure 4.14. The equatorial p_x and p_y orbitals do not have the correct symmetry for interaction with the d_{xz} orbitals. However, there are π*-antibonding interactions between the d_{xz} and equatorial p_z orbitals (shown) and axial p_x orbitals (not shown). The bonding interactions cause the dispersion of the p bands and are not being considered. The overlap between the d_{xz} and equatorial p_z orbitals in Figure 4.14 is poor, so these orbitals interact weakly. The d_{xz}–d_{xz} interaction at X is both π- and δ-bonding. As at Γ, these are much stronger than the p–d interactions, as well as the eight remaining d–d interactions. The d_{xz} band is thus lower in energy, relative to Γ.

Another symmetry point in the first Brillouin zone for the fcc lattice is W ($k_x = 2\pi/a$, $k_y = \pi/a$, $k_z = 0$). At this point, $\lambda = a$ along x and $2a$ along y. The Bloch orbitals for this k-point are shown in Figure 4. 15. The p–d interactions are the same as those at X. However, the net d–d interaction is *less* bonding than it is at X. The d_{xz} band is higher in energy at W, relative to X. It is also slightly higher than at Γ, due to a greater number of antibonding p–d interactions.

We can follow this same procedure to investigate the dispersion of one of the e_g-block bands, for example, the d_{x2-y2} band. The interactions at Γ, X, and W are shown in Figure 4.16. At point $\Gamma(k_x = k_y = k_z = 0)$, $\lambda = 2\pi/0 = 8$, all d–d interactions are bonding. The net interaction between the d_{x2-y2} and p_y orbitals is

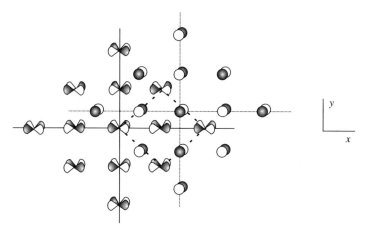

Figure 4.15 A schematic illustration of the superposition of the d_{xz} and p_z Bloch sums at W viewed down the [001] direction. The net d–d interaction is π-bonding, while the p–d interactions are antibonding.

nonbonding, as is the d_{x2-y2}–p_x interaction (not shown). The p_z orbitals are not of the correct symmetry for interaction.

At X ($k_x = 2\pi/a$, $k_y = k_z = 0$), $\lambda = a$ along the x-axis and 8 along the y-axis. The *net d–d* interaction is roughly nonbonding, while the d–p_y and d–p_z interactions are not symmetry-allowed. The four d–p_x interactions, however, are antibonding. Thus, the d_{x2-y2} band is raised in energy at X, relative to Γ.

At W ($k_x = 2\pi/a$, $k_y = \pi/a$, $k_z = 0$), $\lambda = a$ along the x-axis and $2a$ along the y-axis. The net d–d interaction is antibonding. None of the d–p interactions are

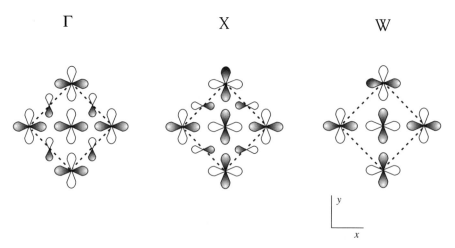

Figure 4.16 The d_{x2-y2}–d_{x2-y2} and p–d interactions at Γ, X, and W, in the transition metal monoxides with the rock-salt structure, viewed down the [001] direction.

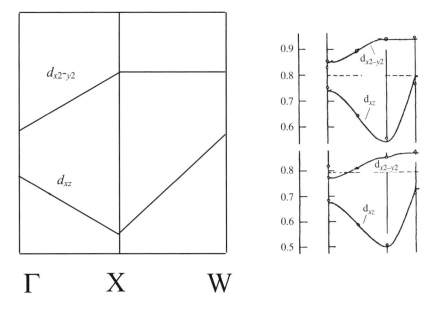

Figure 4.17 Dispersion curves for the d_{x2-y2} and d_{xz} bands in the transition metal monoxides. Upper right: calculated band dispersions for TiO. The Fermi level is indicated by the dashed line. Bottom right: calculated band dispersions for VO.

symmetry-allowed. In moving from W to X, the d_{x2-y2} band may be either nondispersive or slightly raised in energy. The actual bandwidth will depend on the strength of the M $3d$–M $3d$ interactions in comparison to the M $3d$–O $2p$ interactions.

The top curve on the left-hand side of Figure 4.17 shows our qualitative dispersion curves for the d_{x2-y2} band and d_{xz} band from Γ to X to W. Note that, although we have no numerical values for the energies, we have placed the d_{xz} band lower than the d_{x2-y2} band. That is, the "zero of energy" for each band is not equal. A cubic crystal field would be expected to have this effect in a solid, just as in a molecule. The top curves in each diagram on the right-hand side show the corresponding calculated dispersion curves in TiO and VO.

4.5.5 Corollary

With the conceptual approach, it is possible to acquire a reasonably accurate picture of the *shape* of the dispersion curves, which helps us understand exactly what it is we are looking at when we inspect an actual band-structure diagram. In the previous sections, we examined the dispersion of the curves in the conduction band. The width of the conduction band directly influences the magnitude of the electrical conductivity. Electrons in wide bands have higher velocities and mobilities than those in narrow bands. Hence, wide-band metals have higher electrical conductivities. Had we chosen to do so, the dispersion curves making up the valence band could have been evaluated as well. The most striking feature of the band structures of ionic compounds (e.g., KCl) is the narrow bandwidth of the valence band. With the electron transfer from cation to anion comes contraction of the lower-lying filled cation orbitals (due to a higher effective nuclear charge). For example, K^+ ($3s^2 3p^6 3d^{10}$) has spatially smaller orbitals than K ($3s^2 3p^6 3d^{10} 4s^1$). Thus, the overlap integral between orbitals on nearest neighbors (e.g., K^+ $3d$ and Cl^- $3s$ and $3p$), β, is small, which leads to very narrow valence bands.

We should be careful not to overplay the significance of the results from the last three sections, however. Our treatment has *not* provided any numerical estimates for the bandwidth, the extent of band filling (occupancy), or the magnitude of the band gap, all of which are extremely important properties with regard to electronic transport behavior. This is discussed in Section 5.3.1. Knowledge of the bandwidth is especially critical with transition-metal compounds. It is demonstrated in Chapter 6 that nonmetallic behavior will be observed in a metal, if the one-electron bandwidth, W, is less than the Coulomb interaction energy, U, between two electrons at the same bonding site of a crystal orbital. For that matter, in regard to predicting the type of electrical behavior, one has to be careful not to place excessive credence on *actual* electronic structure calculations that invoke the independent electron approximation. One-electron band theory predicts metallic behavior in all of the transition metal monoxides, although it is only observed in the case of TiO! The other oxides, NiO, CoO, MnO, FeO, and VO, are all insulating, despite the fact that the Fermi level falls in a partially filled band. In the insulating phases, the Coulomb interaction energy is over 4 eV, whereas the bandwidths have been found to be approximately 3 eV, that is, $U > W$.

4.6   TOTAL ENERGY TIGHT-BINDING CALCULATIONS

Although the tight-binding (TB) method was originally intended for explaining electronic spectra and structure, Chadi showed it can also be applied to total energy calculations (Chadi, 1978). These calculations can be used for determining bulk properties, such as lattice constants, cohesive energies, and elastic constants in solids, as well as surface structure. Because of its ease of implementation, low computational workload for large systems, and relatively good reliability, the TB scheme is a good compromise between *ab initio* simulations and model-potential calculations (Masuda-Jindo, 2001). A detailed discussion of this topic is beyond the scope of this book. For this, the reader is referred to specialized texts, such as (Ohno et al., 1999). A few words, however, are in order here.

The total energy of a solid can be written as

$$E_{tot} = E_{ee} + E_{ie} + E_{ii} \tag{4.57}$$

where E_{ee} is the electron–electron interaction energy, E_{ie} is the ion–electron energy, and E_{ii} is the ion–ion interaction energy. We can regroup these contributions into two terms. First, the band energy, E_{band}, is given by the sum over occupied orbital energies (up to the Fermi level), Σ_{ε_i}, derived from the diagonalization of the electronic Hamiltonian. In the TB scheme, this term is equal to $E_{ie} + 2E_{ee}$, where the factor 2 comes from double counting the Coulomb repulsion energy between electrons. The band-energy term is attractive, in that valence-band energy levels are lower in energy than the atomic orbitals from which they are derived, and it is thus responsible for the cohesion of a solid.

The second term in our new total energy expression is a short-range repulsive two-particle interaction and contains a correction for double counting the electrons in the band energy. It is equal to $E_{\text{rep}} = E_{ii} - E_{ee}$. Symbolically, we can thus write the new total energy expression as:

$$E_{\text{tot}} = E_{\text{band}} + E_{\text{rep}} \tag{4.58}$$

Some important vibrational data, which allow us to estimate some mechanical properties of solids, can be determined from the repulsive term. For example, in Chapter 9 it is shown that the Young's modulus (an elastic constant) of a solid can be approximated by treating the interatomic bonds as springs. The restoring force for small displacements of the atoms from their equilibrium positions (the stiffness of the spring) is given by the first derivative of the potential energy between a pair of ions, U, with respect to the interatomic distance, dU/dr. Of course, one should take the derivative of Eq. 4.58, the *total* energy, in order to include the contribution of the band term to the force, as well. One can even deduce IR and Raman spectra for small clusters of atoms from TB total energy minimizations, as well as ground-state geometries and binding energies per atom.

REFERENCES

Albright, T. A.; Burdett, J. K.; Whango, M.-H. 1985, *Orbital Interactions in Chemistry*, Wiley-Interscience, New York.

Blasé, X; Benedict, L. X.; Shirley, E. L.; Louie, S. G. 1994, *Phys. Rev. Lett.*, *72*, 1878.

Bloch, F. 1928, *Z. Physik.*, *52*, 555.

Canadell, E.; Whangbo, M.-H. 1991, *Chem. Rev.*, *91*, 965.

Chadi, D. J. 1978, *Phys. Rev. Lett.*, *41*, 1062.

Cohen, R. E.; Mehl, M. J.; Papaconstantopoulos, D. A. 1994, *Phys. Rev.*, *B50*, 14694.

Cotton, F. A. 1990, *Chemical Applications of Group Theory*, 3rd ed., John Wiley & Sons, New York.

Hamada, N.; Sawada, S.; Oshiyama, A. 1992, *Phys. Rev. Lett.*, *68*, 1579.

Harrison, W. A. 1989, *Electronic Structure and the Properties of Solids*, Dover Publications, Inc. New York.

Löwdin, P. O. 1950, *J. Chem. Phys.*, *18*, 365.

Masuda-Jindo, K. 2001, *Mat. Trans.*, *42*, 979.

Mattheiss, L. F. 1969, *Phys. Rev.*, *181*, 987.

Mattheiss, L. F. 1972, *Phys. Rev.*, *B5*, 290.

Mehl, M. J.; Papaconstantopoulos, D. A. 1996, *Phys. Rev.*, *B54*, 4519.

Ohno, K.; Esfarjani, K.; Kawazoe, Y. 1999, *Computational Materials Science: From Ab Initio to Monte Carlo Methods*, Springer Verlag, Berlin.

Rousseau, R.; Palacín, M. R.; Gómez-Romero, P.; Canadell, E. 1996, *Inorg. Chem.*, *35*, 1179.

Slater, J. C.; Koster, G. F. 1954, *Phys. Rev.*, *94*, 1498.

Saito, R.; Fujita, M.; Dresselhaus, G.; Dresselhaus, M. S. 1992, *Appl. Phys. Lett.*, *60*, 2204.

Whangbo, M.-H.; Canadell, E. 1989, *Acc. Chem. Res.*, *22*, 375.

CHAPTER 5

Transport Properties

The phenomena investigated in this chapter include thermal conduction, electrical conduction, and atomic/ionic diffusion in solids. These are transport processes, which can be described by the set of linear phenomenological equations given in Table 5.1. These equations relate a flux (diffusion of atoms, electrons, etc.) to a driving force (electric field, thermal gradients, etc.). Note that we need not rely on the details of any microscopic mechanism in order to *describe* macroscopic transport behavior, although we do need to learn about mathematical quantities called *tensors*. The phenomenological equations can be regarded as *continuum-level descriptions*. Of primary interest to us in this chapter, however, will be the underlying *atomistic-level descriptions*, which *explain* the behavior using principles of atomic and electron dynamics.

5.1 AN INTRODUCTION TO TENSORS

Single crystals are generally not isotropic. We can expect, then, that the physical properties of single crystals will be anisotropic, that is, dependent on the direction in which they are measured. Table 5.1 shows that the *magnitudes* of the fluxes and driving forces describing the transport properties are linearly proportional. However, the *directions* of the flux and driving force are parallel only for a cubic crystal or polycrystal with a random crystallite orientation. It is necessary to use *tensors* to explain anisotropic transport properties in the most precise manner.

The terms "scalar," "vector," and "tensor" were used in the 1840s by the Irish mathematician Sir William Rowan Hamilton (1805–1865) in his lectures on quaternions, or complex numbers. A quaternion consists of both a real part (a scalar) and a vector (possessing magnitude and direction). Hamilton referred to the magnitude or "modulus" of a quaternion as its tensor. Hence, the tensor of a quaternion is a scalar quantity. Although the modulus of a quaternion is still a tensor, the meaning of the word "tensor" has evolved since its introduction, and the word was apparently first used, in the present sense, in 1887 by Woldemar Voigt

Principles of Inorganic Materials Design   By John N. Lalena and David A. Cleary
ISBN 0-471-43418-3   Copyright © 2005 John Wiley & Sons, Inc.

TABLE 5.1   Physical Properties Relating Vector Fluxes and Their Driving Forces

Flux	Driving Force	Physical Property Tensor	Equation
Heat flow, q	Temperature gradient, ∇T	Thermal conductivity, κ	$q = -\kappa \nabla T$
Current density, j	Electric field, E	Electrical conductivity, σ	$j = \sigma E$
Diffusional flux, J	Concentration gradient, ∇c	Diffusion constant, D	$J = -D\nabla c$

(1850–1919) in describing a set of relations between space and time intervals to derive the Doppler shift (Voigt, 1887).

A tensor is an object with many components that look and act like components of ordinary vectors. The number of components necessary to describe a tensor is given by p^n, where p is the number of dimensions in space and n is called the *rank*. For example, a zero-rank tensor is a scalar, which has $3^0 = 1$ component, regardless of the value of p. A first-rank tensor is a vector; it has three components in 3D space (3^1) that are the projections of the vector along the axes of some reference frame, for example, the mutually perpendicular axes of a Cartesian coordinate system. Although the magnitude and direction of a physical quantity, intuitively, do not depend on our arbitrary choice of a reference frame, a vector is defined by specifying its components from projections onto the individual axes of the reference system. Thus, a vector can be defined by the way these components change, or *transform*, as the reference system is changed by a rotation or reflection. This is called a transformation law. For example, a vector becomes the negative of itself if the reference frame is rotated 180°, while a scalar is invariant to coordinate system changes.

Second-rank tensors, like the transport properties discussed in this chapter, relate two first-rank tensors, or vectors. Thus, a second-rank tensor representing a physical property has nine components (3^2), usually written in [3 × 3] matrix-like notation. Each component is associated with two axes: one from the set of some reference frame and one from the material frame. Three equations, each containing three terms on the right-hand side, are needed to describe a second-rank tensor exactly. For a general second-rank tensor τ that relates two vectors, p and q, in a coordinate system with axes x_1, x_2, x_3, we have:

$$p_1 = \tau_{11}q_1 + \tau_{12}q_2 + \tau_{13}q_3$$
$$p_2 = \tau_{21}q_1 + \tau_{22}q_2 + \tau_{23}q_3 \qquad (5.1)$$
$$p_3 = \tau_{31}q_1 + \tau_{32}q_2 + \tau_{33}q_3$$

The tensor is written as:

$$\begin{pmatrix} \tau_{11} & \tau_{12} & \tau_{13} \\ \tau_{21} & \tau_{22} & \tau_{23} \\ \tau_{31} & \tau_{32} & \tau_{33} \end{pmatrix} \qquad (5.2)$$

Note from Eq. 5.1 that each component of p is related to all three components of q. Thus, each component of the tensor is associated with a pair of axes. For example, τ_{32} gives the component of p parallel to x_3 when q is parallel to x_2. In general, the number of indices assigned to a tensor component is equal to the rank of the tensor. We will come across higher-order tensors in later chapters. Tensors of all ranks, like vectors, are defined by their transformation laws. For our purposes, we need not consider these.

Some simplifications can be made. Transport phenomena are processes whereby systems transition from a state of nonequilibrium to a state of equilibrium. Thus, they fall within the realm of irreversible or nonequilibrium thermodynamics. *Onsager's theorem* (Onsager, 1931a, 1931b), which is central to nonequilibrium thermodynamics, dictates that as a consequence of time-reversible symmetry, the off-diagonal elements of a transport property tensor are symmetrical, that is, $\tau_{ij} = \tau_{ji}$. This is known as a *reciprocal relation*. The Norwegian physical chemist Lars Onsager (1903–1976) was awarded the 1968 Nobel Prize in Chemistry for reciprocal relations. Equation 5.2 can thus be rewritten as:

$$\begin{pmatrix} \tau_{11} & \tau_{12} & \tau_{13} \\ \tau_{12} & \tau_{22} & \tau_{23} \\ \tau_{13} & \tau_{23} & \tau_{33} \end{pmatrix} \tag{5.3}$$

where there are now only six independent components.

Symmetrical tensors can also be diagonalized. For second-rank tensors, three mutually perpendicular unit vectors can be found that define three *principal axes*, such that if these axes are used as coordinate axes, the matrices are diagonal:

$$\begin{pmatrix} \tau_{11} & 0 & 0 \\ 0 & \tau_{22} & 0 \\ 0 & 0 & \tau_{33} \end{pmatrix} \tag{5.4}$$

Because of this further simplification, only three independent quantities in a symmetrical second-rank tensor are needed to define the magnitudes of the principal components. The other three components (from the initial six) are still needed to specify the directions of the axes with respect to the original coordinate system.

In the case of physical properties, crystal symmetry imposes even more restrictions on the number of independent components. This was deduced by the German physicist Franz Ernst Neumann (1798–1895) in the nineteenth century and is now known as *Neumann's principle*: The symmetry elements of any physical property of a crystal must include [at least] all the symmetry elements of the point group of the crystal (Neumann, 1833). Put another way, any type of symmetry exhibited by the point group of the crystal is also exhibited by every physical property of the crystal. Thus, according to Neumann's principle a tensor representing a physical property must be invariant with regard to every symmetry operation of the given crystal class. Tensors that must conform to the crystal symmetry are called *matter tensors*.

Because of Neumann's principle, the orientation of the principal axes of a matter tensor must also be consistent with the crystal symmetry. The principal axes of crystals with orthogonal crystallographic axes will be parallel to the crystallographic axes. In the monoclinic system, y is parallel to [010], while x and z are in (010). For triclinc crystals, there are no fixed relations between the principal and crystallographic axes, and so there are no restrictions on the directions of the principal axes. The effects of crystal symmetry on symmetrical second-rank matter tensors are given by Eqs. 5.5–5.9.

$$\text{Cubic crystals, polycrystals } \tau = \begin{pmatrix} \tau_{11} & 0 & 0 \\ 0 & \tau_{11} & 0 \\ 0 & 0 & \tau_{11} \end{pmatrix} \quad (5.5)$$

$$\text{Tetragonal, trigonal, hexagonal crystals } \tau = \begin{pmatrix} \tau_{11} & 0 & 0 \\ 0 & \tau_{11} & 0 \\ 0 & 0 & \tau_{33} \end{pmatrix} \quad (5.6)$$

$$\text{Orthorhombic crystals } \tau = \begin{pmatrix} \tau_{11} & 0 & 0 \\ 0 & \tau_{22} & 0 \\ 0 & 0 & \tau_{33} \end{pmatrix} \quad (5.7)$$

$$\text{Monoclinic crystals } \tau = \begin{pmatrix} \tau_{11} & 0 & \tau_{31} \\ 0 & \tau_{22} & 0 \\ \tau_{31} & 0 & \tau_{33} \end{pmatrix} \quad (5.8)$$

$$\text{Triclinic crystals } \tau = \begin{pmatrix} \tau_{11} & \tau_{12} & \tau_{31} \\ \tau_{12} & \tau_{22} & \tau_{23} \\ \tau_{31} & \tau_{23} & \tau_{33} \end{pmatrix} \quad (5.9)$$

The diagonal elements in Eqs. 5.5–5.9 follow from the indistinguishability of the axes in those crystal classes. For example, the [100], [010], and [001] directions are indistinguishable in the cubic lattice, but are distinguishable in the orthorhombic, monoclinic, and trigonal lattice. The off-diagonal elements in the monoclinic and triclinic crystals give the additional components necessary to specify the tensor. It can be seen that a cubic single crystal has isotropic transport properties. This is also true for polycrystals with a random crystallite orientation (e.g., powders), regardless of the crystal class. Thus, a single scalar quantity is sufficient for describing the conductivity in these materials.

John Nye (Courtesy H. H. Wills Physics Laboratory, University of Bristol. © University of Bristol. Reproduced with permission.)

JOHN FREDERICK NYE

(b. 1923) is a Fellow of the Royal Society and Emeritus Professor of Physics at the University of Bristol, England. He earned his Ph.D. from Cambridge in 1948 where, with Sir Lawrence Bragg and Egon Orowan, he helped to develop the bubble-raft model of a metal and showed how the photoelastic effect could be used to study arrays of dislocations in crystals. While in the mineralogy department at Cambridge University (1949–1951), and later at Bell Telephone Laboratories (1952–1953), Nye wrote *Physical Properties of Crystals: Their Representation by Tensors and Matrices*, which became the definitive textbook. He joined the physics department at the University of Bristol in 1953, becoming emeritus in 1988. Nye has applied physics to glaciology, serving as president of both the International Glaciological Society and of the International Commission of Snow and Ice. In 1974, Nye discovered the existence of dislocations in propagating wavefronts, and in 1983 their analogues, polarization singularities, in electromagnetic waves. These discoveries have developed into a new branch of optics, called singular optics, described in his book *Natural Focusing and Fine Structure of Light* (1999). Nye was elected a Fellow of the Royal Society in 1976.

(*Source*: J. F. Nye, personal communication, July 17, 2003.)

5.2 THERMAL CONDUCTIVITY

The phenomological equation for heat flow is Fourier's law, by the mathematician Jean Baptiste Joseph Fourier (1768–1830). It appeared in his 1811 work, *Théorie analytique de la chaleur* ("The Analytic Theory of Heat"). Fourier's theory of heat conduction entirely abandoned the caloric hypothesis, which had dominated eighteenth century ideas about heat. In Fourier's heat-flow equation, the flow of heat (heat flux), q, is written as:

$$q = -\kappa \nabla T \tag{5.10}$$

where κ is a positive quantity called the thermal conductivity and ∇T is the temperature gradient. Actually, thermal conductivity is a second-rank tensor, since the heat flux and temperature gradient are vectors. Heat flow within the system is in the direction of greatest temperature fall, but is not required to be exclusively parallel to the temperature gradient. For example, if a temperature gradient is set up along one axis of a Cartesian coordinate system, transverse heat flow may be measured parallel to the other two axes. Therefore, the equations representing the heat flow along the three axes are:

$$
\begin{aligned}
q_x &= \kappa_{11}(dT/dx) \quad \kappa_{12}(dT/dy) \quad \kappa_{13}(dT/dz) \\
q_y &= \kappa_{12}(dT/dx) \quad \kappa_{22}(dT/dy) \quad \kappa_{23}(dT/dz) \\
q_z &= \kappa_{13}(dT/dx) \quad \kappa_{23}(dT/dy) \quad \kappa_{33}(dT/dz)
\end{aligned}
\tag{5.11}
$$

where each scalar flux is parallel to an axis of a Cartesian coordinate system.

As discussed in Section 5.1, the number of independent components needing to be specified is reduced by the crystal symmetry. For example, if the x-axis is taken as the [100] direction, the y-axis as the [010], and the z-axis as the [001], for a cubic crystal or a polycrystalline sample with a random crystallite orientation we have:

$$
\kappa = \begin{pmatrix} \kappa_{11} & 0 & 0 \\ 0 & \kappa_{11} & 0 \\ 0 & 0 & \kappa_{11} \end{pmatrix}
\tag{5.12}
$$

in which case we have reduced the thermal conductivity tensor to a single independent component, the equivalent of a scalar quantity. Henceforth, for simplicity during our discussions of the underlying physical basis for transport phenomena, we will consider only cubic crystals or nontextured polycrystalline solids, for which case a single scalar quantity is sufficient.

5.2.1 The Free-Electron Contribution

The thermal conductivity of a metal or alloy consists of two components, a phonon contribution (a phonon is a quantum of acoustic energy, which possesses wave–particle duality), κ_{ph}, and an electronic contribution (free electrons moving through the crystal also carry thermal energy), κ_{el}. In pure metals, κ_{el} is the dominant contribution to the total thermal conduction. This free-electron contribution to the thermal conductivity is given by the gas-kinetic formula as:

$$\kappa_{el} = \pi^2 n k_b^2 T \tau / 3 m_e \tag{5.13}$$

where n is the electron density (number of free electrons per cm^3), τ is the average time an electron travels between collisions, m_e is the electron rest mass, T is the absolute temperature, and k_b is the Boltzmann constant. The parameter τ can be calculated from:

$$\tau = \lambda_{el}/V_F \qquad (5.14)$$

in which λ_{el} is the mean free path and V_F is the electron velocity at the Fermi surface. The denominator of Eq. 5.14 is related to the Fermi energy, ε_F, by:

$$V_F = (2\varepsilon_F/m_e)^{1/2} \qquad (5.15)$$

The fact that the thermal conductivity in a pure metal is dominated by the free-electron contribution was illustrated in 1853 by Wiedmann and Franz, who showed that κ_{el} and the electrical conductivity, σ_{el}, are proportionally related (Wiedemann and Franz, 1853):

$$\kappa_{el}/\sigma_{el} = LT \qquad (5.16)$$

Solving for L, and substituting both the expression for κ_{el} (Eq. 5.13) and the Drude expression for σ_{el} (Eq. 5.20), gives

$$L = \pi^2 k_B^2/3q^2 \qquad (5.17)$$

The Lorentz number L has the theoretical value $2.45 \times 10^{-8}\,\mathrm{W\Omega\,K^{-2}}$. If the experimental value determined for a metal is close to or equal to this, it can be assumed the electronic contribution dominates the thermal conductivity. At room temperature, experimental values for L range from $2.28 \times 10^{-8}\,\mathrm{W\Omega\,K^{-2}}$ for silver (the best conductor) to $3.41 \times 10^{-8}\,\mathrm{W\Omega\,K^{-2}}$ for bismuth (a poor conductor).

The electronic contribution to the thermal conductivity is reduced by electron-scattering events, including electron–electron, electron–phonon, and electron–defect scattering. Since they also limit the electrical conductivity, we will discuss electron-scattering mechanisms in Section 5.3.

5.2.2 The Phonon Contribution

The thermal conductivity of a pure metal is lowered by alloying, whether the alloy formed is a single-phase (solid solution) or multiphase mixture. There are several reasons for this. First, electrons are scattered by crystal imperfections and solute atoms (electron–defect scattering). Second, a substantial portion of the thermal conductivity in alloys, in contrast to that of pure metals, is by phonons, κ_{ph} (phonons are the sole contribution in electrically insulating solids) and phonons may also be scattered. Finally, electron–phonon interactions limit both κ_{el} and κ_{ph}.

So, what exactly are phonons? Phonons are quantized vibrational excitations (waves) moving through a solid, due to coupled atomic displacements. The idea of a lattice vibration evolved from work on the theory of the specific heat of solids by

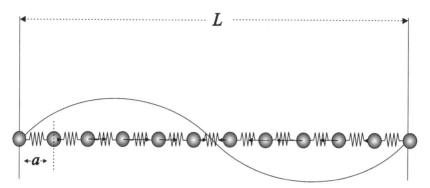

Figure 5.1 A 1D chain of atoms, of length L and lattice spacing a, connected by springs that obey Hooke's law. With the end-atoms fixed, only standing waves of atomic displacement are possible.

Einstein, Debye, Born, and von Kármán (Einstein, 1906, 1911; Debye, 1912; Born and von Karman, 1912, 1913). Let us examine a 1D chain of atoms, of finite length L, such as the one illustrated in Figure 5.1. In this figure, the chemical bonds (of equilibrium length a) are represented by springs. We assume that the interatomic potential is harmonic (i.e., the springs obey Hooke's law). Because the atoms are all connected, the displacement of just one atom gives rise to a vibrational wave, involving all the other atoms, which propagates down the chain. The speed of propagation of the vibrational excitation is the speed of sound down the chain.

For atoms constrained to move along one dimension, of course, only long-itudinal waves composed of compressional and rarefactional atomic displacements are possible. In general, a chain of N atoms interacting in accordance with Hooke's law has N degrees of freedom and acts like N independent harmonic oscillators with N independent *normal modes of vibration*. By analogy, a 3D harmonic solid has $3N$ normal modes of vibration (with both longitudinal waves and transverse waves possible). For simplicity in the present discussion, we chose fixed-boundary conditions in which the amplitudes of the atomic displacements are zero at the surface of the crystal, that is, the atoms at the surface are at fixed positions. Note that this does not properly account for the number of degrees of freedom. Nonetheless, with these boundary conditions, an integer number of phonon half-wavelengths must fit along the length of the chain in Figure 5.1 (or along each dimension of a 3D crystal). For the monatomic chain, the allowed frequencies are given by the relation:

$$\omega_k = 2(K/M)^{1/2}|\sin(ka/2)| \qquad (5.18)$$

where K is the force constant, M is the atomic mass, and $\boldsymbol{k}$ is the wave vector related to the wavelength, λ, of the wave by $|\boldsymbol{k}| = 2\pi/\lambda = n\pi/a (n = \pm1, \pm2, \ldots)$.

There are N allowed $\boldsymbol{k}$-states in the first Brillouin zone ($-\pi/a$ to π/a) corresponding to the independent normal modes of the system. The mode frequency and energy vary in a periodic way with the wave vector, which gives rise to a dispersion curve and a density of states, or number of possible modes (vibrational states) per frequency range. The wave function corresponding to one of these normal modes for the monatomic chain of Figure 5.1 is shown in that figure, with the necessary atomic displacements indicated by arrows. We now define a phonon as a particle-like entity (representing the wave) that can exist in one of these allowed states. The phonon occupation number at thermal equilibrium, or number of phonons in a given state (i.e., with a particular frequency, ω_k, or wave vector, k), follows Bose–Einstein statistics.

If a temperature gradient is now set up across the sample, a nonequilibrium distribution of phonons, called a wave packet (or pulse), is produced. The system reacts by attempting to restore the equilibrium distribution. That is, a thermal current occurs only if the phonon occupation number has a nonequilibrium value. The heat flux is, in fact, equal to the sum of the energies of the phonons in the distribution multiplied by the phonon group velocity divided by the crystal volume. Heat conduction is reduced by any event that reduces the phonon mean-free path or that causes a net change in momentum of the phonons. Phonons can propagate through a defect-free array of harmonic oscillators very easily, for, when phonons collide with one another in such an array, the total energy and momentum of the wave packet is conserved. An infinite defect-free harmonic crystal would possess an infinite thermal conductivity, because there would be no limit to the mean-free path of the phonons.

Real crystals are not defect free and real interatomic forces are anharmonic, that is, the interatomic potential is nonparabolic for all but small displacements. Anharmonicity gives rise to phonon–phonon scattering. First, however, we shall consider phonon–defect scattering. Point defects (including impurities and even different isotopes), dislocations, and grain boundaries in polycrystalline samples all cause elastic phonon scattering at low temperatures where only long-wavelength phonons are excited. The scattering shortens the phonon mean free path and limits thermal conductivity. The easiest way to think of phonon–defect scattering is to compare it to the way a long-wavelength light is Rayleigh scattered by small particles (where the particle radius is much less than the phonon wavelength). The scattering cross section is proportional to the radius of the crystal imperfection.

For impurity atoms with masses lighter than those of the other atoms in the lattice, a spatially localized vibrational mode called an *impurity exciton* is genera-ted, which has a frequency above that of the maximum allowed for propagating waves. The amplitude of the vibration decays to zero not far from the impurity. For heavier atoms, the amplitude of the mode is enhanced at a particular frequency. In noncrystalline solids, which lack translational periodicity (e.g., glasses), only those thermal excitations in a certain frequency range become localized, namely, in the band tails of the density-of-states (the number of vibrational states over a given frequency range). These impurity modes and excitations actually are what cause the phonons to scatter. That is, a phonon is scattered by the alteration in the lattice caused by the exciton. Thus, phonon–defect scattering is really due to exciton–phonon interactions.

With regards to the second feature of real crystals mentioned earlier, there are different types of anharmonicity-induced phonon–phonon scattering events that may occur. However, only those events that result in a total momentum change can produce resistance to the flow of heat. A special type, in which there is a net phonon momentum change (reversal), is the three-phonon scattering event called the *Umklapp process*. In this process, two phonons combine to give a third phonon propagating in the reverse direction.

The expression for the thermal conductivity in a single crystal, due to a *single* phonon mode, from the kinetic theory of gases, is:

$$\kappa_{ph} = \frac{1}{3} C \Lambda v \tag{5.19}$$

where C is the specific heat, Λ is the phonon mean-free path, and v is the phonon group velocity (the velocity at which heat is transported) (Elliot, 1998). Of course, one must take account of the contribution from all the phonon modes. Since the level of analysis required is somewhat above that assumed in this book, we merely quote the expression (Srivastava, 1990):

$$\kappa_{ph} = [h^2/(3Vk_BT^2)] \sum_{q,s} c_s^2(\boldsymbol{q})\omega^2(\boldsymbol{q}s)\tau(\boldsymbol{q}s)n(\boldsymbol{q}s)(n(\boldsymbol{q}s)+1) \tag{5.20}$$

where h is Planck's constant divided by 2π; V is the crystal volume; T is temperature; k_B is Boltzmann's constant; ω is the phonon frequency; c_s is the *wave packet*, or phonon group velocity; τ is the effective relaxation time; n is the Bose–Einstein distribution function; and $\boldsymbol{q}$ and s are the phonon wave vector and polarization index, respectively.

Equation 5.20 is a rather formidable expression. From the experimentalist's standpoint, it will be beneficial to point out some simple, but useful, criteria. First, we note that thermal conductivity is *not* an additive property. It is generally not possible to predict the thermal conductivity of an alloy or compound from the known thermal conductivities of the substituent pure elements. For example, the thermal conductivity of polycrystalline silver and bismuth are, respectively, 429 and 8 W/mK. Yet, the two-phase alloy Bi-50 wt % Ag has a thermal conductivity of only 13.5 W/mK. Often, alloys have thermal conductivities about equal to those of the metal oxides, indicating that a large portion of the total conductivity in alloys is due to phonons. Hence, our limited ability to predict thermal conductivities for alloys (and compounds) is mostly due to the difficulty in determining the κ_{ph} contribution. One of the best-suited methods of obtaining estimates for the phonon contribution is molecular dynamics. However, it is computationally intensive and most effective for modeling systems of limited size.

Based on approximate solutions to Eq. 5.20, four simple criteria for choosing high-conductivity single crystal materials have been established (Srivastava, 2001): (1) low atomic mass, (2) strong, highly covalent, interatomic bonding, (3) simple crystal structure, and (4) low anharmonicty. A systematic evaluation (Slack et al., 1987) has revealed that most of the high thermal conductivity (>100 W m K^{-1}) ceramics are compounds of the light elements that possess a diamond-like crystal structure (e.g., BN, SiC, BeO, BP, AlN, BeS, GaN, Si, AlP, GaP). Conversely, it should be expected that low thermal-conductivity materials generally possess complex crystal structures, especially those exhibiting low dimensionality, or have constituents with high atomic masses.

5.3 ELECTRONIC CONDUCTIVITY

The phenomenological equation for electrical conduction is Ohm's law, which first appeared in *Die galvanische kette mathematisch bearbeitet* ("The Galvanic Circuit Investigated Mathematically"), the 1827 treatise on the theory of electricity by the Bavarian mathematician Georg Simon Ohm (1789–1854). Ohm discovered that the current through most materials is directly proportional to the potential difference applied across the material. An equivalent form of Ohm's law is:

$$j = \sigma E \tag{5.21}$$

where j is the current density, σ is the electrical conductivity tensor, and E is the electric field. For simplicity, in our discussion of the physical basis for electrical conductivity we presume that σ is a scalar; that is, we are only considering isotropic media such as cubic crystals or polycrystalline samples. Likewise, the theory of current flow in heterostructures (e.g., through semiconductor $p - n$ junctions) belongs in the realm of electrical engineering or semiconductor physics, and it is not discussed in this text.

In the Drude model for an electron gas, it is predicted that the electronic conductivity, σ (in units of Ω^{-1} m^{-1}), is proportional to a quantity known as the electron relaxation time, which characterizes the decay of the drift velocity upon removal of the electric field. The conductivity is expressed as:

$$\sigma = nq^2\tau/m_e \tag{5.22}$$

where n is the electron density (m^{-3}), q is the electron charge (coulombs), m_e is the electron mass (kg), and τ is the relaxation time (seconds). Grouping some of the terms into a single parameter for the mobility, gives a well-known alternative expression for the conductivity:

$$\sigma = nq\mu \tag{5.23}$$

in which $\mu = q\tau/m_e$. The mobility is defined as $\mu_e = v/E$, where v is the steady-state velocity of the particle in the direction of an electric field E. In semiconductors, there may be a contribution from both electrons and holes. Thus, the expression for the electrical conductivity becomes:

$$\sigma = nq\mu_e + pq\mu_h \tag{5.24}$$

where p is the hole density (m^{-3}) and μ_h is the hole mobility. For the Drude model to yield Ohm's law (Eq. 5.21), one must neglect the influence of the applied electric field on the drift velocity of the charge carriers. That is, the relaxation time has to be considered independent of the applied field strength.

Example 5.1   The mean atomic mass of potassium is 39.10 amu, its density is 0.86×10^3 kg m^{-3}, and its electron configuration is [Ar]$4s^1$. What is n, the number of valence electrons per unit volume? If the electrical conductivity is 0.143×10^8 Ω^{-1} m^{-1}, what is τ, the relaxation time between collisions?

Solution   The density of K $= 0.86 \times 10^3$ kg m$^{-3} = 0.86$ g cm^{-3}. Dividing this by the mass of one mole, 39.10 g, gives the number of moles of K atoms per cubic centimeter:

$$0.86/39.10 = 0.022 \, \text{mole cm}^{-3}$$

Multiplying this by Avogadro's number yields the number of K atoms per cubic centimeter:

$$6.0223 \times 10^{23} \times 0.022 = 1.3 \times 10^{22} \, \text{atoms per cm}^3$$

Since there is one $4s$ valence electron per K atom, the electron density, n, is also equal to 1.3×10^{22} valence electrons per cm^3, or 1.3×10^{28} per m^3. Drude's formula can now be used to calculate the relaxation time between collisions:

$$\tau = m_e \sigma_e / nq^2$$
$$\tau = (9.11 \times 10^{-31} \, \text{kg})(0.143 \times 10^8 \, \Omega^{-1} \, \text{m}^{-1})/(1.3 \times 10^{28} \, \text{m}^{-3})(-1.6 \times 10^{-19} \, \text{C})^2$$
$$\tau = 3.9 \times 10^{-14} \, \text{seconds}$$

It is not immediately obvious that the units of τ should be seconds. However, this can be confirmed by making the following substitutions:

$$\Omega = V/A = V/(C/s), \quad V = J/C, J = \text{kg m}^2 \, \text{s}^{-2}$$

Equation 5.22 predicts that electronic conductivity is dependent on the electron relaxation time. However, it suggests no physical mechanisms responsible for controlling this parameter. Since electrons exhibit wave–particle duality, we might suspect scattering events to play a part. In a perfect crystal, the atoms of the lattice scatter electrons coherently so that the mean-free path of an electron is infinite. However, in real crystals there exist different types of electron scattering processes that can limit the electron mean-free path and, hence, conductivity. These include the collision of an electron with other electrons (electron–electron scattering), lattice vibrations, or phonons (electron–phonon scattering), and impurities (electron–impurity scattering).

Electron–electron scattering is generally negligible due to the Pauli exclusion principle. Nonetheless, it is important in the transition metals, where s-conduction elecrons are scattered by d-conduction electrons. Electron–phonon scattering is the dominant contribution to the electrical resistivity in most materials at temperatures well above absolute zero. Note that both these scattering processes must be *inelastic* in order to cause a reversal in the electron momentum and increase the electrical resistance. At low temperatures, by contrast, *elastic* electron–impurity scattering is the dominant mechanism responsible for reducing electrical conductivity. Anderson showed that multiple elastic scattering in highly disordered media stops electron propagation by destructive interference effects (Anderson, 1958). It has also been shown that even in a weakly disordered medium, multiple elastic scattering can be sufficient to reduce the conductivity (Bergmann, 1983, 1984). Elastic electron–impurity scattering is further discussed in Section 6.2.

5.3.1 Band-Structure Considerations

We now examine how both the density and mobility of charge carriers in metals and band semiconductors (i.e., those in which electrons are not localized by disorder or correlation) are influenced by particular features of the electronic structure, namely band dispersion and band filling. Taking mobility first, let us briefly revisit the topic of *band dispersion*. Charge carriers in narrow bands have a lower mobility because they are "almost" localized in atomic orbitals that overlap very poorly. Although this may seem intuitive enough, we can build a more convincing case simply by looking at some fundamental physical relationships.

Consider the meaning of the wave vector k, which is equal to π/a. Since $\lambda = 2a$, k must also be equal to $2\pi/\lambda$. The DeBroglie relationship states that $\lambda = h/p$, where p is the electron momentum. Therefore, $p = hk/2\pi = \hbar k$. The velocity, v, is related to the momentum through $v = \hbar k/m$. However, we also know that the electron energy $E(k)$ is equal to $\hbar^2 k^2/2m$. Hence, $dE(k)/dk = \hbar^2 k/m = v\hbar$. Since $dE(k)/dk$ is the slope of the tangent line to $E(k)$, a conduction electron in a wide band (greater slope) has a higher velocity and higher mobility than one in a narrow band. Consequently, it follows from Eq. 5.23 that a wide-band metal, that is, a metal with wide bands in its band-structure diagram, will have a higher electrical conductivity than a narrow-band metal.

Turning our attention now to the other factor, charge-carrier density, requires us to examine *band filling*. In the absence of a strong correlation or disorder, band filling determines the type of electrical behavior (insulator, semiconductor, metal) observed in a solid. When spin degeneracy is included, each crystal orbital can hold two electrons per bonding site, for a total of $2N$ electrons, where N is equal to the number of atoms in the crystal. The lowest crystal orbitals are filled first, followed by the next lowest, and so on, in a completely analogous manner, as is done in a MO energy-level diagram. Completely filled low-lying crystal orbitals make up the valence band. The conduction band consists of the higher-energy crystal orbitals that are either vacant or partially filled.

Considering only the ground-state band in the tight-binding approximation, Bloch's picture of the electron wave function in a periodic lattice was successful at explaining how electrons are free to move in a metal and conduct an electrical current.

Unfortunately, immediately after Bloch's work, the existence of insulators became puzzling! The currently accepted explanation for the difference between insulators and metals was first proposed by the British physicist Sir Alan Herries Wilson (1906–1995) while working with Bloch and Heisenberg at Leipzig. Recall how the sign of k corresponds to the direction of motion of the electron. Application of an electric field causes a shift, Δk, of the entire electron distribution in an energy band, which accelerates the electrons, increasing their momentum in the direction of the field. In a half-filled band (all electrons have the same spin), there are available k-states immediately above the highest filled ground state that can become occupied. Recall that for a system of N atoms, the energy separation between states becomes infinitesimal as N approaches infinity. Hence, the field can cause a change in the distribution of the electrons and produce a *net* current flow in one direction. However, when all the k-states in an energy band are filled (two electrons, paired with opposite spin, per bonding site), and a sizable energy gap separates this band from the next highest band, the application of an electric field cannot bring about a shift in the k-distribution of the electrons (Wilson, 1931).

The highest filled ground state is termed the *Fermi level*, E_F. The energy of this level is the *Fermi energy*. All states above the Fermi level are empty. Thus, the criterion for metallicity in the Bloch/Wilson band picture can be stated as follows. If the Fermi level lies in a partially filled band of delocalized states, band theory predicts metallic behavior. The partially filled band could be either the conduction band or the valence band, since in both cases the Bloch wave functions composing the band are delocalized throughout the crystal.

This may seem counterintuitive at first, since it might appear as though we are saying the presence of delocalized electrons is not a sufficient condition for electronic conduction. Perhaps, the confusion can be cleared up with a more chemical picture. If one recalls the molecular orbital description of benzene, a similar situation is found. Even though the π-bonding and antibonding molecular orbitals in benzene are delocalized throughout the entire molecule, the benzene molecule is not metallic-like. Benzene, rather, can be considered a molecular resonant tunneling transistor. Single molecules of benzene sandwiched between two gold electrodes can only transport a current, upon application of a potential difference, via an excited state in which an electron is excited from a filled π-bonding molecular orbital to a vacant π^*-antibonding molecular orbital, where it can be accelerated by the electric field. At low voltages, the π^* state is the sole contributor, while at higher voltages, the π states also participate (Pantelides et al., 2002). Molecular electronics, which is based on the goal of using single molecules as active devices, is a concept that has been around for three decades (Aviram and Ratner, 1974). Unfortunately, it is not within the scope of this textbook.

Band theory predicts insulating behavior if there is a wide energy gap (say, $E_g > 3$ eV) separating the top of a completely filled valence band and the bottom of an empty conduction band. No net current flow can occur in one direction or the other within the filled valence band and the wide band gap prevents electrons from being thermally excited into singly occupied states in the conduction band. Since energy varies with k, the bottom of the conduction band can occur at the same point (direct-gap) or at a different point (indirect-gap) than the top of the valence band. Band-structure diagrams show only k values of one sign, positive or negative. In insulators, the chemical potential of the highest-energy electron (Fermi energy) falls within the band gap where there is a zero density of states.

Intrinsic semiconductors, by comparison, have small band gaps ($E_g < 3$ eV), which allow electrons to be thermally excited from the valence band to the conduction band. The excitation of an electron from the valence band produces a "hole" or positive charge carrier (it behaves like an electron with a positive charge) in the valence band so that current flow in a semiconductor is comprised of both electron flow and hole flow. The concept of a "hole" was first developed in 1931 by Heisenberg (concurrent with Wilson's work), based on the work of Peierls (Heisenberg, 1931). However, the experimental proof of "hole flow" is made by measuring the so-called Hall effect, which was discovered by E. H. Hall in 1879 (Section 5.3.2). *Extrinsic* semiconductors are those in which the carrier concentration, either holes or electrons, are controlled by intentionally added impurities, called *dopants*. The dopants are termed *shallow* impurities because their energy levels lie within the band gap close to one or other of the bands. Because of thermal excitation, *n*-type dopants (*donors*) are able to donate electrons to the conduction band, and *p*-type dopants (*acceptors*) can accept electrons from the valence band, the result of which is equivalent to the introduction of holes in the valence band. At low temperatures, a special type of electrical transport known as *impurity conduction* proceeds. This topic is discussed in Section 6.3.

Although scattering processes in both semiconductors and metals increase with rising temperature, thereby decreasing the mobility of the carriers, the scattering is more than offset in a semiconductor by an increase in the charge-carrier concentration. Thus, the electrical resistivity of a semiconductor *decreases* with increasing temperature, $d\rho/dT < 0$. In a metal, the resistivity has the opposite temperature dependency: $d\rho/dT > 0$.

A final possibility is that of a *semimetal*. In this case, there is a zero density of states at the Fermi level, but no band gap. Semimetals differ from semiconductors in that their resistivities have a metallic-like temperature dependency. All of these behaviors are illustrated schematically in Figure 5.2, which shows the simplified band picture and the corresponding density of states (DOS) for each case.

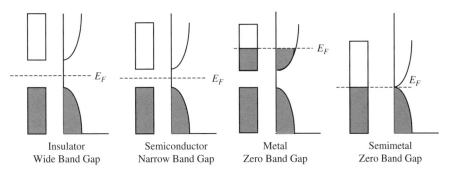

Figure 5.2 Schematic density-of-states diagrams illustrating the different classes of electrical behavior.

Alan Wilson (Courtesy of The Royal Society. © Godfrey Argent. Reproduced with permission.)

SIR ALAN HERRIES WILSON

(1906–1995) earned a B.S. degree in mathematics from Cambridge University in 1926. He stayed for a Research Studentship in Quantum Mechanics under R. H. Fowler (who had already supervised, among others, Paul Dirac, Douglas Hartree, and John E. Lennard-Jones), earning an M.A. in applied mathematics in 1929. Wilson was later awarded a Rockefeller Traveling Fellowship, which afforded him the opportunity to work in Leipzig with Werner Heisenberg. It was at Leipzig where Wilson explained the differences between metals, semiconductors, and insulators. Wilson became a Lecturer in Mathematics at Cambridge in 1933. In 1936, he published the well-known book *The Theory of Metals*, and, in 1939, *Semi-conductors and Metals*. In the late 1930s, Wilson became interested in nuclear physics. His work in that field contributed to the understanding of the importance of unitarity in the modern theory of hadron interactions. During World War II, he worked on the British atomic bomb project. After the war, there was pressure for academics to join industry and Wilson was passed over as Fowler's successor at Cambridge. When asked to become head of research and development at Courtaulds, Ltd., a large British textile company, Wilson reluctantly accepted the position. He was to spend the rest of his career as a business executive. In his spare time, he continued work in solid-state physics and he published an advanced book on thermodynamics and statistical mechanics in 1957. He left Courtaulds in 1962 and joined Glaxo where he became Chairman in 1963, retiring in 1973. Wilson was elected a Fellow of the Royal Society in 1942. He was knighted in 1961.

(*Source*: E. H. Sondheimer, *Biographical Memoirs of the Fellows of the Royal Society of London*, Vol 45, **1999**, pp. 547–562.)

5.3.2 Electronic Conduction in the Presence of Other External Gradients

Thermoelectric Effects A thermoelectric device is a closed circuit comprised of two dissimilar materials joined together, which converts thermal energy from a temperature gradient (the junctions being maintained at different temperatures) into electrical energy. This is known as the *Seebeck effect*. Alternatively, electrical energy can be converted into a temperature gradient, referred to as the *Peltier effect*. The thermoelectric effects may be derived starting from equations representing the rate of heat evolution in each of the two materials in the couple (see Nye, 1957). The result is an expression for the thermoelectric power of the couple, called the *Seebeck coefficient*, in terms of the absolute thermoelectric powers for each of the two materials, which can be written as

$$S = E/\Delta T = -(\Sigma^{(a)} - \Sigma^{(b)})/e \qquad (5.25)$$

where S is the Seebeck coefficient, E is the potential difference, or net electromotive force generated, ΔT is the temperature gradient, e is the charge of the electron, and $\Sigma^{(i)}$ is the absolute thermoelectric power of component i. It should be noted that $T(\Sigma^{(a)} - \Sigma^{(b)})/e$ is the *Peltier coefficient*, normally denoted as Π, and can be considered the rate of heat evolution per unit area of interface. Hence, the Seebeck coefficient is related to the Peltier coefficient through:

$$S = -\Pi/T \qquad (5.26)$$

This expression is known as the *second Thomson relation*. Couples exhibiting a high Seebeck coefficient are potentially useful as power-generation devices, whereas couples with a high Peltier coefficient can be utilized as heat pumps in small-scale solid-state refrigeration applications.

Couples comprised of two dissimilar elemental metals have low Seebeck coefficients, since the absolute thermopowers of metals are in the microvolt per degree Celsius range (superconductors have *zero* absolute thermopowers). This makes them unsuitable for use in most thermoelectric applications, with the exception of thermocouples. Many elemental and compound semiconductors, by contrast, have absolute thermopowers in the millivolt per degree Celsius range. Thermoelectric materials must also have a relatively high electrical conductivity, in order to suppress the deleterious effects of resistive heating, and a low thermal conductivity, to reduce the tendency of the material to dissipate the established temperature gradients (Winder et al., 1996).

Thermoelectric materials can be distinguished based on their fractions of Carnot efficiency attainable in a cooling device. This coversion efficiency is called the thermoelectric power figure of merit, Z, and is given by the expression:

$$Z = S^2 \sigma T/\kappa \qquad (5.27)$$

where S is the Seebeck coefficient ($\mu V/K$), σ is the electrical conductivity (S/cm), and κ is the thermal conductivity (W/cmK) (Barnard, 1972; Meng et al., 2000). The figure of merit for a material must be at least 3×10^{-4} K^{-1}, in order for it to be useful as a heat pump in refrigeration. The figure of merit is heavily dependent on the carrier concentrations, which typically are 10^{14}–10^{21} carriers/cm^3 in semiconductors and 10^{22} carriers/cm^3 in metals. The greatest Z values are obtained for the range 10^{18}–10^{21} carriers/cm^3, which substantiates the earlier claim that the best thermoelectric devices are comprised of semiconductors.

The thermoelectric properties of many different semiconductors have been investigated. Various oxides have been found to have relatively high figures of merit. For example, the layered $NaCo_2O_4$ (Terasaki et al., 1997) with Cu substitution has a Z value of 8.8×10^{-4} K^{-1} (Yakabe et al., 1997). However, the sodium content is very difficult to control, owing to the element's high volatility (Nishiyama et al., 1999). The values of the figure of merit for most other oxides, to date, are in the range 0.59×10^{-4} K^{-1} to 2.4×10^{-4} K^{-1}.

All the oxides thus far investigated have Z values considerably smaller than those of the chalcogenide thermoelectric materials, which are as typically as high as 1.0 K^{-1}, in the case of suitably doped antimony bismuth telluride. Bismuth telluride (Bi_2Te_3), antimony telluride (Sb_2Te_3), and bismuth selenide (Bi_2Se_3) have a nine-layer structure, composed of close-packed Te(Se) anions with Bi(Sb) cations occupying two-thirds of the octahedral holes. Very recently, cubic $AgPb_{10}SbTe_{12}$ and $AgPb_{18}SbTe_{20}$ were reported as having figures of merit of about 2.2×10^{-4} K^{-1} at 800 K (Hsu et al., 2004). Other intermetallic compounds that have been investigated include: $Ba_4In_8Sb_{16}$ (Kim et al., 1999), with $(In_8Sb_{16})^{8-}$ layers separated by Ba^{2+} ions; $Ba_3Bi_{6.67}Sb_{13}$ and its variants, consisting of a 3D $(Bi_{6.67}Se_{13})^{6-}$ anionic network with open channels containing Ba^{2+} cations; and the clathrate compound $Cs_8Zn_4Sn_{12}$ (Nolas et al., 1999).

Galvanomagnetic Effects and Magnetotransport Properties When a current-carrying sample is placed in the presence of orthogonal (perpendicular) magnetic and electric fields (e.g., the x and z directions), the current is deflected by the Lorentz force (due to the magnetic field) in a mutually perpendicular third direction (e.g., the y direction). The associated charge buildup on the $+y$ and $-y$ faces of the sample normal to the y direction creates a secondary electric field across those faces (Hall, 1879). This field is termed the *Hall field*. In steady state, it is balanced by the opposing Lorentz force, so that no current flows in that direction. Hall-effect measurements are used to determine the sign of the charge carriers and the carrier concentration. In ferromagnets, a spontaneous (anomalous) Hall current is observed in the presence of an electric field alone. The current is transverse to the field.

The deflection of the mobile electrons by the Lorentz force also increases the electrical resistivity. This phenomenon, termed *positive* magnetoresistivity, is often observed in metals with anisotropic Fermi surfaces. On the other hand, *negative* magnetoresistivity, that is, a decrease in resistivity with the application of a magnetic field, can occur when a field-induced ferromagnetic alignment of spins (electrons) reduces the scattering of the charge carriers. A detailed discussion of this topic is postponed until Chapter 7 on magnetic properties.

5.4 ATOMIC TRANSPORT

Diffusion is a process in which the transport of matter through a substance occurs. *Self-diffusion* refers to atoms diffusing among others of the same kind (e.g., in pure metals). *Interdiffusion* is the diffusion of two dissimilar substances (a diffusion couple) into one another. *Impurity diffusion* refers to the transport of dilute solute atoms in a host solvent.

In solids, diffusion is several orders of magnitude slower than in liquids or gases. Nonetheless, diffusional processes are important to study because they are basic to our understanding of how solid–solid, solid–vapor, and solid–liquid reactions proceed. Atomic transport processes may also be categorized into those that are concentration-gradient induced (possible in all types of solids) and those that are electric-field induced (occurring only in ionic crystals), which gives rise to an electrical current. Temperature gradients may also give rise to diffusion, in the absence of a concentration gradient (the Soret effect), but this is not discussed here. Another phenomenon, called *electromigration*, is the mass transport resulting from momentum transfer from conduction electrons to atoms. It is only significant at high current densities in metals and is likewise not discussed in this textbook.

The governing phenomenological equation for ionic conduction is, as in electronic conduction, Ohm's law (Eq. 5.21). Concentration-gradient-induced processes, on the other hand, follow *Fick's laws of diffusion*, derived by Adolf Eugen Fick (1829–1901) in 1855 (Fick, 1855). We discuss the latter case first.

5.4.1 Atomic Diffusion

Fick's first law is written as:

$$\boldsymbol{J}_i = -D_i \nabla n_i \tag{5.28}$$

where D is known as the *diffusivity* or *diffusion coefficient* (units of cm^2/s), $\boldsymbol{J}_i$ is the *net* diffusional flux (the number of particles crossing a unit area), and ∇n_i is the concentration gradient of species of type i. Three equations represent the diffusional fluxes along the principal axes exist and are entirely analogous to Eq. 5.11 for heat flow. It should be noted that ∇n_i might refer to a concentration gradient involving solute atoms, vacancies, or interstitials. Equation 5.28 states that the flux is proportional to the concentration gradient (the higher the value for D_i, the larger $\boldsymbol{J}_i$ for the same ∇n_i), and that flow will cease only when the concentration is uniform. It was later proposed by Einstein (Einstein, 1905) that the force acting on a diffusing atom or ion is, in fact, the negative gradient of the chemical potential, which is dependent on not only concentration, but temperature and pressure as well. However, for our purposes, we shall consider T and P to be uniform, in which case the concentration gradient will determine the flow.

Fick's first law represents steady-state diffusion. The concentration profile (the concentration as a function of location) is assumed constant with respect to time. In general, however, concentration profiles do change with time. In order to describe these non-steady-state diffusion processes use is made of *Fick's second law*, which is derived from the first law by combining it with the *continuity equation* ($\partial n_i/\partial t = -[J_{in} - J_{out}] = -\nabla \mathbf{J}_i$):

$$\partial n_i/\partial t = \nabla(D_i \nabla n_i) \tag{5.29}$$

If the material is sufficiently homogeneous, D_i can be considered constant (independent of concentration), so that Eq. 5.29 reduces to:

$$\partial n_i/\partial t = D_i \nabla^2 n_i \tag{5.30}$$

Fick's second law, which is also known simply as the *diffusion equation*, indicates that nonuniform gradients tend to become uniform.

When two dissimilar substances are joined together (forming a diffusion couple), they homogenize by interdiffusion, a concept introduced in the 1930's by the German physical chemist Carl Wagner (1901–1977). Fick's laws imply that each species will have its own intrinsic diffusion coefficient. Interestingly, an apparent bulk flow occurs as a result of the differing intrinsic diffusion coefficients. Specifically, the side of the couple with the fastest diffuser shortens while the side with the slowest diffuser lengthens. Smigelskas and Kirkendall demonstrated that fine molybdenum wire could be used as a marker in the Cu-Zn system to provide an alternative description of interdiffuional processes between substances with dissimilar diffusion coefficients (Smigelskas and Kirkendall, 1947). A brass (Cu-Zn) bar, wound with molybdenum wire, was plated with copper metal. The specimen was annealed in a series of steps, in which the movements of the molybdenum wires were recorded. The inert markers had moved from the interface towards the brass end of the specimen, which contained the fastest diffuser – zinc. This is now called the Kirkendall effect. A similar marker experiment had actually been performed by Hartley a year earlier while studying the diffusion of acetone in cellulose acetate (Hartley, 1946), but most metallurgists were not familiar with this work (Darken and Gurry, 1953).

Lawrence Stamper Darken (1909–1978) subsequently showed how, in such a marker experiment, values for the intrinsic diffusion coefficients (i.e. $D = N_{Zn}D_{Cu}$ and $N_{Cu}D_{Zn}$) could be obtained from a measurement of the marker velocity and a single diffusion coefficient, called the *interdiffusion coefficient* (Darken, 1948). This quantity is sometimes called the *mutual, or chemical, diffusion* coefficient and it is a more useful quantity than the more fundamental intrinsic diffusion coefficients from the standpoint of obtaining analytical solutions to real engineering diffusion problems. Interdiffusion, for example, is of obvious importance to the study of the chemical reaction kinetics. Indeed, studies have shown that interdiffusion is often the rate-controlling step in solid-state reactions.

When there is a linear relationship between the diffusivity and chemical potential gradient, the principle of microscopic reversibility is assumed to hold.

A more general statement of this principle is that the mechanical equations of motion of the individual particles are invariant under the reversal of the time from t to $-t$. Hence, the nonequilibrium system can be divided up into volume elements for which the state functions have the equilibrium values. Local equilibrium was, in fact, presumed by the German physical chemist Carl Wagner in his early work relating the diffusivity of the rate-determining species with the reaction rate constant for diffusion-controlled reactions, years before classical irreversible thermodynamics was a generally accepted field.

So far, the equations of mass flow have been at the continuum-level. Our primary objective is to examine the microscopic, or atomistic, description. Wagner and Schottky showed that diffusion is mediated by point defects. Diffusion of atoms or ions can occur by at least three possible mechanisms, as shown schematically in Figure 5.3. In some cases, transport proceeds primarily by the *vacancy mechanism*, where an atom jumps into an adjacent, energetically equivalent vacant lattice site. The vacancy mechanism is generally much slower than the interstitial mechanism discussed below. Nonetheless, it is believed responsible for *self-diffusion* in pure metals and for most alloys (Shewmon, 1989).

When thermal oscillations become large enough, an atom already in an interstitial site or one that migrates there, can move among energetically equivalent

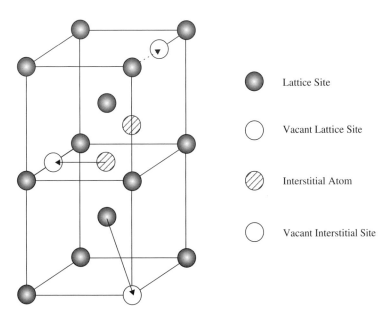

Lattice Site

Vacant Lattice Site

Interstitial Atom

Vacant Interstitial Site

Figure 5.3 Schematic illustrating the different mechanisms for atomic diffusion in a bcc lattice. In the vacancy mechanism, an atom in a lattice site jumps to an adjacent vacant lattice site. In the interstitial mechanism, an interstitial atom jumps into an adjacent vacant interstitial site. In the intersitialcy mechanism, an interstitial atom pushes an atom residing in a lattice site into an adjacent vacant insterstitial site and occupies the displaced atom's site.

vacant interstitial sites without permanently displacing the solvent (lattice) atoms. This is termed the *direct interstitial mechanism,* and is known to occur in interstitial alloys and in some substitutional alloys when the substitutional atoms spend a large fraction of their time on interstitial sites. Although it is found empirically that only solute atoms with a radius less than 0.59 that of the solvent atom dissolve interstitially, diffusion studies have shown that solute atoms with radii up to 0.85 that of the solvent can spend enough time in interstitial sites for the interstitial mechanism to dominate solute transport (Shewmon, 1989).

Atoms larger than this would produce excessively large structural distortions if they were to diffuse by the direct interstitial mechanism. Hence, in these cases diffusion tends to occur by what is known as the *interstitialcy mechanism.* In this process, the large atom that initially moves into an interstitial position displaces one of its nearest neighbors into an interstitial position and takes the displaced atom's place in the lattice. This is the dominant diffusion process for the silver ion in AgBr, where an octahedral site Ag^+ ion moves into a tetrahedral site, then displaces a neighboring Ag^+ ion into a tetrahedral site and takes the displaced Ag^+ ion's formal position in the lattice.

It is possible to obtain expressions for the diffusivity for a particular case, if one knows the microscopic mechanism of diffusion. However, it will be instructive to first derive an expression for D using a probabilistic approach for which no detailed mechanism is assumed: the theory of *random walk.* A species (atom or ion) starting at its original position, makes n jumps and ends up at a final position that is related to the original position by a vector, designated as R_n:

$$R_n = \sum_{i=1}^{n} \lambda_i \tag{5.31}$$

in which λ_i are the vectors representing the various jumps. To obtain the *average* value of R_n (more exactly R_n^2), it is necessary to consider many atoms, each of which takes n jumps. If we assume that each jump is independent of the jump preceding it, and that positive and negative directions are equally probable (completely uncorrelated and random jumps, as in the case of self-diffusion in pure metals, for example), the following is true:

$$\langle R_n^2 \rangle^{1/2} = \lambda \sqrt{n} \tag{5.32}$$

Equation 5.32 states that the root-mean-square displacement is proportional to the square root of the number of jumps. For very large values of n, the net displacement of any one atom is extremely small compared to the total distance it travels. It turns out that the diffusion coefficient is related to this root-mean-square displacement. It was shown independently by Albert Einstein (1879–1955) and Marian von Smoluchowski (1872–1917) that, for Brownian motion of small particles suspended in a liquid, the root-mean-square displacement, $\langle R_n^2 \rangle^{1/2}$, is equal to $\sqrt{(2Dt)}$, where t is the time (Einstein, 1905; von Smoluchowski, 1906).

In the case of 3D diffusion, the following expression is valid for isotropic media (equivalent principal axes along which diffusion takes place), such as cubic crystals:

$$\langle R_n^2 \rangle = 6Dt \qquad (5.33)$$

Note that in Eq. 5.33 the mean-square displacement is used, rather than the root-mean-square displacement. For a 1D random walk, the mean-square displacement is given by $2Dt$, and for a 2D random walk, $4Dt$. Since the jump distance (a vector) is λ, if we now define the jump frequency, $\Gamma = n/t$ (the average number of jumps per unit time), then on combining Eq. 5.32 and 5.33 we get:

$$D = (1/6)\Gamma\lambda^2 \qquad (5.34)$$

It must be stressed that this result holds only for uncorrelated jumps. It is possible to include a correlation factor, and the interested reader is referred to the book by Elliot for this (Elliot, 1998).

The jump vector, λ, will obviously depend on the mechanism and the structure. For example, an atom diffusing through the octahedral interstitial sublattice in an fcc metal, with lattice spacing a (Figure 5.4), must jump the distance between interstitial sites, $\lambda = a/\sqrt{2}$. This is, of course, the same distance that an atom diffusing by the vacancy mechanism must jump. It will be recalled that for every atom in a close-packed structure, there are two tetrahedral interstitial sites and one octahedral interstitial site. The reader might ask if the distances between the tetrahedral sites are the same.

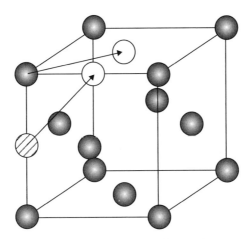

Figure 5.4 Simple geometry shows that the jump distance between octahedral interstitial sites and between lattice sites in an fcc metal, with unit cell edge a, is $a/\sqrt{2}$.

Diffusion is a thermally activated process. It is expected that mass transport will proceed more rapidly at elevated temperatures. Empirically, the Arrhenius equation is found to hold:

$$D = A \exp(-Q/RT) \tag{5.35}$$

where T is the absolute temperature (K), R is the gas constant (8.31 J/mol K), A is a the preexponential constant (sometimes called the frequency factor and denoted as D_0) in units of m^2/s, and Q is the activation energy (J/mol). Both A and Q vary with composition, but are independent of temperature. Usually, values for these parameters are obtained by plotting $\ln D$ versus $1/T$. The slope of the plot yields Q/R and the intercept at $1/T = 0$ is $\ln A$.

Much effort has gone into developing theoretical expressions for A, which must reflect the temperature independence. The origin of this behavior can be seen by referring back to Figure 5.4. An appreciable local distortion of the lattice must take place, for example, before an interstitial jump can occur. Energy must be supplied to the lattice to cause closed-packed atoms to move apart and let the interstitial through. The difficulty with which this is accomplished constitutes the activation barrier to interstitial diffusion, and it can be used to estimate A in Eq. 5.35. Thus, Γ in Eq. 5.34 has been described by:

$$\Gamma = v_0 z [\exp(-\Delta G_m \ddagger /RT)] \tag{5.36}$$

where v_0 is the vibrational frequency of an atom in the lattice (the frequency of attempts an atom makes), z is the coordination number (the number of distinct sites that interstitial atoms can reach in a single jump), and $\Delta G_m \ddagger$ is the free energy of activation for migration (Zener, 1951). Combining Eqs. 5.34, 5.35, and 5.36, and realizing that $\Delta G_m \ddagger = \Delta H_m \ddagger - T\Delta S_m \ddagger$, one obtains:

$$D = [\lambda^2 v_0 z/6] \exp(\Delta S_m \ddagger /R) \exp(-\Delta H_m \ddagger /RT) \tag{5.37}$$

In Eq. 5.37, $[\lambda^2 v_0 z/6]\exp(\Delta S_m\ddagger/R)$ is equal to A in Eq. 5.35.

In the vacancy mechanism, the activation barrier to diffusion is generally much smaller. This is because a lot of energy is not required to displace the surrounding atoms, as in the interstitial mechanism. However, diffusion may be impeded somewhat due to there being a finite number of vacancies. Each atom must wait its turn for a vacancy to become available. Accounting for this, together with the energy required to form the vacancies, Eq. 5.37 for the interstitial mechanism is modified to become:

$$D = [\lambda^2 v_0] \exp[(\Delta S_v + \Delta S_m) \ddagger /R] \exp[(-\Delta H_v - \Delta H_m) \ddagger /RT] \tag{5.38}$$

In this expression, $[\lambda^2 v_0]\exp[(\Delta S_v + \Delta S_m)\ddagger/R]$ is equal to A in Eq. 5.35. Note the conspicuous absence of the geometric constant $z/6$ in Eq. 5.38. For the vacancy mechanism, the diffusivity is given by an expression (equivalent to Eq. 5.34) of the form: $D = \lambda^2 N_v w$, where N_v is the vacancy concentration, and w is the jump frequency of an atom into an adjacent vacancy. Equation 5.38 is a consequence of the thermodynamic description for the equilibrium concentration of vacancies (for a derivation, see Shewmon, 1989):

Equations 5.37 and 5.38 represent *volume* diffusion effects. Surface diffusion and grain boundary diffusion may also be important, under certain circumstances. Recall (Section 1.2.5) that grain boundaries are less dense than the grains, and so are easy diffusion paths. Hence, grain boundary diffusion coefficients are generally much greater than volume diffusion coefficients. However, for a unit sample area, the overall flux is the sum of that through the grain plus that through the boundary. Consequently, grain boundary diffusion is expected to be of importance only for very small grain sizes (higher fraction of boundary area) or at low temperatures.

5.4.2 Ionic Conduction

The phenomenological equation for electrical conduction in an ionic solid is the same as that for electronic conduction, that is, Eq. 5.21. However, the universal expression for electrical conductivity (Eq. 5.23) must be modified to include the transport contribution by all the species, including cations, anions, and electrons:

$$\sigma_{total} = \sigma_{el} + \sum_i n_i q_i \mu_i \qquad (5.39)$$

Materials in which both ionic and electronic conduction occur are termed *mixed conductors*. Examples include $LiNiO_2$ and $LiCoO_2$,. In most ceramics, however, σ_{el} is negligible at normal temperatures.

A more general expression for the ionic conductivity term in Eq. 5.39 is:

$$\sigma = B_i n_i (z_i q)^2 \qquad (5.40)$$

where now B_i is the mobility of species i, and z_i is the charge on the species. If the substitutions $B = \mu/q$ and $z_i = \pm 1$ are made, Eq. 5.40 is found to be equivalent to Eq. 5.23 for electronic (hole) conduction.

In physical chemistry, it is customary to speak in terms of the fraction of the conductivity carried by a particular species. The fraction of the total current carried by the ith species is called the *transference number*, or transfer number t_i. The total of all the transference numbers, including that of the electrons in the case of mixed conductors, must equal one. In many ceramics, the conduction occurs predominantly via the movement of one type of ion. For example, in the alkali halides, $t_{cation} \sim 1$, whereas in CeO_2, BaF_2, and $PbCl_2$, $t_{anion} \sim 1$.

For any particular solid, the relative activation barriers for the available mechanisms determine whether the anions or cations are responsible for the ionic conduction. For example, in yttria-stabilized ZrO_2, with the formula $Zr_{1-x} Y_x O_{2-(x/2)}$ aliovalent substitution of Zr^{4+} by Y^{3+} generates a large number of oxygen vacancies, giving rise to a mechanism for oxide-ion conduction. Indeed, it is found that the O^{2-} anions diffuse about six orders of magnitude faster than the cations.

The microscopic mechanisms for ionic conduction are the same as those for atomic diffusion, namely, the vacancy and interstitial models discussed in the previous section. In fact, the diffusivity can be related to the conductivity via the *Nernst–Einstein equation*:

$$D_i = B_i k_B T \tag{5.41}$$

where k_B is the Boltzmann constant, B_i is the mobility of i ions, and D_i is their diffusivity. Dividing Eq. 5.40 by Eq. 5.41, shows that:

$$\sigma_i/D_i = n_i(z_i q)^2/k_B T \tag{5.42}$$

where D_i is given by Eq. 5.37 (interstitial mechanism) or Eq. 5.36 (vacancy mechanism).

Like diffusion, ionic conduction is a thermally activated process. Low activation barriers are therefore achieved in the same manner. Geometric features such as open channels result in larger diffusivities (easier ion movement) because this lowers the magnitude of the ΔH_m terms in Eqs. 5.37 and 5.38. For example, in β-alumina, the sodium ions are located in sparsely populated layers positioned between spinel blocks. Accordingly, they diffuse through these channels easily due to the large number of vacancies present.

REFERENCES

Anderson, P. W. 1958, *Phys. Rev.*, *109*, 1492.

Aviram, A.; Ratner, M. A. 1974, *Chem. Phys. Lett.*, *29*, 277.

Barnard, R. D. 1972, *Thermoelectricity in Metals and Alloys*, Taylor & Francis, London.

Bergmann, G. 1983, *Phys. Rev. B*, *28*, 2914.

Bergmann, G. 1984, *Phys. Rep.*, *107*, 1.

Born, M.; von Kármán, Th. 1912, *Z. Phys.*, *13*, 297.

Born, M.; von Kármán, Th. 1913, *Z. Phys.*, *14*, 15.

Darken, L. 1948, *Trans. AIME*, *175*, 184.

Darken L. S.; Gurry, R. W. 1953, *Physical Chemistry of Metals*, McGraw-Hill Book Company, New York.

Debye, P. 1912, *Ann. Phys.*, *39*, 789.

Einstein, A. 1905, *Ann. Phys.*, *17*, 549.

Einstein, A. 1906, *Ann. Phys.*, *22*, 180, 800.

Einstein, A. 1911, *Ann. Phys.*, *34*, 170.

Elliot, S. 1998, *The Physics and Chemistry of Solids*, John Wiley & Sons, Chichester, UK.

Fick, A. 1855, *Pogg. Ann. Phys. Chem.*, *94*, 59.

Frenkel, J. 1926, *Z. Phys.*, *35*, 652.

Hall, E. H. 1879, *Am. J. Math.*, *2*, 287.

Hartley, G. S. 1946, *Trans. Faraday Soc.*, *42*, 6.

Heisenberg, W. 1931, *Ann. Phys.*, *10*, 888.

Hsu, H. F.; Loo, S.; Guo, F.; Chen, W.; Dyck, J. S.; Uher, C.; Hogan, T.; Polychroniadis, E. K.; Kanatzidis; M. G. 2004, *Science*, *303*, 818.

Kim, S-J.; Hu, S.; Uher, C.; Kanatzidis, M. G. 1999, *Chem. Mater.*, *11*, 3154.

Meng, J. F.; Polvani, D. A.; Jones, C. D. W.; DiSalvo, F. J.; Fei, Y.; Badding, J. V. 2000, *Chem. Mater.*, *12*, 197.

Neumann, F. E. *Pogendonff Ann. Phys.*, 1833, *27*, 240.

Nishiyama, S.; Ushijima, T.; Asakura, K.; Hattori, T. 1999, *Key Eng. Mater.*, *67*, 169.

Nolas, G. S.; Weakly, T. J. R.; Cohn, J. L. 1999, *Chem. Mater.*, *11*, 2470.

Nye, J. F. 1957, *Physical Properties of Crystals: Their Representation by Tensors and Matrices* Oxford University Press, London.

Onsager, L. 1931a, *Phys. Rev.*, *37*, 405.

Onsager, L. 1931b, *Phys. Rev.*, *38*, 2265.

Pantelides, S. T.; Di Ventra, M.; Lang, N. D.; Rashkeev, S. N. 2002, *IEEE Trans. Nano.*, *1*, 86.

Schottky, W.; Wagner C. 1930, *Z. Phys. Chem.*, *11B*, 163.

Seitz, F. 1940, *The Modern Theory of Solids*, McGraw-Hill Book Company, New York.

Shewmon, P. 1989, *Diffusion in Solids*, The Minerals, Metals, and Materials Society, Warrendale, PA.

Slack, G. A.; Tanzilli, R. A.; Pohl, R. O.; Vandersande, J. W. 1987, *J. Phys.Chem. Solids*, *48*, 641.

Smigelskas, A.; Kirkendall, E. 1947, *Trans. AIME*, *171*, 130.

Srivastava, G. P. 1990, *The Physics of Phonons*, Adam Hilger, Bristol, UK.

Srivastava, G. P. 2001, *MRS Bull.*, *26*, 445.

Terasaki, I.; Sasago, Y.; Uchinokura, K. 1997, *Phys. Rev. B*, *56*, R12685.

Voigt, W. 1887, "Über das Dopplersche Prinzip," *Nachr. Königlichen Ges. Wiss. Georg-Augusts-Univ. Göttingen*, *2*, 41–51.

Von Smoluchowski, M. 1906, *Ann. Phys.*, *21*, 756.

Watari, K.; Shinde, L.; 2001, *MRS Bull.*, *26*, 440.

Wiedemann, G.; Franz, W. 1853, *Ann. Phys.*, *89*, 497.

Wilson, A. H. 1931, *Proc. R. Soc. London*, *133*, 458.

Winder, E. J.; Ellis, A. B.; Lisensky, G. C. 1996, *J. Chem. Ed.*, *73*, 940.

Yakabe, H.; Kikuchi, K.; Terasaki, I.; Sasago, Y.; Uchinokura, K. 1997, *Proc. Int. Conf. Thermoelectr.*, 523.

Zener, C. 1951, *J. Appl. Phys.*, *22*, 372.

CHAPTER 6

Metal–Nonmetal Transitions

We have seen that, in the Bloch–Wilson band picture, insulators are materials with a completely filled valence band and an empty conduction band, separated by a sizable band gap. Semiconductors are similar, but have smaller band gaps. Thermal excitation of charge carriers across the band gap ensures that the electrical conductivity of a band insulator or semiconductor will increase with increasing temperature. In metals, electron scattering mechanisms decrease the electrical conductivity as the temperature is raised. Commonly, the reciprocal quantity resistivity, ρ, is used. An experimental criterion for metallic behavior (and semimetals) is thus that $d\rho/dT > 0$, while for insulators and semiconductors $d\rho/dT < 0$. A material need not possess a low electrical resistivity (e.g., copper has an electrical resistivity of about 10^{-6} Ω cm) to be regarded as metallic; rather its resistivity must merely have a positive temperature coefficient. Materials that meet this criterion for metallicity but that, nonetheless, exhibit high resistivities, are termed *marginal metals*.

It is possible to induce metallic transport behavior in solids with band gaps by doping the conduction band with charge carriers. This is appropriately termed a *filling-control metal–nonmetal transition*, since one is filling a formerly vacant conduction band with charge carriers. We expand on that topic later in this chapter. First, however, we need to discuss other kinds of insulators, in order to more fully develop the concept of the metal–nonmetal (M–NM) transition.

About one decade after the development of band theory, de Boer and Verwey reported that many transition metal oxides with partially filled bands, which band theory predicted to be metallic, were poor conductors and some even insulating (de Boer and Verwey, 1937). Rudolf Ernst Peierls (1907–1995) first pointed out the possible importance of electron correlation in controlling the electrical behavior of these oxides (Peierls, 1937). Electron correlation is the term applied to the interaction between electrons, via Coulomb's law.

Somewhat later, Mott intuitively predicted a M–NM transition resulting from a strong Coulomb attraction between the conduction electrons and the positively charged ion cores, which are screened from any one conduction electron by all the other conduction electrons (Mott, 1949). A conduction electron, diffusing under an

Principles of Inorganic Materials Design   By John N. Lalena and David A. Cleary
ISBN 0-471-43418-3   Copyright © 2005 John Wiley & Sons, Inc.

electric field, encounters other electrons as it moves among singly occupied sites. Hubbard then introduced a model in which a band gap opens in the conduction band because of the Coulomb repulsion between two electrons at the same site (Hubbard, 1963, 1964a, 1964b). Thus, by this definition the Mott–Hubbard insulator is only realizable when the conduction band is at integer filling, but only partially filled. That is only possible with multicomponent systems, that is, compounds.

The opening of a band gap within the conduction band obviously leads to insulating behavior. However, we must emphasize the difference between the origin of gaps in correlated systems and systems with noninteracting electrons. In a Bloch–Wilson insulator, electrons in the valence band are delocalized in the sense that the crystal orbitals are extended throughout the volume of the crystal. As a particle, the electron has an equal probability of existing in any cell of the crystal. However, the Pauli principle prohibits an electron from moving into an already completely filled crystal orbital. There is also a sizable energy gap (>3 eV) separating the top of the filled valence band and the bottom of the empty conduction band. Hence, no energetically near degenerate states (immediately above the Fermi level) exist in which an electron can be accelerated by an electric field. By contrast, in a Mott–Hubbard insulator, the Coulomb repulsion between two conduction electrons (with antiparallel spin) at the same bonding site is strong enough to keep these electrons away from one another and spatially localized in individual atomic orbitals, rather than delocalized Bloch functions. A narrow bandwidth and concomitant band gap results. The localized conduction electrons are rather like the electrons in the very narrow valence band of an ionic crystal (e.g., KCl). Because of this picture of localized wave functions, such insulators are also sometimes called Heitler–London-type insulators (Section 2.2.2).

The intrasite Coulomb gap is completely unaccounted for in Hartree–Fock theory (e.g., tight-binding calculations), since electron correlation is neglected in the independent-electron approximation. It is underestimated in the LDA to the DFT. Simple LDA–DFT calculations are also generally poor at reproducing the metallic state near the Mott transition. There are several reasons for this. First, it is difficult to treat the nonuniform electron densities of correlated systems with localized wave functions. Second, there is the necessity in density functional formalisms of introducing a self-interaction correction term for the Coulomb interaction energy of an electron in a given eigenstate with itself (not needed in the Hartree–Fock approximation!), and that is fitted to the results for a uniform electron density.

Electrons in periodic solids can also be localized by disorder. This is a rather different mechanism, which involves multiple elastic scattering of the conduction electrons by the impurities. When the disorder and, hence, scattering is strong enough, the electrons can no longer propagate through the solid. This is known as Anderson localization. However, unlike band insulators and Mott insulators, no band gap is opened in Anderson insulators. There remains a finite DOS at the Fermi level, but the electrons are localized on individual sites, as in Mott insulators, rather than existing as itinerant, or delocalized, particles. Anderson argued that in such systems, transport is by nondiffusive, phonon-assisted hopping between localized centers (Anderson, 1958).

6.1 CORRELATED SYSTEMS

Mott originally considered an array of monovalent metal ions on a lattice, in which the interatomic distance, d, can be varied. Very small interatomic separations correspond to the condensed crystalline phase. Because the free-electron-like bands are half-filled in the case of ions with a single valence electron, one-electron band theory predicts metallic behavior. However, it predicts that the array will be metallic, *regardless* of the interatomic separation. Clearly, this cannot be true given that, in the opposite extreme, isolated atoms are electrically insulating.

For an isolated atom, the single valence electron feels a strong Coulomb attraction to the ion core, $V = -e^2/r$, where e is the proton charge and r is is distance separating the electron and ion core. In the solid state, the Coulomb attraction of any one valence electron to its ion core is reduced because it is "screened" from it by all the other electrons of the free-electron gas. The potential now becomes $V = (-e^2/r)\exp(-\xi r)$, where ξ is the screening parameter. Mott argued that when the potential is just strong enough to trap an electron, a transition from the metallic to the nonmetallic state occurs. The Mott criterion for the occurrence of a M–NM transition was found to be:

$$n_c^{1/3} a_0^* = 0.25 \tag{6.1}$$

where n_c is the critical electron density, and a_0^* is the effective Bohr radius.

The screening parameter, ξ, increases with the increasing magnitude of the tight-binding hopping integral, which, in turn, increases with the decreasing internuclear separation between neighboring atoms. In the context of the tight-binding or LCAO method, rather than considering the screened potential, one usually speaks in terms of the bandwidth, which also increases with decreasing internuclear separation (due to stronger overlap between adjacent orbitals). Hence, metallic behavior would be expected for internuclear distances smaller than some critical value. A large internuclear distance results in more localized atomic-like wave functions for the electrons (i.e., the Heitler–London model), as opposed to extended Bloch functions, since the hopping integral and the overlap integral are very small in this case. Thus, Mott recognized that there should be a sharp transition to the metallic state at some critical bandwidth: $W = 2zb$ (Eq. 4.38), where z is the coordination number and b is the hopping integral (Mott, 1956, 1958).

The concepts described here are particularly important for the d-electrons, since they do not range as far from the nucleus as s- or p-electrons. The d-electrons in solids are best described with localized atomic orbitals if the magnitudes of the hopping and overlap integrals between neighboring atoms are very small (i.e., narrow-band systems). An increase in the extent of d-electron localization is expected as one moves to the right in a period, and for smaller principal quantum numbers because of a contraction in the spatial extension of the d-atomic orbitals. Valence s and p orbitals are always best described by Bloch functions, while $4f$-electrons are localized and $5f$ are intermediate. For heavy elements, however, one may also need to consider s and p orbital contraction due to both increased electrostatic attraction toward the nucleus (the lanthanide and actide contraction) and direct relativistic effects. The latter, in fact, causes a d and f orbital expansion, termed the indirect relativistic effect, due to the increased shielding of those orbitals from the nucleus.

A semiempirical approach for characterizing the transition region separating the localized (Heitler–London) and itinerant (Bloch) regimes, in terms of the critical internuclear distance, was developed by Goodenough (Goodenough, 1963; 1965; 1966; 1967; 1971). This method employs the tight-binding hopping integral (transfer integral), b, which is proportional to the overlap integral between neighboring orbitals. The approach is useful for isostructural series of transition-metal compounds, since b can be related to the interatomic separation. In the $ANiO_3$ perovskites (A = Pr, Nd, Sm, Eu), for example, the M–NM transition is accompanied by a very slight structural change involving coupled tilts of the NiO_6 octahedra. In the ABO_3 perovskites, BO_6 octahedral tilting is in response to nonoptimal values of the geometrical tolerance factor, t (Eq. 2.20), which places the B–O bonds under compression and the A–O bonds under tension, or vice versa (Goodenough and Zhou, 1998). Tilting relieves the stresses by changing the B–O–B angles, which govern the transfer interaction. The smaller the tolerance factor (for $t < 1$), the greater the deviation of the B–O–B angle from 180°. This reduces the one-electron d-bandwidth and localizes the conduction electrons. For $ANiO_3$, only when A is lanthanum is the oxide metallic. The other oxides in the series, for which the t's are smaller ($t < 1$), are insulators. The maximum Ni–O–Ni angle deviation is $\sim -0.5°$. The structural transition is thus a subtle one. Nonetheless, it may be considered the driving force for the M–NM transition, because the changes to the Ni–O bond length affect the d-bandwidth.

The crucial step in our present understanding the M–NM transition in systems with interacting electrons was Hubbard's introduction of a model that included dynamic correlation effects–the short-range Coulomb repulsion between electrons with antiparallel spin at the same bonding site (Hubbard, 1963; 1964a; 1964b). Long-range (*inter*site) electron–electron interactions were neglected. The inclusion of dynamic correlation effects in narrow-band systems, such as transition-metal compounds, form the basis for what is now known as the Mott–Hubbard M–NM transition.

Nevill Mott (Courtesy of AIP Emilio Segrè Visual Archives. © The Nobel Foundation. Reproduced with permission.)

SIR NEVILL FRANCIS MOTT

(1905–1996) received his bachelor's degree (1927) and master's degree (1930) from the University of Cambridge. He became a professor of theoretical physics at the University of Bristol in 1933 and returned to Cambridge in 1954 as Cavendish Professor of Physics, a post from which he retired in 1971. In his early career, Mott worked on nuclear physics; atomic collision theory; the hardness of metals; the electronic structures of metals and alloys; and latent image formation in photographic emulsions. In the 1940s, he postulated how a crystalline array of hydrogen-like electron donors could be made insulating by increasing the inter-atomic separation. Mott was to devote the rest of his career to the study of the metal–nonmetal transition. He was a co-recipient of the 1977 Nobel Prize in Physics for his work on the magnetic and electronic properties of noncrystalline substances. Mott was elected a Fellow of the Royal Society in 1936. He was knighted in 1962.

(*Source*: *Nobel Lectures*, Physics 1971–1980, World Scientific Publishing Company, Singapore.)

6.1.1 The Mott–Hubbard Insulating State

The Hubbard picture is the most celebrated and simplest model of the Mott insulator. It is comprised of a tight-binding Hamiltonian, written in the second quantization formalism. Second quantization is the name given to the quantum field theory procedure by which one moves from dealing with a set of particles to a field. Quantum field theory is the study of the quantum mechanical interaction of elementary particles with fields. Quantum field theory is a notoriously difficult subject, and we dare not attempt to go beyond the level of merely quoting equations in this textbook. The Hubbard Hamiltonian is:

$$\mathcal{H} = t_{ij} \sum_{i,j\sigma} C_{i\sigma}^{\dagger} C_{j\sigma} + U \sum_{i} n_i \uparrow n_i \downarrow \tag{6.2}$$

Equation 6.2 contains just two parameters: t, an intersite hopping energy for an electron jump from one site to an adjacent site (hybridization), and U, which represents the Coulomb repulsion of two antiparallel spin electrons at the *same* site. It is possible to include interactions between electrons on adjacent sites and, in fact, 30 years earlier Shubin and Wonsowsky of the Russian school put forth a model of interacting electrons that accounts for intersite interactions between adjacent sites (Shubin and Wonsowsky, 1934; Izyumov, 1995). Nowadays, the inclusion of next-nearest-neighbor interactions is called the *extended Hubbard model*. However, for our purposes, we need not consider this. The Hubbard Hamiltonian can be solved exactly only in the 1D case and requires numerical techniques for higher dimensions. Nonetheless, it often adequately describes tendencies and qualitative behavior.

The first term in Eq. 6.2 represents the electron kinetic energy. It contains the Fermi creation operator $(C_{i\sigma}^{\dagger})$ and annihilation operator $(C_{j\sigma})$ for an electron or hole, at sites i and j with spin $\sigma = \uparrow$ or $\downarrow$. In quantum field theory, Hamiltonians are written in terms of Fermi creation and annihilation operators, which add and subtract particles from multiparticle states, just as the ladder operators (raising and lowering operators) for the quantum harmonic oscillator add and subtract energy quanta. Multiparticle wave functions are specified in terms of single-particle *occupation numbers*. The second term in Eq. 6.2 is the potential energy term. The number of electrons at the site is given by n_i, and the Coulomb repulsion between two electrons at the *same* site, U, is given by:

$$U = \int \int [e^2/|r_{12}|]|\varphi(r_1)|^2|\varphi(r_2)|^2 dr_1 dr_2 \qquad (6.3)$$

The distance between two electrons at a given site is given as r_{12}. The electron wave function for one of the electrons is given as $\varphi(r_1)$, and the wave function for the second electron, with antiparallel spin, is $\varphi(r_2)$. Equation 6.3 is called the *Hubbard intra-atomic energy*, and it is not accounted for in conventional band theory, in which the independent electron approximation is invoked. Finally, it should also be noted that the Coulomb repulsion interaction was introduced earlier in the Anderson model describing a magnetic impurity coupled to a conduction band (Anderson, 1961). In fact, it has been shown that the Hubbard Hamiltonian reduces to the Anderson model in the limit of infinite-dimensional (Hilbert) space (Izyumov, 1995). Hence, Eq. 6.3 is sometimes referred to as the Anderson–Hubbard repulsion term.

The Hubbard intra-atomic energy is one of three critical quantities that determine whether a solid will be an insulator or a conductor, the other two being band filling and bandwidth. The Hubbard energy is the energy cost of placing two electrons on the same site. In order to see this, consider two atoms separated by a distance a. In the limit $a \rightarrow 8$ (i.e., two isolated atoms), the excitation energy required for electron transfer from an atomic orbital on one of the atoms to an atomic orbital on the other atom, already containing an electron, is defined as:

$$U = (I_{el} - \chi) \qquad (6.4)$$

where I_{el} is the ionization energy, and χ is the electron affinity (the minus sign signifies that a small portion of I_{el} is recovered when the electron is added to a neutral atom). In a solid, the situation is analogous, but U is given by Eq. 6.3, instead of Eq. 6.4. The Coulomb repulsion due to Eq. 6.3 is typically on the order of 1 eV or less. This is much smaller in magnitude than the energy gaps between the valence and conduction bands in covalent or ionic insulators (>3 eV). For wide-band metals, such a small Coulomb repulsion term is insufficient to drive the system to the insulating state. However, in narrow-band systems, it may.

As alluded to in Section 4.5.5, the transition-metal monoxides with the rock-salt structure are the archtypical examples of correlated systems. Of these oxides, only TiO is metallic. The others, NiO, CoO, MnO, FeO, and VO, are all insulating, despite the fact that the Fermi level falls in a partially filled band (in the independent electron picture). Direct electron transfer between two of the transition-metal cations (in the rock-salt structure, d–d interactions are important due to the proximity of the cations; see Section 4.5.3), say, manganese, is equivalent to the disproportionation reaction:

$$Mn^{2+} + Mn^{2+} \rightarrow Mn^+ + Mn^{3+}$$
$$d^5 + d^5 \rightarrow d^6 + d^4$$

(6.5)

It costs energy U to transfer an electron from one Mn^{2+} into the already occupied d orbital of another Mn^{2+} ion ($d^5 \rightarrow d^6$), thereby, opening a Coulomb gap. The d band is split into two equivalent subbands, shifted symmetrically relative to the Fermi level: a filled, singly occupied lower Hubbard band (LHB) and an empty upper Hubbard band (UHB), separated by a band gap of the order of U. A schematic representation of this type of electronic structure is shown in Figure 6.1a. The LHB represents low-energy configurations with no doubly occupied sites, that is $Mn^{2+} \rightarrow Mn^{3+}$. The UHB is the high-energy configuration corresponding to doubly occupied sites, that is, $Mn^{2+} \rightarrow Mn^+$. In general, for a Mott–Hubbard insulator, charge excitations across the gap correspond to: $d^n + d^n \rightarrow d^{n+1} + d^{n-1}$. In Hubbard's original approximation, the system is an insulator for *any* value of U, however small. In what is known as the Hubbard-3 approximation (the models are named in the order of Hubbard's three papers), band splitting occurs only when U is sufficiently large.

In fact, the Hubbard-3 approximation describes the occurrence of a transition from insulating to metallic behavior, as first predicted by Mott (Mott, 1961), when $U \sim W$, where W represents the one-electron bandwidth. Hence, the bandwidth is another quantity that determines the type of electronic behavior exhibited by a

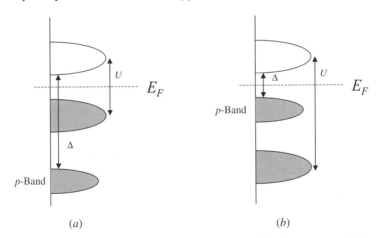

(a)   (b)

Figure 6.1 (a) The band gap is determined by the d–d electron correlation in the Mott–Hubbard insulator, where $\Delta > U$. (b) By contrast, the band gap is determined by the charge-transfer excitation energy in the charge-transfer insulator, where $U > \Delta$.

solid. The critical value of the ratio $(W/U)_c$, corresponding to the Mott M–NM transition, is:

$$(W/U)_c \approx 1 \tag{6.6}$$

with numerical calculations showing the actual value to be 1.15 (Mott, 1990). Although there is some disagreement in the literature over the exact value of the ratio at which a M–NM transition occurs, the gist of the matter is that a correlated system is expected to become metallic (the Hubbard band gap closes) when U becomes smaller than the bandwidth. In terms of the simple picture described earlier, this may be attributed to a large portion of the energy penalty for transferring an electron to an occupied atomic orbital being recovered. The amount of energy regained corresponds to the amount that the crystal orbitals are stabilized, relative to the free atomic orbitals, as reflected in the bandwidth.

John Hubbard (Courtesy AIP Emilio Segrè Visual Archives. © IBM. Reproduced with permission.)

JOHN HUBBARD

(1931–1980) received his B.S. and Ph.D. degrees from Imperial College, University of London, in 1955 and 1958, respectively. Hubbard spent most of his 25-year career at the Atomic Energy Research Establishment in Harwell, England, as head of the solid-state theory group. He is best known for a series of papers, published in 1963–1964, on dynamic correlation effects in narrow-band solids. The terms "Hubbard–Hamiltonian" and "Mott–Hubbard metal-insulator transition" became part of the jargon of condensed matter physics. Walter Kohn once described Hubbard's contributions as "the basis of much of our present thinking about the electronic structure of large classes of magnetic metals and insulators." Hubbard also conducted research on the nature of gaseous plasmas in nuclear fusion reactors, the efficiency of centrifuges in isotope separation, and developed the functional integral method many-body technique, as well as a first-principles theory of the magnetism of iron. In 1976, Hubbard joined the IBM Research Laboratory in San Jose, California where he remained until his premature death at the age of 49.

(*Source: Physics Today*, Vol. 34, No. 4, 1981. Reprinted with permission from *Physics Today*. Copyright 1981, American Institute of Physics.)

6.1.2 Charge-Transfer Insulators

Unlike the Mott–Hubbard insulator MnO described earlier, the band gap in the iso-structural oxide NiO is much smaller than expected from intrasite Coulomb repulsion. Fujimori and Minami showed that this is due to the location of the NiO oxygen $2p$ band—*between* the lower and upper Hubbard subbands (Fujimori and Minami, 1984). This occurrence can be rationalized by considering the energy level of the d band as one moves from Sc to Zn in the first transition series.

As one moves to the right in a period, the d orbital energy drops due to an increasing effective nuclear charge. Hence, the energy of the nickel $3d$ (conduction) band in NiO decreases to a level close to the oxygen $2p$ (valence) band, which allows Ni $3d$–O $2p$ hybridization. Although the d band still splits from on-site Coulomb repulsion, the O $2p$ band now lies closer to the UHB than the LHB does. The magnitude of the band gap is thus determined by the charge-transfer energy, δ, associated with the process: $d^n \rightarrow d^{n+1}L^1$, where L^1 represents a hole in the ligand O $2p$ band. In other words, the band gap is now δ, not U_{dd}, as illustrated in Figure 6.1b. Zaanen, Sawatzky, and Allen (1985) named these types of systems charge-transfer insulators.

Whether the band gap is of the charge-transfer type or Mott–Hubbard type is determined by the relative magnitudes of Δ and U. If $U < \Delta$, the band gap is determined by the d–d Coulomb repulsion energy. When $\Delta < U$, the charge-transfer excitation energy determines the band gap. Crude estimates for U can be obtained using Eq. 6.4. However, neglecting hybridization effects in this manner gives an "atomic" value, resulting in the values for the solid state being overestimated. The Coulomb repulsion energy is primarily a function of the spatial extension of the orbital. The charge-transfer energy, by contrast, depends on the effective nuclear charge. Both of these parameters, in turn, are dependent on the transition metal's atomic number and oxidation state, as well as hybridization effects and the ligand electronegativity.

A systematic semiempirical study of the core-level photoemission spectra of a wide range of $3d$ transition-metal compounds has been carried out (Bocquet et al., 1992, 1996). The values for U and Δ obtained from a simplified CI cluster-model analysis are demonstrated in Figure 6.2. As can be inferred from the graphs, the heavier $3d$ transition-metal compounds shown in the figure are expected to be charge-transfer insulators, whereas the compounds of the lighter metals are generally expected to be of the Mott–Hubbard type.

6.1.3 Marginal Metals

It was pointed out in the introductory paragraphs of this chapter that an experimental criterion for metallicity is the observation of a positive temperature coefficient to the electrical resistivity. The so-called "bad" or "marginal" metals are those that meet this criterion, but in which the value for the resistivity is relatively high ($\rho > 10^{-2}$ Ω cm). Many transition metal oxides behave in this manner, while others (e.g., ReO_3 and RuO_2) have very low electrical resistivities, like conventional metals ($\rho < 10^{-4}$ ohm cm). Consider the Ruddlesden–Popper ruthenates. Both strong Ru $4d$–O $2p$ hybridization and weaker intrasite correlation effects compared

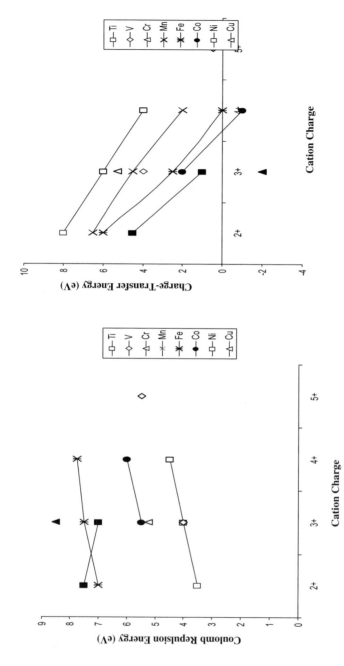

Figure 6.2 The on-site d–d Coulomb repulsion energy (left) and the charge-transfer excitation energy (right) for the $3d$ transition metal oxides. Plots are from data by Bocquet et al. (1992, 1996).

to the $3d$ transition metals are expected because of the greater spatial extension of the $4d$ atomic orbitals. Nonetheless, U is sufficiently strong to open a band gap in the ruthenates. Optical conductivity measurements indicate that the one-electron bandwidth in the $n = 1$ Ruddlesden–Popper insulator Ca_2RuO_4 is $\sim$1 eV, while the on-site Coulomb interaction is estimated to be in the range 1.1–1.5 eV (Puchkov and Shen, 2000). Since $U > W$, correlation prevents metallicity. However, in the isostructural Sr_2RuO_4, the conductivity in the ab plane of the perovskite layers is metallic with a low resistivity ($\rho \sim 10^{-4}$ Ω cm), and the out-of-plane transport along the c-axis is marginally metallic at low temperatures ($\rho \sim 10^{-2}$ Ω cm) and semiconducting at high temperatures (Figure 1.15). Below 1 K, Sr_2RuO_4 exhibits superconductivity.

For the $n = 3$ Ruddlesden–Popper phase $Ca_3Ru_2O_7$, the bandwidth is expected to be slightly larger, while the on-site Coulomb interaction is presumed to be essentially unchanged from that of Ca_2RuO_4. Likewise, $Ca_3Ru_2O_7$ and $Sr_3Ru_2O_7$ are marginally metallic in the ab planes of the perovskite blocks ($\rho \sim 10^{-2}$ Ω cm), but also exhibit a very slight metallic conductivity in the perpendicular out-of-plane direction along the c-axis (Puchkov and Shen, 2000). The anisotropy is smaller than in the $n = 1$ Ruddlesden–Popper phases. It is a general observation that the anisotropy in transport properties of such layered structures decreases with increases in the number of layers, n.

It is believed that electron correlation plays an important role with the anomalously high resistivity exhibited in marginal metals. Unfortunately, although the Mott–Hubbard model adequately explains behavior on the insulating side of the M–NM transition, it does so on the metallic side only if the system is far from the transition. Electron dynamics of systems in which U is only slightly less than W (i.e., metallic systems close to the M–NM transition), are not well described by a simple itinerant or localized picture. The study of systems with "almost" localized electrons is still an area under intense investigation within the condensed-matter physics community. A dynamical mean field theory (DMFT) has been developed for the Hubbard model, which enables one to describe both the insulating state and the metallic state, at least for weak correlation. However, solution of the mean-field equations is not trivial, and this topic is not be discussed here.

6.2   ANDERSON LOCALIZATION

We have seen in the previous section that the ratio of the on-site electron–electron Coulomb repulsion and the one-electron bandwidth are a critical parameters. The Mott–Hubbard insulating state is observed when $U > W$, that is, with narrow-band systems like transition-metal compounds. Disorder is another condition that localizes charge carriers. In crystalline solids, there are different types of disorder. One type arises from the random placement of impurity atoms in lattice sites or interstitial sites. The term *Anderson localization* is applied to systems in which the carriers are localized by such disorder. Anderson localization is important in a wide range of materials, from phosphorus-doped silicon to the perovskite oxide strontium-doped lanthanum vanadate, $La_{1-x}Sr_xVO_3$.

In a crystalline solid, the presence of strong disorder results in destructive quantum interference effects from multiple elastic scattering, which is sufficient to stop propagation of the conduction electrons. Inelastic scattering (Section 5.3) also increases electrical resistivity, but does not cause a phase transition. Anderson's original model is the tight-binding Hamiltonian on a d-dimensional hypercubic lattice with random site energies, characterized by a probability distribution. On a hypercubic lattice, each site has $2d$ nearest neighbors and $2d(d-1)$ next nearest neighbors.

In the second quantization language, the tight-binding Hamiltonian is written as:

$$\mathcal{H} = \sum_{i,\sigma} \varepsilon_{i\sigma} n_{i\sigma} + t_{ij} \sum_{ij,\sigma} C_{i\sigma}^{\dagger} C_{j\sigma} \qquad (6.7)$$

In Eq. 6.7, t_{ij} is the nearest-neighbor hopping matrix (assumed constant), the $C_{i\sigma}^{\dagger}$ and $C_{j\sigma}$ terms are creation and annihilation operators at the lattice sites i and j, and $n_{i\sigma} = C_{i\sigma}^{\dagger} C_{i\sigma}$. The site energies, $\varepsilon_{i\sigma}$, are taken to be randomly distributed in the interval $[-B/2, B/2]$, where B is the width characterizing the probability distribution function. Random site energies correspond to varying potential well depths, or *diagonal disorder* (diagonal elements of the Hamiltonian matrix) and typically arise from the presence of random substitutional impurities in an otherwise periodic solid. When there is variation in the nearest-neighbor distances, this gives rise to fluctuations in the hopping elements, t_{ij}. This is *off-diagonal disorder* and is the type present in truly amorphous solids, such as glasses and polymers. Off-diagonal disorder is sometimes called *Lifshitz* disorder (Lifshitz, 1964; 1965).

Consider the simplest possible case, a monatomic crystalline solid. The potential at each lattice site is represented in the Kronig–Penney model by a single square well (Kronig and Penney, 1931). For a perfect monatomic crystalline array (Figure 6.3a), all the potential wells are of the same depth and are the same distance apart. However, the random introduction of impurity centers in the crystalline lattice produces variation in the well depths (diagonal, or Anderson, disorder), as illustrated in Figure 6.3b. Amorphous substances and glasses exhibit variations in nearest-neighbor distances. This off-diagonal disorder essentially produces unevenly spaced potential wells (Figure 6.3c).

It is found that off-diagonal disorder is not very effective at inducing electron localization. For example, most metallic glasses do not show a M–NM transition, but only moderate changes in resistivity with temperature (Mott, 1990). By contrast, the multiple elastic scattering from the many fluctuations in site energies, arising from strong diagonal disorder, is sufficient to reduce the mean free path traveled by the electrons between scattering events to a value comparable to the electron wavelength. Destructive quantum interference between different scattering paths ensues, localizing the electron wave functions. It must be noted that this localized electron wave function does not exist solely on a single isolated atomic center, but rather exponentially decays over several atomic centers, referred to as the *localization length*.

In all substances, at high temperatures, the electrical resistivity is dominated by *inelastic* scattering of the electrons by phonons, and other electrons. As classic

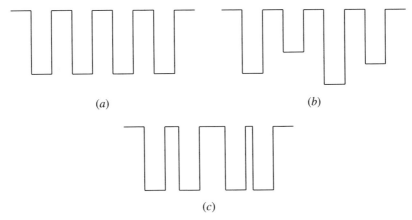

(a)   (b)

(c)

Figure 6.3 (a) In the Kronnig–Penney model, the potential at each lattice site in a monatomic crystalline solid is of the same depth. (b) The random introduction of impurity atoms in the crystalline lattice produces variation in the well depths, known as diagonal, or Anderson, disorder. (c) Amorphous (noncrystalline) substances have unevenly spaced potential wells, or off-diagonal disorder.

particles, the electrons travel on trajectories that resemble random walks, but their apparent motion is diffusive over large length scales, because there is enough constructive interference to allow propagation to continue. Ohm's law holds and with increasing numbers of inelastic scattering events, a decrease in the conductance can be directly observed.

By contrast, we can think of Anderson localization as occurring at some critical number of *elastic* scattering events, in which the free electrons of a crystalline solid experience a phase transition from the diffusive (metallic) regime to the nondiffusive (insulating) state. In this insulating state, nondiffusive transport occurs via a thermally activated quantum mechanical hopping of localized states from site to site. There is one very important aspect of Anderson localization that distinguishes it from band-gap insulators: *In the latter case, nonmetallic transport is due to the lack of electronic states at the Fermi level, whereas with Anderson localization, there remains a finite DOS at E_F.*

The critical disorder strength needed to localize *all* the states via the Anderson model is:

$$(W/B)_{complete} \approx 2 \tag{6.8}$$

although a value as low as 1.4 has also been calculated (Mott, 1990). To treat cases of weak disorder, Mott introduced the concept of a mobility edge, E_c, which resides in the band tail and separates localized from nonlocalized states (Mott, 1966). The system is nonmetallic for all values of the Fermi energy that fall below E_c into the localized regime, as illustrated in Figure 6.4.

Percolation theory is helpful for analyzing disorder-induced M–NM transitions (recall the classic percolation model used to describe grain boundary transport

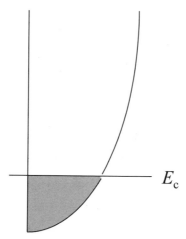

Figure 6.4   A schematic density-of-states curve showing localized states below a critical energy, E_c, in the conduction-band tail. The conduction electrons are localized unless the Fermi energy is above E_c.

phenomena in Chapter 1). In this model, the M–NM transition corresponds to the percolation threshold. Perhaps the most important result comes from the very influential work by Abrahams (Abrahams et al., 1979), based on scaling arguments from quantum percolation theory. This is the prediction that no percolation occurs in a 1D or 2D system with nonzero-disorder concentration at 0 K in the absence of a magnetic field. It has been confirmed in a mathematically rigorous way that all states will be localized in the case of disordered 1D transport systems (i.e., chain structures).

In 2D systems (e.g., layered structures with alternating conducting and insulating sheets), the effect of disorder is controversial. Theoretically, 2D systems with any amount of disorder are expected to behave as insulators at any temperature. Such behavior has been confirmed for many compounds. For example, in the phase $Na_{2-x+y}Ca_{x/2}La_2Ti_3O_{10}$ 2D metallic transport should be possible within the perovskite slabs. This is because electrons have been donated to the formally empty conduction band by a two-step sequence. The first step consists of alloying the alkali metal sites in the NaO rock-salt layers of the parent phase $Na_2La_2Ti_3O_{10}$ with Ca^{2+}, thereby introducing vacancies (yielding $Na_{2-x}Ca_{x/2}La_2Ti_3O_{10}$, where $0.48 < x < 1.66$) This is followed by intercalation of additional sodium atoms into the vacancies (yielding $Na_{2-x+y}Ca_{x/2}La_2Ti_3O_{10}$), which chemically reduces a portion of $Ti^{4+}(d^0)$ cations in the octahedral layers to the Ti^{3+} (d^1) state and donates electrons to the conduction band. However, the random electric fields generated by the distribution of impurities (Ca^{2+}) in the rock-salt layer are primarily responsible for the nonmetallic behavior in this system, possibly with weak correlation effects (Lalena et al., 2000). By contrast, in the analogous 3D transport system, the mixed-valent perovskite $La_{1-x}TiO_3$ (Ti^{3+}/Ti^{4+}), metallic conduction is observed at $x = 0.25$.

Recently, the prediction that metallicity is impossible in disordered 2D systems appears to have been contradicted. A metallic regime has been observed at zero applied field, down to the lowest accessible temperatures, in $GaAs/Ga_{1-x}Al_xAs$ heterostructures with a high density of InAs quantum dots incorporated just 3 nm below the heterointerface Ribeiro et al., 1999). Some have argued that this is an artifact, and that other effects mask the M–NM transition. The possibility of metallicity in disordered 2D transport systems is another hotly debated topic within the physics community.

Philip Anderson (Coutesy Department of Physics, Princeton University. © Princeton University. Reproduced with permission.)

PHILIP WARREN ANDERSON

(b. 1923) received his Ph.D. in theoretical physics under John H. Van Vleck in 1949 at Harvard University. From 1949 to 1984, he was associated with Bell Laboratories and was chairman of the theoretical physics department there from 1959 to 1961. From 1967 to 1975, he was a Visiting Professor of Theoretical Physics at the University of Cambridge. In 1975, he joined the Physics Department at Princeton University, becoming emeritus in 1997. Anderson was the first to estimate the magnitude of the antiferromagnetic superexchange interaction proposed by Kramers, and to point out the importance of the cation–anion–cation bond angle. In addition to basic magnetics, Anderson has made seminal contributions to our understanding of the physics of disordered media, such as the spin glass state and the localization of noninteracting electrons. For this work, he was a co-recipient of the 1977 Nobel Prize in Physics. He has worked almost exclusively in recent years on high-temperature superconductivity. Anderson was on the U.S. National Academies of Sciences Council from 1976 to 1979, and he was awarded the U.S. National Medal of Science in 1983.

(*Source: Nobel Lectures, Physics 1971–1980*, World Scientific Publishing Company, Singapore.)

6.3 EXPERIMENTALLY DISTINGUISHING ELECTRON CORRELATION FROM DISORDER

The main effect of both types of electron localization, of course, is a crossover from metallic to nonmetallic behavior (a M–NM transition). But is there a way of experimentally distinguishing between the effects of electron–electron Coulomb repulsion and disorder, and if so, how? In cases where only one or the other type of localization is present, this task is made much simpler. The Anderson transition, for example, is predicted to be continuous. That is, the zero-temperature electrical conductivity should drop to zero continuously as the impurity concentration is increased. By contrast, Mott predicted that electron–correlation effects lead to a first-order, or discontinuous transition. The conductivity should show a discontinuous drop to zero with increasing impurity concentration. Unfortunately, experimental verification of a true first-order Mott transition remains elusive.

Disorder and correlation are both often present in a system. One then has the more difficult task of ascertaining which is the *dominant* mechanism. As might be expected, the most useful experimental approaches to this problem involve magnetic and conductivity measurements. In the paramagnetic regime at high temperatures, the magnetic susceptibility, χ, of a sample with a nonzero density of states at the Fermi level (e.g., metals and Anderson localized states) is temperature-independent. This is known as *Pauli paramagnetism* (see Example 6.1 and Figure 6.6a), since it results from the Pauli exclusion principle. The Pauli susceptibility is given by:

$$3Nz_e\mu_B^2/2E_F(0) \tag{6.9}$$

where Nz_e is the number of atoms times the number of electrons per atom in the band at the Fermi energy, E_F, and μ_B is the Bohr magneton, 9.27×10^{-24} J/T (Pauli, 1927). For narrow bands and high temperatures, however, there may be a slight temperature dependency to the Pauli susceptibility (Goodenough, 1963).

Disorder cannot be the sole mechanism of electron localization when there also exists experimental evidence suggesting the presence of a band gap. Direct methods for detecting the presence of a band gap include optical conductivity and photoemission spectroscopy. However, magnetic characterization is also very useful here as well. This is because electron correlation induces the exchange interactions responsible for spontaneous magnetization in insulators. Goodenough has emphasized the use of magnetic criteria as a means of characterizing the transition region between the itinerant and localized regimes. The validity of this approach can be seen by considering the *Heisenberg Hamiltonian*:

$$\mathcal{H}_{ex} = -\sum J_{ij} S_i \cdot S_j \tag{6.10}$$

This is the expression for the exchange interaction between localized magnetic moments. In Eq. 6.10, J is the *exchange integral*, given by:

$$J = -2b^2/4S^2 U \tag{6.11}$$

where S is the spin quantum number of the system (Anderson, 1959). In a mean field approximation, the exchange integral can be incorporated into a dimensionless constant (the Weiss constant) and the Heisenberg Hamiltonian can be rewritten in terms of this constant and the magnetization of the sample. Thus, Eq. 6.9 for the Pauli susceptibility of itinerant electrons, and Eq. 6.11 relating $\mathcal{J}$ to the on-site Coulomb energy and the tight-binding transfer integral, establishes how magnetic data can be used to characterize the itinerant-to-localized transition region.

Systems exhibiting both strong disorder and electron correlation, so-called *disordered Mott–Hubbard insulators*, are difficult to evaluate. The description of electronic states in the presence of both disorder and correlation is still an unresolved issue in condensed-matter physics. Whether disorder or the correlation is the predominant factor in controlling transport properties in a material depends on a complex temperature/composition-dependent three-way interplay between U, B, and W. Many materials exhibit the behavior illustrated in the electronic phase diagram shown in Figure 6.5. For cases of strong disorder with weak correlation, a temperature-independent (Pauli) paramagnetic susceptibility, $\chi(T)$, is usually observed at high temperatures because of the finite density of states at E_F. However, this may cross over to a Curie- or Curie–Weiss-type behavior (linear dependence of χ on T) at low temperatures, as illustrated later in Figure 6.6a of Example 6.1. The crossover from Pauli-type to Curie-type behavior has been shown to be due to

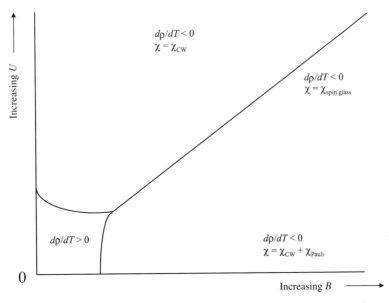

Figure 6.5 An electronic phase diagram illustrating the location of a metallic regime, at low U (Coulomb repulsion energy) and B (disorder), where $d\rho/dT > 0$. With increasing U and/or B, the system becomes nonmetallic ($d\rho/dT < 0$), but may still exhibit Pauli paramagnetism if there is no band gap ($U \ll B$).

correlation, which gives rise to the occurrence of singly occupied states below the Fermi energy, E_F (Yamaguchi et al., 1979). A Curie law results whenever the probability of finding two electrons in a localized state is less than the probability of finding one (because of Coulomb repulsion). If electron correlation is much stronger than disorder ($U \gg B$), the magnetic susceptibility curve typically has the general appearance of an antiferromagnetic insulator.

Another characteristic feature of disordered systems is the variable range hopping (VRH) mechanism to the electrical conduction, which is observed on the non-metallic side of the M–NM transition at low temperatures. This arises from the hopping of charge carriers between localized states, or impurity centers. Hence, the phenomenon is also known as *impurity conduction*. Experimentally, it is indicated by characteristic temperature dependency to the direct current (dc) conductivity. For 3D systems with noninteracting electrons, the logarithm of the conductivity and $t^{-1/4}$ are linearly related, in accordance with the equation given by Mott (Mott, 1968)

$$\sigma = A \, \exp[-(T_0/T)^n] \quad n = 1/4 \tag{6.12}$$

The exponent n, which is known as the hopping index, is actually equal to $1/(d+1)$ where d is the dimensionality. Hence, for 2D systems, $n = 1/3$. *Strictly speaking, Eq. 6.12 holds only when the material is near the M–NM transition and at sufficiently low temperatures.* At high temperatures, conduction proceeds by thermal excitation of electrons from donors into the conduction band, or injection of holes from acceptors into the valence band.

The Anderson transition is described in terms of noninteracting electrons. For real materials, electron–electron interactions cannot be ignored, even when disorder is the primary localization mechanism. In other words, with Anderson localized states, correlation effects are normally present to some extent. Of course, the reverse is not true; correlated systems may be completely free of disorder. Electrical conduction in insulators with band gaps is also thermally activated. Hence, it can be difficult to determine the value of n in strongly correlated disordered systems unambiguously. The hopping index, however, is often found to be equal to 1/2 at low temperatures, with 1/4 still being observed at higher temperatures.

For highly correlated systems, a T^2 dependence to the resistivity is frequently observed in the metallic regime. Notable exceptions to this rule are the high-temperature superconducting cuprates, in which conduction occurs within CuO_2 layers. Electron correlation is believed to be important to the normal state (nonsuperconducting) properties. In fact, all the known high-temperature superconducting cuprates are compositionally located near the Mott insulating phase. Both hole-doped superconductors (e.g., $La_{2-x}Sr_xCuO_4$, $YBa_2Cu_3O_{7-y}$) and electron-doped superconductors (e.g., $Nd_{2-x}Ce_xCuO_4$) are known, in which the doping induces a change from insulating to metallic behavior with a superconducting phase being observed at low temperatures. However, a T linear dependence to the resistivity, rather than a t^2 dependence, is widely observed in the nonsuperconducting metallic state above the critical temperature, T_c, in these materials.

Lengthy review articles on disorder-induced M–NM transitions with minor contributions by electron correlation are available (Lee and Ramakrishnin, 1985; Belitz and Kirkpatrick, 1994). An extensive review article on correlated systems has also been published recently (Imada et al., 1998).

6.4   TUNING THE METAL–NONMETAL TRANSITION

There are two basic approaches for tuning the composition of a phase to induce metallicity. One is the *filling control* approach, in which the chemical composition of an insulator or semiconductor (either a band type or correlated type) is varied in such a way as to increase the charge-carrier (hole or electron) concentration. The sodium tungsten bronzes, introduced in Chapter 2, and La-doped perovskite $SrTiO_3$ are examples of filling control. Both WO_3 and $SrTiO_3$ are band insulators. However, as sodium is intercalated into the interstitials of WO_3, giving Na_xWO_3, or when La^{3+} is substituted for Sr^{2+} to give $La_{1-x}Sr_xTiO_3$, a portion of the transition-metal cations is chemically reduced and mixed valency is introduced. The extra electrons are donated in each case to a conduction band that was formally empty. Hole doping may be used to bring about metallicity, too, as in $La_{1-x}Sr_xMO_3$ ($M =$ Mn, Fe, Co).

Introducing substitutional donors directly on the cation sublattice can sometimes also induce metallic behavior. For example, the perovskite oxide $LaCoO_3$ is a band insulator at 0 K (i.e., it contains a filled valence band and an empty conduction band separated by a band gap) since Co^{3+} is in the low-spin d^6 (t_{2g}^6) ground state. Upon increasing the temperature to approximately 90 K, the Co^{3+} ion is excited into a high-spin state. Nonetheless, the oxide remains in the insulating state, due to correlation effects. Upon further increasing the temperature to about 550 K, the oxide becomes conducting. Metallic behavior can be induced at much lower temperatures, however, by electron doping $LaCoO_3$ with Ni^{3+} ($t_{2g}^6 e_g^1$) to give $LaNi_{1-x}Co_xO_3$ (Raychaundhuri et al., 1994).

Despite its success in $LaNi_{1-x}Co_xO_3$, the strategy of alloying the transition metal site is not often attempted because of the danger of reducing the already narrow d bandwidth, which promotes electron localization. Transition-metal compounds generally tend to have narrow d bands (small W), which becomes pronounced if dopant atoms are not near in energy (due to a different effective nuclear charge or orbital radius) to those of the host. For example, low concentrations of impurities may have energy levels *within* the band gap and effectively behave as isolated impurity centers. With heavier doping, appreciable overlap between the orbitals on adjacent impurities may occur with the formation of an *impurity band*. However, nonthermally activated metallic conduction would be predicted in the Bloch/Wilson band picture only if the impurity band and conduction band overlap. This is believed to be the mechanism for the M–NM transition in phosphorus-doped silicon.

The filling-control approach has even been applied to some nanophase materials. For example, the onset of metallicity has been observed in individual alkali metal-doped single-walled zigzag carbon nanotubes. Zigzag nanotubes are semiconductors with a band gap around 0.6 eV. Using tubes that are (presumably) open on each end, it has been observed that upon vapor-phase intercalation of potassium into the interior of the nanotube, electrons are donated to the empty conduction band, thereby raising the Fermi level and inducing metallic behavior (Bockrath, 1999).

A second way of inducing metallicity is known as *bandwidth control*. Usually, the lattice parameters of a phase are varied while maintaining the original structure. Thus, U remains essentially the same, while W is made larger. This may be accomplished by the application of pressure or the introduction of substitutional dopants. The former is commonly exhibited in V_2O_3, while the latter is successful in $Ni_2S_{2-x}Se_x$, where the Se doping widens the S $2p$ bandwidth, closing the charge-transfer gap. There has been disagreement in the literature as to whether NiS_2 is a charge-transfer or Mott–Hubbard insulator. It has been speculate that at low temperatures, the compound is of the Mott–Hubbard type and, at higher temperatures, of the charge-transfer type (Honig and Spalek, 1998). As with filling control, alloying of the transition metal cation sublattice is not usually successful at driving a system metallic.

In the ABO_3 perovskites, bandwidth control may be accomplished by changing the ionic radius of the A-site (12-coordinate) cation. Essentially, this mimics the effect of pressure, which also effectively changes the $B–O–B$ angle (see the discussion in Section 6.1). The intrasite correlation energy usually also changes somewhat during bandwidth control, in accordance with the expectations outlined earlier. However, the method primarily relies on increasing the hybridization, in order to obtain a wider bandwidth favorable for metallic behavior. In some cases, structure-type changes may occur, for example, as with the transformation of a cubic perovskite to a rhombohedral or orthorhombic structure.

Example 6.1   The M–NM transition has been studied in powder samples (prepared by the ceramic method) of the perovskite oxide series $La_{1-x}TiO_3$ (Table 6.1) with $0 \leq x \leq 0.33$ (MacEachern et al., 1994). Figure 6.6 contains the curves of the temperature dependencies to the magnetic susceptibility (Figure 6.6a) and electrical resistivity (Figure 6.6b) for different compositions, as well as the cell parameters for the phases: (a) For what value of x does metallic behavior first appear? (b) Is electron localization primarily due to disorder or correlation? (c) Speculate as to the nature of the M–NM transition. Does varying the chemical composition (the value of x) primarily result in filling control or bandwidth control?

TABLE 6.1   Cell Parameters

Composition	$a(\text{Å})$	$b(\text{Å})$	$c(\text{Å})$	$vol(\text{Å}^3)$
$La_{0.70}TiO_3$	5.464	7.777	5.512	234.22
$La_{0.75}TiO_3$	5.541	7.793	5.528	238.70
$La_{0.80}TiO_3$	5.557	7.817	5.532	240.30
$La_{0.88}TiO_3$	5.582	7.882	5.559	244.58
$La_{0.92}TiO_3$	5.606	7.914	5.584	247.74

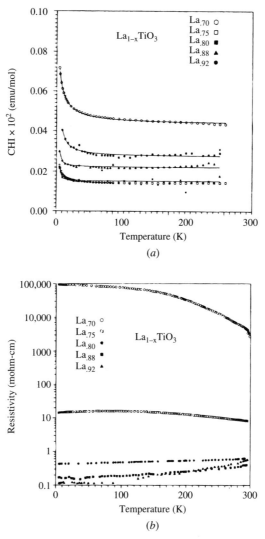

Figure 6.6 (From MacEachern et al. 1994, *Chem. Mater.*, 6, 2092. Reproduced by permission of the American Chemical Society.)

Solution   For $x = 0.30$, $d\rho/dT$ is negative over the entire temperature range, indicating insulating behavior. For $x = 0.25$, $d\rho/dT$ is negative at high temperatures, but changes sign, becoming positive (*i.e.*, the resistivity begins to increase with increasing temperature) around 150 K and remains positive for all temperatures below this point. Hence, the M–NM transition appears at the composition $x = 0.25$.

(a)   For these materials, there is evidence for both disorder and electron correlation. The temperature-independent component to the magnetic susceptibility at high temperatures is too great in magnitude for Van Vleck paramagnetism (see Section 7.2) but, rather, is Pauli-like paramagnetism, indicating the presence of a finite density of states at the Fermi level (Anderson localization). Furthermore, plots of $\ln \sigma$ vs. $T^{-1/4}$ were reported as being linear, indicating variable range hopping. However, there is also a Curie–Weiss-like behavior at temperatures below 50 K in the two lower curves of Figure 6.6a and below about 150 K in the two upper curves, which is indicative of electron correlation. Since Ti is to the left of the first transition period, we would expect relatively strong electron correlation in the Ti $3d$ orbitals. In fact, a linear dependency of the resistivity to T^2 (plots of ρ vs. T^2 were linear) was also reported in the metallic phases, supporting the presence of correlation effects. Hence, for $La_{1-x}TiO_3$, both disorder and correlation are probably important. These phases are best considered *disordered Mott–Hubbard insulators*.

(b) The unit cell volume expands with increasing lanthanum content. However, the metallic behavior appears with the larger unit cells. This is contradictory to bandwidth control, in which greater atomic orbital overlap would be expected to overcome correlation effects with contractions in the unit cell.

For $La_{0.66}TiO_3$, titanium is fully oxidized to Ti^{4+} (d^0). Hence, the conduction band is empty. As the Ti^{3+} (d^1) content increases with increased lanthanum deficiency, however, the conduction band becomes occupied with electrons. Thus, the M–NM transition appears to be under filling control.

6.5 OTHER TYPES OF ELECTRONIC TRANSITIONS

In this chapter we have so far focused on the Mott (Mott–Hubbard) and Anderson transitions. When charge ordering is present, other types of transitions are also possible. A classic example is the mixed-valence spinel Fe_2O_3. There are two types of cation sites in Fe_2O_3, denoted as A and B. The A sites are tetrahedrally coordinated Fe^{3+} ions, and the B sites are a 1:1 mixture of octahedrally coordinated Fe^{2+} ($t_{2g}^4 e_g^2$) and Fe^{3+} ($t_{2g}^3 e_g^2$) ions. Electrical transport is on the B sublattice, and is nonmetallic ($d\rho/dT < 0$) thermally activated hopping below about 300 K. Interestingly, it is found that at 120 K, the resistivity increases by two orders of magnitude, the oxide remaining nonmetallic. This is the *Verwey transition* in Fe_2O_3. Since the system remains nonmetallic, it is sometimes called an insulator–insulator transition.

At the Verwey transition, the mixed-valence ions on the B sublattice become ordered. There is some controversy about the exact structure of the ordered state. Verwey first proposed that the mixed-valence cations are ordered onto alternate B-site layers (Verwey and Haayman, 1941; Verwey et al., 1947). Charge-ordering transitions are also observed in Ti_4O_7, as well as many transition-metal perovskite oxides, such as $La_{1-x}Sr_xFeO_3$, $La_{1-x}Sr_{1+x}MnO_4$, and $La_{2-x}Sr_xNiO_4$ (the latter in which the charge ordering drives a true M–NM transition).

As pointed out in Section 3.4.1, another type of charge ordering, called the *charge-density-wave* (CDW) *state* can open a band gap at the Fermi level in the partially filled band of a metallic conductor. This causes a M–NM transition in quasi-1D systems or a metal–metal transition in quasi-2D systems, as the temperature is lowered. At high temperatures, the metallic state becomes stable, because the electron energy gain is reduced by thermal excitation of electrons across the gap. Most CDW materials contain weakly coupled chains, along which electron conduction takes place, such as the nonmolecular $K_{0.3}MoO_3$ and the molecular compound $K_2Pt(CN)_4Br_{0.3}3H_2O$ (KCP). Perpendicular to the chains, electrical transport is much less easy, giving the quasi-1D character necessary for CDW behavior.

A CDW is a periodic modulation of the conduction-electron density within a material. It is brought about when an applied electric field induces a symmetry-lowering lattice modulation in which the ions cluster periodically. The modulation mechanism involves the coupling of degenerate electron states to a vibrational normal mode of the atom chain, which causes a concomitant modulation in the electron density that lowers the total electronic energy. In 1D systems, this is the classic *Peierls distortion* (Peierls, 1930, 1955). It is analogous to the Jahn–Teller distortion observed in molecules.

Figure 6.7a shows the extended-zone electronic band structure for a 1D crystal–an atom chain with real-space unit cell parameter a and reciprocal lattice vector π/a–containing a half-filled (metallic) band. In this diagram, both values of the wave vector, $\pm k$, are shown. The wave vector is the reciprocal unit cell dimension. The Fermi surface is a pair of points in the first Brillouin zone (Figure 6.7c). When areas on the Fermi surface can be made to coincide by mere translation of a wave vector, q, the Fermi surface is said to be *nested*. The instability of the material toward the Peierls distortion is due to this nesting. In one dimension, nesting is complete and a 1D metal is converted to an insulator because of a Peierls distortion. This is shown in Figure 6.7b, where the real-space unit cell parameter of the distorted lattice is $2a$ and a band gap opens at values of the wave vector equal to half the original values, $\pi/2a$.

The total electronic energy is reduced, because filled states are lowered in energy and empty states are raised, relative to the same states in the undistorted lattice. The distortion is favored as long as the decrease in electronic energy outweighs the increase in elastic strain energy (Elliot, 1998). Note that the Peierls transition does not involve electron correlation effects. In higher dimensions, only sections of the Fermi surface can be translated and superimposed on other portions. The band-gap opening is thus only partial, and a metal–metal transition, rather than a M–NM transition, is observed.

A related phenomenon is the spin-density wave (SDW), in which the Coulomb repulsion between electrons in doubly occupied sites results in a spin modulation, rather than a charge-density modulation. The SDW state leads to a band-gap-opening M–NM transition like the CDW state. However, the lattice is not distorted by a SDW state, so these waves cannot be detected by X-ray diffraction techniques. SDW instabilities can be induced in low-dimensional metals by magnetic fields as well as by lowering the temperature (Greenblatt, 1996). Hence, neutron diffraction

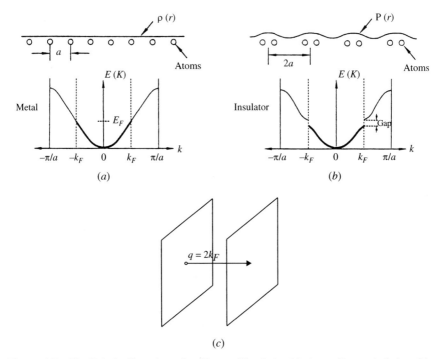

Figure 6.7 The Peierls distortion of a 1D metallic chain. (a) An undistorted chain with a half-filled band at the Fermi level (filled levels shown in bold) has an unmodulated electron density. (b) The Peierls distortion lowers the symmetry of the chain and modulates the electron density, creating a CDW and opening a band gap at the Fermi level. (c) The Fermi surface nesting responsible for the electronic instability.

can be used for observing the existence of a SDW state, since it results in a spatially varying magnetization of the sample.

REFERENCES

Abrahams, E.; Anderson, P. W.; Licciardello, D. C.; Ramakrishnan, T. V. 1979, 1979, *Phys. Rev. Lett.*, *42*, 673.

Anderson, P. W. 1958, *Phys. Rev.*, *109*, 1492.

Anderson, P. W. 1959, *Phys. Rev.*, *115*, 2.

Anderson, P. W. 1961, *Phys. Rev.*, *124*, 41.

Belitz, D.; Kirkpatrick, D. R. 1994, *Rev. Mod. Phys.*, *6*, 261.

Bockrath, M. W. Ph.D. Thesis, University of California, Berkeley, Berkeley, CA 1999.

Bocquet, A. E.; Fujimori, A.; Mizokawa, T.; Saitoh, T.; Namatame, H.; Suga, S.; Kimizuka, N.; Takeda, Y. 1992, *Phys. Rev. B*, *46*, 3771.

Bocquet, A. E.; Mizokawa, T.; Morikawa, K.; Fujimori, A.;Barman, S. R.; Maiti, K. B.; Sarma, D. D.; Tokura, Y.; Onoda, M. 1996, *Phys. Rev. B*, *53*, 1161.

De Boer, J. H.; Verwey, E. J. W. 1937, *Proc. Phys. Soc., London, A*, *49*, 59.

Elliot, S. 1998, *The Physics and Chemistry of Solids*, John Wiley & Sons, Chichester, UK.

Fujimori, A.; Minami, F. 1984, *Phys. Rev. B*, *30*, 957.

Goodenough, J. B. 1963, *Magnetism and the Chemical Bond*, Interscience Publishers, New York.

Goodenough, J. B. 1965, *Bull. Soc. Chim. Fr.*, *4*, 1200.

Goodenough, J. B. 1966, *J. Appl. Phys.*, *37*, 1415.

Goodenough, J. B. 1967, *Czech. J. Phys.*, *B17*, 304.

Goodenough, J. B. 1971, *Prog. Solid State Chem.*, *5*, 149.

Goodenough, J. B.; Zhou, J.-S. 1998, *Chem. Mater.*, *10*, 2980.

Greenblatt, M. 1996, *Acc. Chem. Res.*, 29, 219 and references therein.

Honig, J. M.; Spalek, J. 1998, *Chem. Mater*, *10*, 2910.

Hubbard, J. 1963, *Proc. R. Soc. London*, *A276*, 238.

Hubbard, J. 1964a, *Proc. R. Soc. London*, *A227*, 237.

Hubbard, J. 1964b, *Proc. R. Soc. London*, *A281*, 401.

Imada, M.; Fujimora, A.; Tohura, Y. 1998, *Rev. Mod. Phys.*, *70*, 1039.

Izyumov, Y. A. 1995, *Physics-Uspekhi*, *38*, 385.

Kronig, R. de L.; Penney, W. G. 1931, *Proc. Roy. Soc.*, *130*, 499.

Lalena, J. N.; Falster, A. U.; Simmons, W. B.; Carpenter, E. E.; Wiggins, J.; Hariharan, S.; Wiley, J. B. 2000, *Inorg. Chem.*, *12*, 2418.

Lee, P. A.; Ramakrishnan, T. V. 1985, *Rev. Mod. Phys.*, *57*, 287.

Lifshitz, I. M. 1964, *Adv. Phys.*, *13*, 483.

Lifshitz, I. M. 1965, *Sov. Phys.-Usp.*, *7*, 549.

MacEachern, M. J.; Dabkowaska, H.; Garrett, J. D.; Amow, G.; Gong, W.; Liu, G.; Greedan, J. E. 1994, *Chem. Mater.*, *6*, 2092.

Mott, N. F. 1949, *Proc. Phys. Soc. London Ser.*, *A62*, 416.

Mott, N. F. 1956, *Can. J. Phys.*, *34*, 1356.

Mott, N. 1958, *Nuovo Cimento*, *10*, *Suppl.*, 312.

Mott, N. F. 1961, *Philos. Mag.*, *6*, 287.

Mott, N. F. 1966, *Philos. Mag.*, *13*, 989.

Mott, N. F. 1968, *J. Non-Cryst. Solids*, *1*, 1.

Mott, N. F. 1990, *Metal-Insulator Transitions*, 2nd ed., Taylor & Frances, London.

Pauli, W. 1927, *Z. Phys.*, *41*, 81.

Peierls, R. E. 1930, *Ann. Phys. (Leipzig).*, *4*, 121.

Peierls, R. 1937, *Proc. Phys. Soc. London*, Ser. *A49*, 72.

Peierls, R. E. 1955, *Quantum Theory of Solids*, Clarendon Press, Oxford.

Puchkov, A. V.; Shen, Z-X. 2000, in Hughes, H. P., Starnberg, H. I., editors, *Electron Spectroscopies Applied to Low-Dimensional/Materials*; Kluwer Academic Publishers, Dordrecht, The Netherlands.

Raychaundhuri, A. K.; Rajeev, K. P.; Srikanth, H.; Mahendiran, R. 1994, *Physica B.*, *197*, 124.

Ribeiro, E.; Jäggi, R. D.; Heinzel, T.; Ensslin, K.; Medeiros-Ribeiro, G.; Petroff, P. M. 1999, *Phys. Rev. Lett.*, *82*, 996.

Shubin, S. P.; Wonsowsky, S. V. 1934, *Proc. R. Soc., London, A*, *145*, 159.

Verwey, E. J. W.; Haayman, P. W. 1941, *Physica*, *8*, 979.

Verwey, E. J. W.; Haayman, P. W.; Romeijn, F. C. 1947, *J. Chem. Phys.*, *15*, 181.

Yamaguchi, E.; Aoki, H.; Kamimura, H. 1979, *J. Phys. C: Solid State Phys.*, *12*, 4801.

Zaanen, J.; Sawatzky, G. A.; Allen, J. W. 1985, *Phys. Rev.*, *55*, 418

CHAPTER 7

Magnetic and Dielectric Properties

The phenomenon of magnetism was observed as far back as ancient times. The earliest discovery of the properties of lodestone (Fe_3O_4) was by either the Greeks or Chinese. The relationship between magnetism and electricity was discovered in the nineteenth century. In 1820, the Danish physicist Hans Christian Oersted (1777–1851) demonstrated that bringing a current-carrying wire close to a magnetic compass caused a deflection of the compass needle. This deflection is caused by the magnetic field generated by the electric current. In 1855, still more than 40 years before the discovery of the electron, Scottish physicist James Clerk Maxwell (1831–1879) showed that a few relatively simple mathematical equations could express the interrelation between electricity and magnetism, in terms of the macroscopic fields. For example, it is possible to calculate the magnetic flux density at any distance from a current-carrying wire if the current density is known. Hendrik Antoon Lorentz (1853–1928) later formulated Maxwell's equations in terms of the analogous microscopic fields. Our coverage of magnetic properties in this book does not require a detailed analysis of either Maxwell's or Lorentz's equations. However, we shall need to examine field equations more closely in the next chapter, which treats the optical properties of solids.

Just as magnetic moments are responsible for the magnetic behavior of a material, electric dipole moments are responsible for dielectric properties. In response to an electric field, the electric polarization of a dielectric material changes. Because of the analogies with magnetism, it is logical to briefly discuss dielectric properties in this chapter.

7.1 MACROSCOPIC MAGNETIC BEHAVIOR

It is possible to describe macroscopic magnetic behavior without any reference to its atomic origin. The starting point for discussing the behavior of a material placed within a magnetic field is the *magnetic moment* (also called the magnetic dipole moment, and the electromagnetic moment), μ_m. When a bar magnet is placed inside

Principles of Inorganic Materials Design   By John N. Lalena and David A. Cleary
ISBN 0-471-43418-3   Copyright © 2005 John Wiley & Sons, Inc.

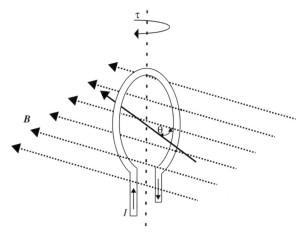

Figure 7.1   The concept of a magnetic moment. When placed in a magnetic field, a current-carrying loop (or a bar magnet) experiences a torque, τ, which tends to align the moment's axis with that of the field.

a field, for example, it experiences a torque (or moment), τ, that tends to align its axis with the direction of the field. The torque increases with the strength of the field and with the separation between the poles of the bar magnet. Similarly, an electrical current flowing in the single loop of a coil produces a magnetic field and has a magnetic moment perpendicular to the plane of the orbit. If such a coil is suspended in an external magnetic field so that it can turn freely, it too will rotate so that its axis tends to become aligned with the external field. An angle is made between the axis of the loop normal and the field direction (Figure 7.1). The magnitude of the magnetic moment is given by the product of the current, I (SI unit, ampere), and the area enclosed by the loop, πr^2 (units, m^2):

$$\mu_m = I\pi r^2 \qquad (7.1)$$

The SI units of magnetic moments are thus A·m^2. The torque experienced by the magnetic moment in the external field is given by the cross product of the magnetic moment and the *internal* magnetic flux density, $\boldsymbol{B}$ (SI units, tesla; or gauss in centimeter-gram-second (cgs) units), which is the sum of the flux due to the external source and that due to the sample itself.

For a macroscopic sample (which need not be a current-carrying coil), comprised of a very large number of microscopic (atomic) magnetic moments, we can define the *magnetization*, $\boldsymbol{M}$, as the net moment per unit volume (SI units A/m; G/cm^3 or electromagnetic units (emu) cm^3 in the cgs system):

$$\sum_i n_i \mu_{m,i} \qquad (7.2)$$

In Eq. 7.2, n_i is the concentration of magnetic moments, μ. The magnetization and the magnetic flux density are related to the magnetic-field intensity $\boldsymbol{H}$ *within* the sample, via the equation:

$$\boldsymbol{B} = \mu_0(\boldsymbol{H} + \boldsymbol{M}) \tag{7.3}$$

where μ_0 is the vacuum permeability ($4\pi \times 10^{-7}$ H/m).

The magnetization is often linearly proportional to the internal field strength, the constant of proportionality being the *magnetic susceptibility*, χ:

$$\boldsymbol{M} = \mu_0\chi\boldsymbol{H} \tag{7.4}$$

Materials with a magnetization of the same polarity as the applied field, and in which $\boldsymbol{M}$ and $\boldsymbol{H}$ are linearly related, are termed *paramagnetic substances*. This behavior results from the susceptibility being independent of the field strength and positive in sign, since paramagnets are attracted to a field. Materials in which the susceptibility is relatively small, field-independent, and negative in sign are termed *diamagnetic substances*. These materials are repelled by a magnetic field. The magnetization and applied field are still linearly related, but they have opposite polarities. *Ferromagnetic substances* are strongly attracted to a magnetic field and have large positive *field-dependent* values of the susceptibility. *Antiferromagnetic substances* have susceptibilities with magnitudes comparable to those of paramagnetic substances, but are field dependent. These behaviors are summarized in Table 7.1.

7.1.1 Magnetization Curves

As discussed earlier, the different types of magnetic behavior are evident in magnetization curves, which show the net magnetization, $\boldsymbol{M}$, of a sample versus applied field strength, $\boldsymbol{H}$. Paramagnetic and diamagnetic materials exhibit a linear relationship between $\boldsymbol{M}$ and $\boldsymbol{H}$ (Figures 7.2a and 7.2b), whereas the magnetization curve for a ferromagnetic substance has the appearance shown in Figure 7.2c. The net magnetization rapidly increases (following the initial curve) until it reaches the saturation magnetization, $\boldsymbol{M}_S$. If the field is reduced to zero, the magnetization does

TABLE 7.1 Susceptibility for Different Types of Magnetic Behavior

Type	Sign	Magnitude (emu)[a]	Field Dependence
Diamagnetic	−	10^{-6}	Independent
Paramagnetic	+	0–10^{-4}	Independent
Ferromagnetic	+	10^{-4}–10^{-2}	Dependent
Antiferromagnetic	+	0–10^{-4}	Dependent

[a]SI units for volume susceptibility are 4π times larger than cgs values given in the table.

Source: Drago, R. 1992, *Physical Methods for Chemists*, Saunders College Publishing, FortWorth, TX.

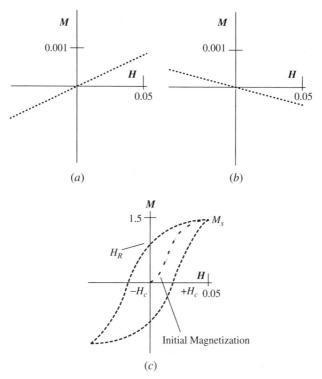

Figure 7.2   Different types of magnetic behavior. (a) A typical M–H curve (both axes in teslas) of a paramagnetic substance. (b) A diamagnetic substance. (c) A ferromagnetic substance (note the change in the scale of the M axis). An isotropic single crystal, or polycrystal with random crystallite orientation, is assumed.

not fall back along the initial curve, but rather follows the *hysteresis loop* in the direction indicated by the arrow. It can be seen from Figure 7.2c that once the material is exposed to a magnetic field, it retains some magnetization (M_R); it is converted to a *permanent magnet*. In order to demagnetize the material, it is necessary to reverse the external field and reach the *coercitive field strength* value, $-H_C$. This behavior is due to the domain structure of ferromagnets, which is discussed in Section 7.4. Note that the M–H curve, in general, will not be the same for different directions in a single crystal.

7.1.2 Susceptibility Curves

The quantity χ is dimensionless (since M and H have the same units), but is usually reported in relation to the sample volume and, hence, is reported as a *volume susceptibility* in cgs units of emu/cm^3, or, in reality, gauss cm^3 per gram. The corresponding value for the volume susceptibility in SI units is 4π times larger than the value in cgs units. The molar susceptibility, χ_M, may be obtained by multiplying χ by the molar volume.

In a single crystal M is, in general, not parallel to H. The susceptibility therefore must be defined by the magnitudes and directions of its principal susceptibilities χ_x, χ_y, χ_z. Hence, χ is a second-rank tensor, which linearly relates the two vectors, M and H (tensors were introduced in Section 5.1). However, we will only consider isotropic solids such as cubic crystals or randomly textured polycrystals, for which the magnetic susceptibility becomes equivalent to a scalar. In a powder with random crystallite orientation, M of the aggregate is in the direction of H, since those components of M transverse to H in the individual crystallites cancel on average. However, the mean value of M for a powder is the mean value of the principal susceptibilities, which are equal in magnitude. Hence, in a powder, $\chi = \frac{1}{3}(\chi_x + \chi_y + \chi_z)$ (Nye, 1957).

An important attribute of the magnetic susceptibility is its temperature dependence. The susceptibility, as a function of temperature, corresponding to each of the three major types of magnetic response is illustrated in Figure 7.3a. Pierre Curie

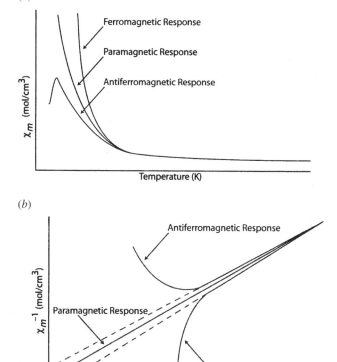

Figure 7.3 (a) Plots of the susceptibility versus temperature, and (b) the inverse susceptibility versus temperature, showing the three major classes of magnetic behavior.

(1859–1906) showed that the paramagnetic susceptibility and temperature are reciprocally related (Curie, 1895):

$$\chi = C/T \tag{7.5}$$

Equation 7.5 is the equation of a line passing through the origin (zero intercept) with χ plotted on the ordinate and $1/T$ plotted on the abscissa. The slope of the line is C, the Curie constant. Equivalently, a plot of $1/\chi$ vs. T yields a straight line with slope $1/C$ (Figure 7.3b). Paul Langevin (1872–1946) later showed that the exact form of the equation for the magnetic susceptibility is:

$$\chi = \mu^2/3kT \tag{7.6}$$

where μ is the magnetic moment and k is Boltzmann's constant (Langevin, 1905).

Although a straight-line plot of Eq. 7.5 might be obtained, a nonzero intercept is sometimes observed. This behavior is due to exchange interactions *between* the elementary magnetic moments composing the sample. The deviation from the Curie law can be treated as a perturbation to Eq. 7.5 by replacing T with a $[T - \theta]$ term, yielding the *Curie-Weiss law*:

$$1/\chi = (1/C)[T - \theta] \tag{7.7}$$

The slope of the line given by Eq. 7.7 is $1/C$, and the intercept of the line with the temperature axis gives the sign and magnitude of θ (the *paramagnetic Curie temperature* also denoted as T_C). A positive value for θ indicates ferromagnetic interactions, while a negative value indicates antiferromagnetic interactions.

In practice, it can be difficult to distinguish a pure paramagnet from a weak ferromagnet solely by examination of a $\chi(T)$ plot. Antiferromagnetic interactions are much more obvious, provided the sample temperature is lowered enough to observe the Néel temperature (T_N), the point at which the susceptibility begins to drop as the temperature is further decreased. It is not possible to fully understand exchange interactions without first discussing the atomic origin of magnetism, a topic we now take up.

7.2 ATOMIC ORIGIN OF PARAMAGNETISM

An electron orbiting an atomic nucleus can be viewed as a circulating electrical current and will have an *orbital magnetic moment*, if it has orbital angular momentum (i.e. *p*-, *d*-, and *f*-electrons). The orbital magnetic moment is perpendicular to the plane of the orbit and parallel to the angular momentum vector, **L**. Actually, the movement of any charged particle will produce a magnetic field. However, the magnetic properties of solids are generally associated with electrons, as the magnetic moment of this subatomic particle is 960 times greater than that of the neutron and 658 times that of the proton. There are, in fact, contributions due to both the orbital motion and spin of the electron.

The orbital contribution to the atomic moment is symbolized as μ_m^L, and is given by:

$$\mu_m^L = -\mu_B L \tag{7.8}$$

where L is the orbital angular momentum eigenstate operator and μ_B is a unit of measurement called the *Bohr magneton* (BM), which is given by $eh/2m_e = 9.274096 \times 10^{-24}$ A·m^2. The BM is identical to the magnetic moment of a single electron moving in a circular path with the Bohr radius. An orbital contribution to the atomic moment is only possible with a net orbital angular momentum hL, which may only occur when there are electrons outside of closed shells. The eigenstate of L is $h[L(L+1)]^{1/2}$ (where L is the total angular-momentum quantum number), but h is already built into the BM in Eq. 7.8. Hence, the magnitude of the orbital contribution to the moment is:

$$\mu_m^L = -\mu_B[L(L+1)]^{1/2} \tag{7.9}$$

Sometimes, the value for μ_B is not included in the calculation of the moment, in which case it must be carried in the units, as expressed by:

$$|\mu_m^L| = [L(L+1)]^{1/2} \, (\text{BM}) \tag{7.10}$$

The electron-spin angular momentum is twice as effective as the orbital angular momentum in giving rise to a magnetic moment. The magnetic moment due to the spin of an electron is given by:

$$\mu_m^s = -\gamma_m hs \tag{7.11}$$

where s is the spin angular momentum vector, h equals $h/2\pi$ (h is Planck's constant), and γ_m is the *gyromagnetic ratio*, which may be expressed as:

$$\gamma_m = ge/2m_e \tag{7.12}$$

In Eq. 7.12, e is the electron charge and m_e is the electron mass. The quantity $eh/2m_e$ is the BM, μ_B, which was already introduced. Equation 7.11 for the magnetic moment due to the intrinsic spin of an electron can thus be rewritten as:

$$\mu_m^s = -g\mu_B s \tag{7.13}$$

The factor g in Eqs. 7.12 and 7.13 is a dimensionless quantity called the electron, or Landé g-factor (or spectroscopic splitting factor) and, for the case of a *free* electron, it equals precisely 2.0023193. However, we will see later that it can have other values when the electron is in condensed matter.

The net spin of an atom containing more than one unpaired electron outside of closed shells is:

$$\mu_m^S = -g\mu_B S \tag{7.14}$$

If the vector S is replaced with its eigenvalue, $h[S(S+1)]^{1/2}$, where S is the total spin quantum number, this yields the magnitude of the *spin-only moment*:

$$\mu_m^S = -g\mu_B[S(S+1)]^{1/2} \tag{7.15}$$

or, equivalently:

$$|\mu_m^S| = g[S(S+1)]^{1/2} \, (\text{BM}) \tag{7.16}$$

We can also write:

$$|\mu_m^S| = g[n/2(n/2+1)]^{1/2} \, (\text{BM}) \tag{7.17}$$

where n is the number of unpaired spins per cation multiplied by the number of cations per formula unit. For example, a free Cr^{3+} ion has three unpaired electrons, so $S = 3/2$ and Eq. 7.16 ($g = 2$) gives $|\mu_m^S| = 3.87$ BM. Alternatively, $n = 3$ and substitution of this into Eq. 7.17 also gives 3.87 BM.

The effects of the crystal (ligand) field must also be considered in determining the value of S for $3d$ transition-metal ions. That is, S will depend on the relative magnitudes of the crystal-field splitting energy, $10Dq$, and the electron-spin pairing energy (on-site Coulomb repulsion energy). Those in which $10Dq$ is the greater of the two quantities will be low spin. If the on-site Coulomb repulsion is larger, a high-spin configuration results. Generally, $4d$ and $5d$ transition metal ions are low spin. By contrast, crystal-field splitting is usually unimportant for rare-earth ions, because their partially filled $4f$ shells lie deep inside the ions, beneath filled $5s$ and $5p$ shells.

Assuming there is no spin-orbit (LS) coupling (Section 7.2.2), the total magnetic moment of a free atom is obtained by combining Eqs. 7.8 and 7.14, the sum of the orbital plus spin moments:

$$\mu_m^{\text{tot}} = -\mu_B(L + 2S) \tag{7.18}$$

or:

$$\mu_m^{\text{tot}} = [L(L+1) + 4S(S+1)]^{1/2}(\text{BM}) \tag{7.19}$$

When g is taken as 2 and L is zero, Eq. 7.19 is equivalent to the spin-only moment (Eq. 7.15). The spin contribution is often the most important in compounds of the first-period transition metals. This is because less than spherically symmetric electric fields generated by the surrounding ions in a solid (the crystal field) either wholly or partially quench the orbital moment. Hence, the orbital contribution can be ignored and we can just use Eq. 7.16. However, this approximation is not as valid in compounds of the rare-earth ($4f$) elements, the transition elements of the third period ($5d$), and, to a lesser extent, those of the second period ($4d$).

7.2.1 Spin-Orbit Coupling

When only the spin contribution to the magnetic moment of an ion of the first-period transition elements is considered, excellent agreement with the experimental values is usually observed. By contrast, the magnetic moments of ions in the second and third transition periods ($4d$, $5d$ metal ions), as well as those of the rare-earth ($4f$) elements, differ significantly from the spin-only values. This is because spin-orbit coupling, which increases with the atomic number, influences the magnetic moment in heavy ions. An orbital contribution to the magnetic moment can be observed only if there is circulation of the electron about an axis and, hence, orbital angular momentum. However, in noncentrosymmetric crystal fields, the electron orbital angular momentum precesses, and its component perpendicular to the plane of its orbit can average to zero, which quenches the angular momentum and orbital magnetic moment.

With certain d electron configurations in cubic crystal fields (tetrahedral and octahedral complexes), the orbital momentum is not completely quenched because of the existence of a rotation axis that permits a half-filled d orbital of the *triply degenerate set* (t_{2g}) to be rotated into a vacant d orbital or another half-filled one (Drago, 1992). There must not be an electron of the same spin already in the orbital. For octahedral complexes, these electron configurations are: t_{2g}^1, t_{2g}^2, $t_{2g}^4 e_g^2$, $t_{2g}^5 e_g^2$. In tetrahedral complexes, orbital angular momentum can result from the electron configurations $e_g^2 t_{2g}^1$, $e_g^2 t_{2g}^2$, $e_g^4 t_{2g}^4$, $e_g^4 t_{2g}^5$.

In general, for cases in which there is a significant orbital angular momentum contribution, the value of the magnetic moment is not given by Eq. 7.19. This is because the spin and orbital momenta couple via the spin-orbit interaction:

$$\mathcal{H}_{SO} = \lambda \boldsymbol{L} \cdot \boldsymbol{S} \qquad (7.20)$$

where $\mathcal{H}_{SO}$ is the spin-orbit Hamiltonian term and λ is the (field-independent) spin-orbit coupling constant (units cm^{-1}), which has a different value for each electronic state. The spin-orbit contribution (1.25–250 meV) to the energy of an electron in a magnetic field is typically an order of magnitude smaller than that due to the crystal field (12.5 meV–1.25 eV), which, in turn, lies about an order of magnitude below the electron kinetic and potential energies (1–10 eV).

In the Russell–Saunders approximation (Russell and Saunders, 1925), appropriate for the $4d$, $5d$, and lighter rare-earth elements, the coupling of $\boldsymbol{L}$ and $\boldsymbol{S}$ gives a total angular momentum quantum number $\boldsymbol{J}$ that takes on values given by $(L + S)$ and $|L - S|$:

$$\boldsymbol{J} = \boldsymbol{L} + \boldsymbol{S} \qquad (7.21)$$

The total magnetic moment, *including spin-orbit coupling*, is then given by:

$$\mu_m^{\text{tot}} = -g_L \mu_B \boldsymbol{J} = -g_L \mu_B [J(J+1)]^{1/2} \qquad (7.22)$$

where g_L is now known as the Landé g-factor, after Alfred Landé (1888–1976) (Landé, 1923). We use the subscript L simply to differentiate this quantity from the free-electron value given earlier. The Landé g-factor may be calculated from:

$$g_L = 1 + \frac{J(J+1) + S(S+1) - L(L+1)}{2J(J+1)} \tag{7.23}$$

and falls within the range $1 \le g_L \le 2$. In using Eqs. 7.21 and 7.23, the rules for determining the ground-state configuration of an isolated atom, in the order they must be followed are:

1. *Hund's first rule*: S has the maximum value consistent with the Pauli exclusion principle.
2. L takes the maximum value consistent with *Hund's second rule*: For states of the same spin multiplicity, the state with the greater orbital angular momentum will be lowest in energy.
3. *Hund's third rule*: $J = |L - S|$ for a less-than-half-filled shell; $J = (L + S)$ for a more-than-half-filled shell; and $J = S$ for a half-filled shell.

7.2.2   Solids

It is now necessary to consider the magnetic behavior of a macroscopic collection of magnetic ions, as opposed to the single-ion magnetic systems discussed in the preceding section. The temperature dependence of the paramagnetic susceptibility of a bulk solid can be derived theoretically by statistical mechanics. The result is:

$$\chi_{\text{mol}} = [Ng_L\mu_B\mathbf{J}/\mathbf{H}]B_J(\alpha) \tag{7.24}$$

where N/V is the number of ions of the αth type per unit volume and $B_J(\alpha)$ is known as the Brillouin function (Brillouin, 1927):

$$B_J(\alpha) = \frac{(J + 1/2)\coth(J + 1/2)\alpha - 1/2\coth\alpha/2}{J} \tag{7.25}$$

$$\alpha = \mathbf{H}g_L\mu_B/kT \tag{7.26}$$

For $\alpha \to 0$, $B_J(\alpha)$ is approximately equal to:

$$B_J(\alpha) \cong J(J + 1)\alpha/3J^2 \tag{7.27}$$

and the *weak-field* molar paramagnetic susceptibility becomes:

$$\chi_{\text{mol}} = [Ng_L^2\mu_B^2/3kT]J(J + 1) \tag{7.28}$$

where N is now Avogadro's number and k is the Boltzmann constant. As stated earlier, the classic analog to Eq. 7.28 (i.e., replacing $g_L^2\mu_B^2J(J + 1)$ with the square of the magnetic moment, μ^2) was first derived by Langevin, while the reciprocal relation between χ and T was first observed experimentally by Curie. In those cases, where there is more than one type of magnetic ion, the individual contributions combine via the relation:

$$\chi_{\text{mol}} = N(\nu_A \chi_A + \nu_B \chi_B) \tag{7.29}$$

where ν_A and ν_B are the numbers of A and B atoms per unit cell. In the special case that $L = 0$, the molar paramagnetic susceptibility is simply:

$$\chi_{\text{mol}} = [Ng^2\mu_B^2/3kT]S(S+1) \tag{7.30}$$

On combining Eqs. 7.28 and 7.5, the Curie constant is found to be:

$$C_{\text{mol}} = [Ng_L^2\mu_B^2/3Vk]J(J+1) \tag{7.31}$$

In an analogous manner, the complete form of the Curie–Weiss equation is:

$$\chi = [Ng_L^2\mu_B^2/3k(T-\theta)]J(J+1) \tag{7.32}$$

A quantity called the *effective magnetic moment* is obtained by measuring the Curie constant. If a substance obeys Curie's law, Eq. 7.5 holds over the paramagnetic region and C_{mol} is given by the slope of the χ_{mol} vs. $1/T$ plot. If $(\mu_{\text{eff}})^2$ is then substituted for $g^2\mu_B^2[J(J+1)]$ in Eq. 7.31, it can be solved for μ_{eff}. The result is:

$$\mu_{\text{eff}} = [8C_{\text{mol}}]^{1/2}\,\mu_B \tag{7.33}$$

Table 7.2 lists the calculated magnetic moments for some ions based on Eqs. 7.16 and 7.22, along with the effective magnetic moment from Eq. 7.33.

TABLE 7.2 Values for the Calculated Magnetic Moment, Calculated Spin-Only Moment, and Effective Magnetic Moment for Some Ground-State 3d Transition-Metal Ions and Rare-Earth Ions

Ion	Number of Electrons	Ground State [JLS]	$g_L[J(J+1)]^{1/2}$	$2[S(S+1)]^{1/2}$	μ_{eff}
V^{4+}	1	[2 3 1]	1.55	1.73	1.80
Cr^{2+}	4	[0 2 2]	0	4.90	4.80
Fe^{2+}	6	[4 2 2]	6.70	4.90	5.40
Ni^{8+}	8	[4 3 1]	5.59	2.83	3.20
Pr^{3+}	2	[4 5 1]	3.62	2.83	3.50
Eu^{3+}	6	[0 3 3]	0	6.90	3.40
Tb^{3+}	8	[6 3 3]	9.72	6.90	9.50
Ho^{3+}	10	[8 6 2]	10.6	4.90	10.4

In addition to the temperature-dependent paramagnetism we have been discussing, a very small (10^{-5}–10^{-4} emu mole^{-1}) temperature-independent paramagnetism is associated with many ions, due to mixing of the electronic ground state with low-lying excited states. This may be observed, for example, with ions that have one less electron than is required to half-fill a shell (Cr^{2+}, high-spin d^4), since the total angular momentum is zero in the ground state, but becomes nonzero upon mixing. This is known as Van Vleck paramagnetism (Van Vleck, 1937).

7.2.3 Diamagnetic Correction

For accurate work, it is also necessary to correct the measured susceptibility for the diamagnetic contribution. The diamagnetic component of the magnetic susceptibility discussed in Section 7.1 is always present, but is generally very small. Diamagnetic substances are those in which all the electron spins are paired, so that the diamagnetic contribution is the only component to the magnetization. Quartz and rock salt are two examples. Diamagnetism arises from a moment that is induced by the external field (even a nonvarying one) and directed opposite to it. That is, the magnetic field produces an EMF, which generates a current that sets up an opposing magnetic field. The fact that even a *steady* magnetic field sets up such a diamagnetic screening current is a quantum effect that is not predicted for a classic system. In the latter case, Lenz's law dictates that only *varying* magnetic fluxes produce opposing fields. The diamagnetic susceptibility can be calculated from empirical data tabulations, such as Table 7.3 (see, for example, reference 1), by Pascal's method:

$$\chi_{\text{dia}} = \lambda + n_i\chi_i \qquad (7.34)$$

where λ is a constitutive correction factor that depends on the type of bonds present, n_i is the number of atoms of each type, and χ_i is the contribution to the susceptibi-

TABLE 7.3 Ionic Diamagnetic Suceptibilities per gram ion (-1×10^{-6} emu gram-ion^{-1}

Ion	Susceptibility	Ion	Susceptibility
Ag^+	24	Nb^{5+}	9
Al^{3+}	2	O^{2-}	12
Ba^{2+}	32	P^{3+}	4
Be^{2+}	0.4	P^{5+}	1
Ga^{3+}	8	Pb^{2+}	28
Ge^{4+}	7	Pb^{4+}	26
In^{3+}	19	Pr^{4+}	17
K^+	13	Re^{7+}	12
La^{3+}	20	Sb^{5+}	14
Mg^{2+}	2	Si^{4+}	1
Mn^{2+}	14	Ta^{5+}	14
Mn^{3+}	10	Ti^{4+}	5
Mo^{6+}	7	W^{6+}	14
Na^+	5	Zn^{4+}	10

From C.J.O'Connor *Prog. Inorg. Chem.* (1982) 29, 203.

lity of each of the constituent atoms. For inorganic solids, χ_{dia} can be approximated to within 10% of their actual values by assuming $\lambda = 0$ (O'Connor, 1982).

7.3 SPONTANEOUS MAGNETIC ORDERING

As has been explained, the presence of unpaired electrons imparts magnetic moments to the atoms, ions, or molecules of materials, causing them to behave like individual tiny magnets. Nevertheless, bulk paramagnetic solids do not exhibit any net magnetization in the absence of an applied magnetic field. Thermal energy is greater than the interaction energies between the individual magnetic moments. Consequently, the orientations of the individual moments stay randomly arranged (Figure 7.4a), as the disordered configuration is the thermodynamic equilibrium

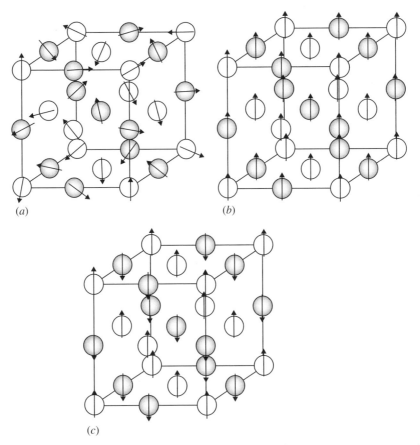

(a)   (b)

(c)

Figure 7.4   (a) In a paramagnetic substance there is no alignment between adjacent magnetic moments in the absence of an applied field. (b) In a ferromagnetic substance adjacent magnetic moments spontaneously align in a parallel fashion at low temperatures. (c) In an antiferromagnetic substance adjacent magnetic moments spontaneously align in an antiparallel fashion at low temperatures.

state. However, at sufficiently low temperatures, a magnetic field is able to overcome thermal disordering effects and force alignment.

By contrast, *spontaneous* magnetic ordering, in the absence of a magnetic field, is possible in some substances, in which adjacent magnetic moments become aligned, either in a parallel (Figure 7.4b) or antiparallel (Figure 7.4c) fashion, below some critical temperature. *Exchange interactions* are responsible for this effect. This, in turn, accounts for the deviation from Curie-law behavior in many paramagnetic materials, as described in the last section. Before we discuss the nature of exchange interactions, we shall first present the phenomenological description of ferromagnetic behavior, which preceded our understanding of its quantum mechanical origin.

The presence of interactions between magnetic moments means that the *effective* field acting on any particular atomic moment in a lattice is *not* identical with the externally applied field. The magnetization actually measured is that obtained by replacing the macroscopic field H in Eq. 7.4 with the *effective* magnetic field, H_{eff}, which may be considered made up of contributions from H and a fictitious internal field due to the neighboring moments. A widely used relationship between H and H_{eff} was derived by Lorentz (Lorentz, 1906), and is given by the expression

$$H_{eff} = H + 4\pi M/3 \qquad (7.35)$$

where M is the sample magnetization and the $4\pi/3$ factor arises from geometrical considerations, namely, an imaginary circumscribed sphere around a given moment. A completely analogous equation to Eq. 7.35 is found for the local electric field and electric polarizability in a dielectric material. Van Vleck pointed out (Van Vleck, 1932) that these relations were actually suspected, and somehat established, by Rudolf Julius Emanuel Clausius (1822–1888) and Ottaviano Fabrizio Mossottii (1791–1863) in the middle of the nineteeth century (Mossottii, 1836; Clausius, 1879)!

If Eq. 7.35 is substituted in Eq. 7.4, the following equation is obtained:

$$M = \mu_0\chi/[1 - (4\pi/3)\mu_0\chi]H \qquad (7.36)$$

If this expression were correct, M would be finite even in a zero applied field. Lorentz's theory thus predicts that all substances obeying the Curie law should be ferromagnetic at sufficiently low temperatures. Although this is not true, Lorentz's theory did predict that spontaneous magnetization in the absence of an externally applied field was possible. Pierre Ernest Weiss (1865–1940) replaced the $4\pi/3$ factor with a large temperature-independent proportionality constant, W, now known as the *Weiss field constant* (Weiss, 1907). This constant is of the order of magnitude 10^4. The product of W and M is the fictitious *Weiss field* (also known as the mean field, internal field, and molecular field). The key assumptions are that this field is proportional to the magnetization, and that it is the field acting on the magnetic moments due to their interactions with the surrounding moments. Thus, the *effective* field acting on the moments is the sum of the Weiss field and the externally applied field:

$$H_{\text{eff}} = H + WM \tag{7.37}$$

It can be shown that the paramagnetic Curie temperature, at which M vanishes, can be obtained from the vector-valued function $M(H)$, when that function is written as two parametric equations (the parameter being α of Eq. 7.26). The solution is obtained graphically (see, for example, Goodenough, 1966, p. 81) and is found to be:

$$\theta = Wg_L\mu_B M_S(J+1)/kT \tag{7.38}$$

The familiar Curie–Weiss law (Eq. 7.7), for the magnetic susceptibility in the paramagnetic regime, is obtained from:

$$M/(H + WM = \mu_0(C/T) \tag{7.39}$$

where it can be seen that the paramagnetic Curie temperature must also equal $\mu_0 CW$.

The Curie–Weiss law adequately explains the high-temperature paramagnetic regime ($T > \theta$) in ferromagnets. Moreover, like Lorentz's equation, the Weiss theory also predicts spontaneous ferromagnetism in the absence of an applied field at sufficiently low temperatures. However, there are some shortfalls. It predicts an exponential temperature-dependence of the magnetization at very low temperatures. In reality, low-temperature spin-wave excitations, which are not properly accounted for in the mean field theory, preclude this behavior and a $T^{3/2}$ power law is observed experimentally. There are also significant deviations of the susceptibility near the Curie point. Because of these reasons, the $M(T)$ curve predicted by the Curie–Weiss law is not necessarily a great fit to the experimentally observed curves.

7.3.1 Exchange Interactions

Werner Heisenberg (1901–1976) showed that the physical origin of the Weiss molecular field is in the exchange integral (Heisenberg, 1928). In magnetically ordered systems, magnetic moments couple with one another quantum mechanically. This coupling is known as the *exchange interaction*, and there are three types: direct exchange, indirect exchange, and superexchange. The Heisenberg exchange Hamiltonian for the *isotropic* interaction between atoms i and j, with *localized* electrons imparting spins S_i and S_j, is written as:

$$\mathcal{H}_{\text{ex}} = -\sum \mathcal{J}_{ij} S_i \cdot S_j \tag{7.40}$$

In Eq. 7.40, $\mathcal{J}_{ij}$ is called the exchange integral. The accepted convention is to use S in Eq. 7.40 to represent the *total* angular momentum, even though this quantity is denoted by J elsewhere. In some older texts, $\mathcal{J}_{ij}$ may be denoted J_{ij}, and one must be cautioned not to confuse it with the total angular-momentum quantum number, J. It has been pointed out that any relationship between the exchange integral and the

Weiss field is only valid at 0 K, since the former considers magnetic coupling in a pairwise manner and the latter results from a mean-field theory (Goodenough, 1963). Finally, it is also essential to understand that Eq. 7.40 is strictly valid only for localized moments (in the context of the Heitler–London model). One might wonder, then, whether the Weiss model is applicable to the ferromagnetic metals, in which the electrons are in delocalized Bloch states, for example, Fe, Co, and Ni. We take this up later (Sections 7.3.2 and 7.4.2).

The sign of the exchange integral determines the type of ordering. A positive integral is, like a positive Weiss constant, associated with *ferromagnetic* coupling. A negative exchange integral is, like a negative Weiss constant, associated with *antiferromagnetic* ordering. In the latter case, an equal number of each type of orientation results in a zero overall magnetic moment for the bulk sample. In the schematic representation of Figure 7.4c, we showed a two-sublattice model, in which each sublattice has a net spin, opposite to the other. In reality, the net spin of each sublattice is, on average, zero.

Antiferromagnetism is the most commonly observed type of magnetic behavior and is even observed over long ranges in materials that order locally ferromagnetically. A third type of ordering is possible in which an unequal number of the opposite orientations result in a nonzero net magnetic moment. Such materials are termed *ferrimagnetic*. In all cases, magnetic ordering does not occur until the order–disorder transition temperature is reached, which is the Curie temperature for ferromagnets and ferrimagnets, and the Néel temperature for antiferromagnets. Above these critical temperatures, paramagnetic behavior is observed due to thermally induced disorder. By contrast, a phenomenon known as *superparamagnetism* may be observed in very small magnetic particles below the ferromagnetic Curie temperature. These materials exhibit a bulklike ferromagnetism only when cooled below a critical blocking temperature. We cover this later.

Direct Exchange and Superexchange Interactions in Magnetic Insulators

The *direct-exchange* interaction is a strong but short-range coupling between magnetic moments that are close enough to have overlapping electron wave functions. In compounds (i.e., sublattice phases), anions separate the cations. Nonetheless, direct interaction between a pair of magnetic cations (not involving the anion) may be possible depending on the cation–anion–cation angle. For example, Anderson first pointed out that when only 90° *M–X–M* angles are present (e.g., rock salt), cation–anion–cation (*superexchange*) interactions are negligibly small in comparison to direct cation–cation interactions (Anderson, 1950). This is illustrated in Figure 7.5, where it can be seen that, in structures with edge-sharing octahedra (and face-sharing octahedra), the *d* atomic orbitals on neighboring octahedral cations are directed toward one another.

Magnetic exchange interactions can be positive (ferromagnetic) or negative (antiferromagnetic) in sign. A set of semiempirical rules, now known as the Goodenough–Kanamori rules (Goodenough, 1955, 1958; Kanamori, 1959), were established in the 1950s, which were highly successful in rationalizing the magnetic properties of a large variety of compounds on a qualitative level. The rules are based on the

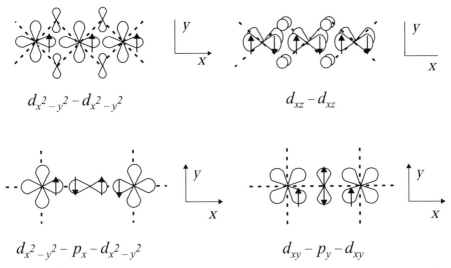

$d_{x^2-y^2} - d_{x^2-y^2}$ $d_{xz} - d_{xz}$

$d_{x^2-y^2} - p_x - d_{x^2-y^2}$ $d_{xy} - p_y - d_{xy}$

Figure 7.5 Exchange interactions are responsible for magnetic ordering. Direct exchange (top) between neighboring d atomic orbitals in a transition-metal compound with edge-sharing octahedra. Superexchange (bottom) between d atomic orbitals via ligand p atomic orbitals in a transition metal compound with vertex-sharing octahedra.

symmetry relations and electron occupancy of the orbitals on neighboring magnetic cations, which can interact via either direct exchange or superexchange. A similar model was proposed earlier by Slater (Slater, 1953) for the superexchange mechanism, but is not discussed here since the signs of the exchange interactions are found to be the same.

The Dutch physicist Hendrik Anthony Kramers (1894–1952), who is better known to chemists for theoretically describing the Raman effect with Heisenberg, first hypothesized the superexchange mechanism, in which magnetic coupling occurs through an intermediary nonmagnetic anion (Kramers, 1934). In addition to the direct exchange interaction previously mentioned, coupling between magnetic cations (M) can occur via atomic orbitals on nonmagnetic intermediary anions (X) having the proper symmetry for overlap with those on the cation. In the case of 180° $M–X–M$ angles (Figure 7.4), the anion $p\sigma$ orbitals are orthogonal to the cation t_{2g} orbitals, but overlap strongly with the e_g orbitals. Therefore, $e_g–p\sigma–e_g$ exchange interactions are strong. Similarly, the anion $p\pi$ orbitals are orthogonal to the cation e_g orbitals. However, the $t_{2g}–p\pi–t_{2g}$ bridge can accommodate exchange interactions, although they are weaker than the $e_g–p\sigma–e_g$ type. Cation–anion–cation superexchange is possible in vertex-sharing octahedra with $M–X–M$ angles as small as 120° (Goodenough, 1960). The guiding principle in determining the sign of the superexchange interaction is that the anion electrons simultaneously form partial-covalent bonds with the cation spin orbitals on opposite sides of the anion.

The Goodenough–Kanamori rules are given in Table 7.4, which lists four types of *inter*-atomic interactions between a pair of cations, denoted as A and B: (a)

TABLE 7.4   The Goodenough–Kanamori Rules for Magnetic Exchange

Electrons per Orbital	Sign of Exchange	Magnitude (Weihe and Güdel)
$[1]_A$–$[1]_B$	Antiferromagnetic	$2b_{ij}^2/[n_A n_B U]$
$[1]_A$–$[0]_B$	Ferromagnetic	$2b_{ij}^2 J_{n_B+1}/[n_A(n_B+1)U^2]$
$[2]_A$–$[1]_B$	Ferromagnetic	$2b_{ij}^2 J_{n_A+1}/[(n_A+1)n_B U^2]$
$[2]_A$–$[0]_B$	Ferro-/antiferromagnetic	$2b_{ij}^2(J_{n_A+1} - J_{n_B+1})/[(n_A+1)(n_B+1)U^2]$

Notes: b_{ij} = interatomic transfer integral; J_{n_A+1}, J_{n_B+1} = *intra*-atomic exchange integrals; n_A, n_B = number of unpaired electrons on cation A or B; U = charge transfer energy $(U \gg b_{ij})$. $1/10 \le J_{intra}/U \le 1/5$.

interaction between half-filled orbitals; (b) between half-filled and empty orbitals; (c) between filled and half-filled orbitals; and (d) between filled and empty orbitals. The Pauli exclusion principle limits a given orbital to one electron of each spin, that is, two electrons of the same spin exclude one another from a common region of space. Thus, type *a* interactions will be antiferromagnetic, while type *c* interactions are ferromagnetic. For coupling between half-filled or filled orbitals on one atom with empty orbitals on another atom (type *b* and type *d* interactions), ferromagnetic exchange is favored when the atom with the empty orbital also contains a nonoverlapping half-filled orbital because of *intra*-atomic exchange, *J* intra, *within* that atom. It has recently been shown that type *d* interactions can be either ferromagnetic (if $n_A < n_B$) or antiferromagnetic (if $n_A > n_B$) (Weihe and Güdel, 1997). As can be deduced from the third column of Table 7.4, when multiple types of interactions are simultaneously present, type *a* (antiferromagnetic) are generally dominant, since they are independent of the J_{intra}/U term (which usually lies in the range $1/10 \le J_{intra}/U \le 1/5$). Of course, it follows from Eq. 7.40 that there is no coupling between a magnetic atom and a nonmagnetic atom.

Use of the Goodenough–Kanamori rules allows one to predict, for simple cases, the net magnetic exchange expected of an *M–M* or *M–X–M* linkage, based on the valence electron configurations of the interacting cations. For example, Table 7.5 gives the sign of the net exchange between high-spin octahedral-site cations with linear *M–X–M* linkages for both direct exchange and superexchange. Note that it is predicted that high-spin d^3–d^5 interactions with *M–X–M* bond angles within the range 135°–150° change from ferromagnetic to antiferromagnetic. This is because, as the *M–X–M* bond angle bends away from 180°, the antiferromagnetic t_{2g}–$p\pi$–t_{2g} interactions (of the type ($[1/2]_A$–$[1/2]_B$) begins to dominate over the ferromagnetic e_g–$p\sigma$–e_g interactions (of the type ($[1/2]_A$–$[0]_B$). Similarly, it is speculated that the unsymmetrical bonding present with *M–X–M* bond angles in the range 125°–150° changes the coupling from antiferromagnetic to ferromagnetic for d^3–d^3 cation configurations (Goodenough, 1963).

The superexchange and direct-exchange mechanisms may compete with one another in structures containing both edge- and vertex-sharing octahedra, such as rock salt-, corundum-, ilmenite-, and rutile-type compounds. The predominant contribution is then determined by the relative magnitudes of the different interactions, which, in turn, depends on the cation–cation separation and electron occupancy of the atomic orbitals.

TABLE 7.5   Magnetic-Exchange Interactions for Octahedral-Site Cations

Type	n_A	n_B	Sign of Excahange
Cat–Cat	0	≤ 10	None
	≤ 5	≤ 5	Antiferromagnetic
	$5 < n < 8$	<8	Ferromagnetic
	$8 \leq n \leq 10$	≤ 10	None
180° Cat–An–Cat	0	≤ 10	None
	≤ 3	≤ 3	Antiferromagnetic
	$4 \leq n \leq 8$	≤ 3	Ferromagnetic
	4	4	Anti-/ferromagnetic
	$4 \leq n \leq 8$	$5 \leq n \leq 8$	Antiferromagnetic
	10	≤ 10	None
135°–150° Cat–An–Cat	0	≤ 10	None
	≤ 3	≤ 3	Antiferromagnetic
	$4 \leq n \leq 8$	≤ 3	Antiferromagnetic
	4	4	Anti-/ferromagnetic
	$4 \leq n \leq 8$	$5 \leq n \leq 8$	Antiferromagnetic
	10	≤ 10	None

Note: n_A, n_B = valence electron configurations of cations A and B.

Source: After Goodenough, J. B. 1960, *Phys. Rev.*, *117*, 1442. Copyright © American Physical Society. Reproduced with permission.

So far, only isotropic or symmetric exchange coupling are been discussed. Moriya proposed that anisotropic superexchange results when the effects of spin-orbit coupling are included in the superexchange formalism (Moriya, 1960). In this case, additional terms have to be included in the exchange Hamiltonian:

$$\mathcal{H}_{ex} = -\sum[\mathcal{J}_{ij}S_i \cdot S_j] + \sum[D_{ij} \cdot (S_i \times S_j) + S_i \cdot \Gamma_{ij} \cdot S_j] \qquad (7.41)$$

where D_{ij} is the Dzialoshinski–Moriya vector constant (Dzialoshinski, 1958) that can be approximated by $\mathcal{J}_{ij}(\Delta g/2)$ and Γ_{ij} is a tensor approximated by $\mathcal{J}_{ij}(\Delta g/2)$. In these expressions, Δg represents the deviation in the gyromagnetic ratio from the free-electron value of two.

Essentially, in anisotropic superexchange an electron or hole hops through an intermediate diamagnetic ion from a ground-state orbital on one magnetic ion to an excited-state orbital on a neighboring magnetic ion via the spin-orbit coupling interaction. This process is found to favor *canted* (tilted) spin configurations. For example, it has been shown that in the parent antiferromagnetic phases of the cuprate superconductors, which contain vertex-sharing CuO_6 octahedra, hopping occurs from a Cu^{2+} $d_{x^2-y^2}$ orbital, through a bridging oxygen p orbital, to an excited-state d_{xy} orbital on the neighboring Cu^{2+} ion (Bonesteel, 1993). The excited state on the adjacent Cu^{2+} ion is accessible via spin-orbit coupling on that ion, which causes the spin to precess as it hops between the adjacent magnetic ions through the intermediate oxygen p orbital. This then appears as a small spin tilting, or canting.

John Goodenough (Courtesy of the College of Engineering, The University of Texas at Austin. © The University of Texas at Austin. Reproduced with permission.)

JOHN BANNISTER GOODENOUGH

(b. 1922) obtained his Ph.D. in solid-state physics from the University of Chicago in 1952 under Clarence Zener. Goodenough then joined the Lincoln Laboratory at M.I.T., where he investigated magnetic properties of transition-metal oxides for magnetic-memory applications. During this period, he wrote the now often-cited book *Magnetism and the Chemical Bond* and the comprehensive review article "Metallic Oxides." Goodenough has contributed greatly to our understanding of the transition from itinerant-to-localized electronic behavior in transition-metal oxides. The Goodenough–Kanamori rules are widely used to predict the signs of magnetic-exchange interactions. From 1976 to 1986, Goodenough was head of the Inorganic Chemistry Laboratory at Oxford University, where he developed the cathode materials currently used in lithium rechargeable batteries, a development for which he received the Japan Prize in 2001. He has also contributed to the development of oxide-ion electrolytes for solid oxide fuel cells. He is now the Virginia H. Cockrell Centennial Chair in Engineering at the University of Texas. Goodenough was elected to the U.S. National Academy of Engineering in 1976, and he is a foreign associate of the French, Spanish, and Indian National Academies.

(*Primary source*: "Interview of John B. Goodenough, March 2001," by Bernadette Bensaude-Vincent and Arne Hessenbruch, 2001, from *The History of Materials Research Project*, funded by the Alfred P. Sloan Foundation and The Dibner Fund. © The Dibner Institute for the History of Science and Technology. Reproduced with permission.)

Indirect-Exchange Interactions Conduction electrons may couple with localized electrons over large distances, where there is little or no direct overlap between the localized electron wave functions. Zener first proposed this model to explain ferromagnetism in transition metals and their alloys (Zener, 1951a, 1951b). The spin of a single electron in the valence s-shell in an isolated atom is always aligned parallel to that of electrons in the inner incomplete d-shell. Zener thus assumed that in the condensed state, since the s-electrons spend more time near the d-electrons, indirect ferromagnetic coupling between the itinerant s-electron and localized d-electron would be even stronger. The properties of the system are obtained by minimizing the free energy with respect to the spin magnetization, or *spin energy*. When the indirect (ferromagnetic) exchange contribution to the spin energy dominates over the antiferromagnetic direct-exchange interaction between the d-shells (i.e., for large interatomic spacing), the system becomes ferromagnetic. This implies that the Curie temperature, T_C, is strongly dependent on the competition between these different spin interactions. For alloys, the Curie temperature may be expected to be dependent on the composition.

Zener's model was later abandoned in favor of the Ruderman–Kittel–Kasuya–Yosida (RKKY) model (see below) because Zener neglected the magnetic contribution due to the conduction electrons themselves (itinerant magnetism) and because he did not account for the oscillation of the spin polarization around the local moments. However, in magnetic semiconductors the Zener and RKKY models become equivalent since these oscillations average to zero (Dietl et al., 1997). Thus, the Zener model has found application in estimating how T_C might change with composition in dilute semiconductor alloys, for example, $Ga_{1-x}Mn_xN$, where $x = 0.06$–0.09 (Dietl et al., 2002).

The currently accepted mechanism for describing magnetic ordering in metallic systems containing unpaired electrons in localized atomic orbitals (particularly rare-earth systems) is the RKKY model (Ruderman and Kittel, 1954; Kasuya, 1956; Yosida, 1957). As mentioned earlier, cations polarize conduction electrons in their vicinity. Two localized electrons *can* interact via this polarized conduction electron density. Thus, in the RKKY model the conduction electrons act as an intermediary, similar to the role of the anions in the superexchange interaction for magnetic insulators. The exchange coefficient has a damped oscillatory nature that switches from positive to negative as the cation separation changes. The sign of the exchange interaction (ferromagnetic or antiferromagnetic) will therefore depend on whether the positions of the cations correspond to peaks or troughs.

7.3.2 Itinerant Ferromagnetism

The discussion of the preceding two sections relied on the presumption that localized moments were present. However, valence s- and p-electrons are *always* best described by Bloch functions, while $4f$ electrons are localized and $5f$ are intermediate. Valence d-electrons, depending on the internuclear distance, may be considered localized (Section 6.1). Our dilemma is that the Heisenberg exchange interaction of Eq. 7.40, which is the physical basis for the Weiss field, is not strictly applicable in the case of delocalized electrons, in spite of the success of the Weiss model.

It will be recalled that, for atoms, we only needed to consider electrons outside of closed shells. Analogously, in solids with delocalized states, only those electrons in partially filled bands (i.e., metals) contribute to ferromagnetic behavior. A proper treatment of metallic materials must include contributions to the spin energy due to the itinerancy of the conduction electrons, that is, their kinetic energy. The British physicist Edmund Clifton Stoner (1899–1968) introduced what is now called the *Stoner criterion* for itinerant ferromagnetism. It is given in the form of the following inequality:

$$I \cdot N(E_F) > 1 \qquad (7.42)$$

In this equation, I is the Stoner parameter, which must be calculated from the energy eigenvalues of the one-electron Bloch states (it can be considered analogous to the Weiss field, or exchange integral), and $N(E_F)$ is one-half the density of states at the Fermi level. When Eq. 7.42 is satisfied, a system is predicted to be unstable in the nonmagnetic state (Stoner, 1938). The primary condition for satisfying the Stoner criterion is a large density-of-states. Hence, the only ferromagnetic metals are narrow *d*-band systems. Stabilization of such a system can be achieved if the band splits into spin-up and spin-down configurations with different energies. This is called *intraband spin polarization*. Conversely, for small values of I and small density-of-states (wide-band metals) the nonmagnetic state is energetically preferred. Magnetic and neutron diffraction data indicate that this model of ferromagnetism is important in iron, nickel, and cobalt. Itinerant ferromagnetism, however, is not restricted to monatomic metals. Many compounds are itinerant ferromagnets including, for example, the layered $LaCrSb_3$ (Raju, et al., 1998) and the pyrochlore $Tl_2Mn_2O_7$, which are discussed later. We should note that the validity of the Stoner criterion does come into question with ultrasmall nanoscale metal particles where the electron energy levels become discrete.

7.3.3 Noncollinear Spin Configurations

In paramagnetic samples, the individual magnetic moments are randomly oriented due to thermal disorder being greater than the magnetic dipole–dipole interactions. Similarly, the spins may not all be oriented in a single direction in the magnetic state of a bulk ferromagnet, or in just two directions within an antiferromagnet. Rather, particular noncollinear spin configurations are observed. There are three distinct phenomena responsible for this magnetic behavior: geometric frustration, magnetic anisotropy, and magnetic domain formation.

Geometric Frustration Some lattices are inherently prone to competing or contradictory exchange coupling. For example, in a 2D triangular lattice, the collinear antiferromagnetic spin arrangement (Figure 7.6a) is not the lowest energy configuration, even though the antiferromagnetic constraint is satisfied globally. Calculations reveal that the noncollinear spin configuration illustrated in Figure 7.6b, in which the vector sum of the three spins (aligned at 120° to each other) on each triangle is zero, is lower in energy. Such magnetic systems are said to be geometrically frustrated.

Alternatively, rather than enter a state of long-range magnetic order at low temperatures via a noncollinear spin orientation, some substances exhibiting

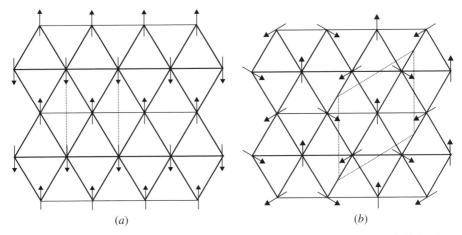

Figure 7.6   In a 2D triangular lattice, (a) the collinear antiferromagnetic structure is higher in energy than (b) the noncollinear antiferromagnetic structure. The unit cells are shown by the dashed lines.

competing magnetic interactions transition to a *spin-glass state*, in which the spins freeze with a random, or paramagnetic, orientation. If the sample is cooled in the presence of an applied field, there may be a partial alignment of the spins and resultant magnetization, which the sample may retain upon removal of the external field. Systems frozen into the spin glass state are also less able to follow fluctuations in an applied oscillating (alternating current (ac)) magnetic field, and the magnetization of the sample may lag behind the field. The spin-glass state thus has several characteristic signatures, which are dependent on the manner in which the magnetic measurements are made (e.g., field cooled, zero-field cooled). The spin-glass transition is interesting theoretically for its relation to critical phenomena.

Spin-glass behavior is often observed in amorphous solids at low temperatures, because the distribution in bond distances and angles gives rise to exchange frustration. However, crystalline substances with diagonal disorder may also exhibit spin-glass behavior. For example, the antiferromagnetic $LiMn_2O_4$ spinel contains a random distribution of Mn^{3+} (d^4) and Mn^{4+} (d^3) cations in the octahedral sites, which give rise to the simultaneous presence of d^3–d^3, d^4–d^4, and d^3–d^4 interactions. The Mn–O–Mn angle in spinel is 90° (edge-sharing MnO_6 octahedra) and the manganese ions reside on a tetrahedral sublattice, which is inherently prone to geometric frustration like the triangular lattice. Furthermore, since Mn^{3+} is a Jahn–Teller ion, this 3D tetrahedral network is distorted. All of these various features culminate in the spin-glass state below 25 K (Jang et al., 1999). It generally has been assumed that both geometric frustration *and* chemical or bond disorder are required in order for the spin-glass state to be observed in crystalline substances, such as in $LiMn_2O_4$. However, in the pyrochlore oxides exhibiting spin-glass behavior, there is no disorder present (Gardner et al., 2001).

While many disordered magnetic systems do display spin-glass behavior, those that are not geometrically frustrated can behave in a fundamentally different way. In

Simpson's amorphous antiferromagnet model, dilute and amorphous antiferromagnets alloyed with nonmagnetic atoms may contain a fraction of the magnetic
ions that are isolated from other magnetic ions by surrounding nonmagnetic ions
(Simpson, 1970). In this case, the exchange or superexchange interactions between
isolated magnetic ions are negligible and the system behaves as a simple paramagnet showing infinite or very large susceptibility near absolute zero, swamping
out any antiferromagnetic behavior at low temperatures. The reciprocal susceptibility versus temperature plot shows a downward curvature with decreasing
temperature. This has been found in some amorphous mixed oxides, such as
P-doped silicon, and In-doped CdS (Walstedt et al., 1979).

Magnetic Anisotropy   In crystals, magnetic anisotropy results when magnetization is easier in some directions than in others. The preferred directions are called
easy directions of magnetization, while the others are termed *hard* magnetization
directions. The energy along a hard direction is greater than along an easy direction
by the *magnetocrystalline anisotropy*. Magnetic anisotropy may originate from
magnetic dipole–dipole interactions or single-ion crystalline–electric-field effects
and spin-orbit coupling, the latter of which can result in anisotropic superexchange.

There is another phenomenon called *anisotropic exchange*, or the *exchange bias
effect*, that arises from the interfacial exchange coupling between a ferromagnetic
layer and an antiferromagnet, whereby a preferred easy direction of magnetization
in the ferromagnetic layer is aligned and *pinned* by the antiferromagnet, so that it
does not reverse in an external magnetic field. The ferromagnetic hysteresis loop
becomes asymmetric and shifted from zero. The exchange bias effect is important
in the read heads, made from magnetic multilayers exhibiting giant magnetoresistance (GMR), already in use in computer hard drives. However, since this effect
normally is associated with magnetic heterostructures (e.g., composites containing
separate ferromagnetic and antiferromagnetic films), as opposed to being an
intrinsic single-crystal property, we do not discuss it further here.

Magnetic Domains   The individual magnetic dipole moments in a ferromagnetic
substance spontaneously align upon reaching a critical temperature (the ferromagnetic Curie temperature, T_C) provided that at some point in its history, the
sample has had previous exposure to an external magnetic field. The requirement of a
previous exposure to an external magnetic field is due to the domain structure of
ferromagnets. There exist regions called domains, *within* which the individual
magnetic moments are aligned parallel below the transition temperature to generate a large magnetic moment given by the vector sum of all the unpaired electrons in
that domain. The domains are *not* identical with the crystalline grains, but may be
smaller or larger than the grain size. *Subgrain* domains were illustrated earlier in
Figure 1.16. At any rate, the different domains within a bulk sample have their
magnetizations pointing in different directions, canceling each other over the
macroscopic extent of the sample, in the absence of an external field.

The boundaries separating domains, called Bloch walls, are transition regions
where the magnetization changes continuously from the value for one domain to the

value for the neighboring domain. The Bloch wall is of higher energy, since there is a cost in the exchange energy for inverting the spins. We can also think of a domain wall as a defect, in an otherwise perfectly magnetically ordered system, in analogy to a grain boundary, which is also of higher energy in comparison to the intragranular region. This domain structure persists until the *saturation magnetization* point is reached, where all the magnetic moments of the entire sample become aligned with the applied field.

As described previously, although spontaneous magnetization occurs on a *microscopic* scale below the ferromagnetic Curie temperature, only a very small net magnetic moment is observed in the zero field for a macroscopic ferromagnetic sample. Upon a first exposure to an applied field, the magnetic moments of each domain align with the field. When magnetization saturation is reached, the material becomes *magnetized*. If the field is removed, the sample remains trapped in a metastable state, retaining a remnant magnetization fixed along the direction associated with minimum energy; it has been converted to a permanent magnet.

At any given temperature above absolute zero, thermal energy acts to restore thermodynamic equilibrium by destroying this magnetic order. However, when the relaxation time is very much longer than the observation time, the magnetization returns to zero only if the field is reversed and reaches the coercitive field strength. Thus, hysteresis is observed when the sample is trapped in the ferromagnetic state over a period much longer than the observation time. This behavior explains the magnetization curve shown in Figure 7.2c. The greater the misalignment among moments, the less the magnetization that remains after the field is removed (*remanence*), and the lower the magnitude of the reverse field required for demagnetization (*coercivity*).

The formation of domains in ferromagnets is attributed to a reduction in the overall magnetostatic energy. The external magnetic field exerted by a material, and its energy density, is decreased if internal local regions (domains) with opposing magnetizations are created, even though there is an energy penalty (increase) associated with the formation of domain walls. Domain structures have also been experimentally confirmed for many antiferromagnets, including NiO, CoO, $CoCl_2$, CoF_2, $MnTe$, and $YBa_2Cu_3O_{6+x}$. Domain structures in antiferromagnets were first suggested by Néel to explain their susceptibility behavior (Néel, 1954). However, as pointed out by Li (Li, 1956), there appears to be no *obvious* compensation for the increase in the free energy expected from the existence of domain walls (Gomonay and Loktev, 2002); that is, the formation of antiferromagnetic domains do not appear to be energetically favorable. The two types of domains found in antiferromagnets are collinear (with antiparallel magnetic vectors in neighboring domains) and orientational (with noncollinear magnetic axes in neighboring domains).

There is a limit to the energy reduction associated with domain formation, since it costs energy to form the domain walls. A single particle comparable in size to this minimum domain size would not break up into domains, but rather exhibit an enhanced magnetic moment. Typical values for the minimum domain size are in the range 10 nm to 100 nm (Sorenson, 2001). Single domains exhibit very large magnetic moments (thousands of Bohr magnetons) and, hence, the largest coercivities.

Another small-particle phenomenon, *superparamagnetism*, may be observed in even smaller ferromagnetic particles (1–10 nm) above a critical blocking temperature, T_B, which is below the ferromagnetic Curie temperature. In fact, any domain may contain subdomain superparamagnetic clusters of spins that act like single paramagnetic ions with large magnetic moments. The magnetic states in these samples are very sensitive to thermal fluctuations, as they have very short thermal relaxation times. The particles remain in the paramagnetic state (zero coercivity and zero remnant magnetization) at temperatures below the ferromagnetic Curie temperature, but exhibit very large total magnetic moments because each superparamagnetic cluster acts like a single independent domain. However, when cooled further, to temperatures below the blocking temperature, superparamagnetic clusters experience a very long thermal relaxation time, and the system exhibits ferromagnetic behavior with hysteresis. The ultimate goal in magnetic data storage is the use of one single-domain particle (above—but as close as possible to—the superparamagnetic state) per data bit.

7.4 MAGNETOTRANSPORT PROPERTIES

The presence of a magnetic field causes the mobile electrons in a conducting sample to be deflected by the Lorentz force, thus increasing the electrical resistivity. This phenomenon, termed *positive* magnetoresistivity, is often observed in metals with anisotropic Fermi surfaces. On the other hand, *negative* magnetoresistivity, that is, a decrease in resistivity with the application of a magnetic field, can occur when a field-induced ferromagnetic alignment of spins (electrons) reduces electron scattering.

This effect has been observed in many mixed-valent manganite perovskite oxides in the family $Ln_{1-x}A_xMnO_3$, where $0.2 \leq x \leq 0.5$, Ln = lanthanide or Bi, and A = Sr, Ca, Ba, Pb (Maignan et al., 1996; Urushibara et al., 1995; Von Helmolt et al., 1995; Zuotao and Yufang, 1996). These particular materials have a paramagnetic insulating phase at high temperatures and a metallic ferromagnetic low-temperature phase. Because the observed drop in resistivity is very large (sometimes as large as $10^6\%$), the effect has been termed *colossal magnetoresistance* (CMR), which is to be differentiated from the giant magnetoresistance in magnetic multilayers. Usually, the change in resistivity occurs at a temperature slightly lower than T_C, as illustrated in Figure 7.7. CMR is also exhibited by some compounds that do not appear to be mixed-valent. This group includes: the metallic pyrochlore oxide ($Tl_2Mn_2O_7$) (Shimikawa et al., 1996); the chromium spinels (ACr_2S_4: A = Fe, Cu, Cd) (Ramirez et al., 1997a, 1997b); some Zintl phases (e.g., $Eu_{14}MnBi_{11}$) (Chan et al., 1997a, 1997b, 1998); and the sulfide $BaFe_2S_3$ (Serpil Gönen et al., 2000). Although CMR is often described as a field-induced NM–M transition, in $Tl_2Mn_2O_7$, the high temperature phase is also metallic.

7.4.1 The Double-Exchange Mechanism

In the case of the mixed-valence manganites with the perovskite structure, Zener's double-exchange mechanism has been used to explain the

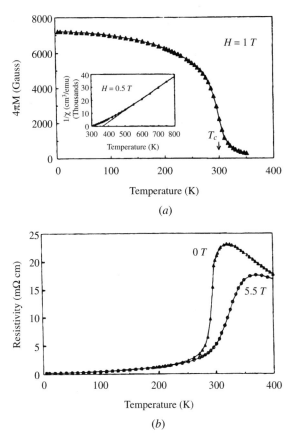

Figure 7.7 Example of experimental data on a CMR material. The ferromagnetic Curie temperature for $La_{0.65}(CaPb)_{0.35}MnO_3$ is slightly higher than the NM–M transition temperature. (After Pickett and Singh, 1996. Copyright © American Institute of Physics. Reproduced with permission.)

CMR effect (Zener, 1951c). In this model, a strong Hund's coupling is presumed to exist between the itinerant e_g electron originating on a high-spin d^4 Mn^{3+} cation (electron configuration = $t_{2g}^3 e_g^1$) and the localized t_{2g} spins on a neighboring d^3 Mn^{4+} cation (electron configuration = t_{2g}^3) through the intermediary oxygen atom. Because the exchange interaction, Δ_{ex}, is greater than the crystal-field splitting, $10Dq$ (Goodenough, 1971), incoherent carrier hopping occurs through the intermediary oxygen atom (i.e., through the Mn^{3+}–O–Mn^{4+} bonds) at no energy cost, and without a change in the itinerant carrier's spin. However, at the ferromagnetic transition temperature, the parallel alignment of spins allows a transition from incoherent hopping to coherent electrical conduction.

Clarence Zener (Courtesy of AIP Emilio Segrè Visual Archives.)

CLARENCE MELVIN ZENER

(1905–1993) received his Ph.D. in physics under Edwin C. Kemble at Harvard in 1929. Zener held teaching, research, and administrative positions in both academia and industry. He was at several American universities, including Washington University in St. Louis (1935–1937), the City College of New York (1937–1940), Washington State University (1940–1942), the University of Chicago (1945–1951), Texas A&M University (1966–1968), where he was the first Dean of the College of Science, and Carnegie Mellon University (1968–1993). Zener was also a physicist at the Watertown Arsenal from 1942 to1945. From 1951 to 1965 he was at Westinghouse, as Director of Research and, later, Director of Science. Zener's scientific accomplishments were as wide-ranging as his posts. He discovered the effect of heavily doping silicon diodes, causing them to exhibit a controlled reverse-bias breakdown and enabling their use as voltage regulators. These devices are now called Zener diodes. He developed a mean field model to explain the ferromagnetism in transition metals and their alloys, and was the first to write an explicit description for the effects of alloying on the magnetic Gibbs energy. Zener proposed the double-exchange mechanism for the magnetotransport properties of mixed-valence manganites. He contributed to the theories of elasticity and fracture mechanics of polycrystals, devised theoretical expressions for diffusion coefficients, and even conceived a design for an energy plant that utilizes oceanic temperature gradients. Zener was elected to the U.S. National Academy of Sciences in 1959.

(*Primary Source*: "On the Occasion of the 80th Birthday Celebration for Clarence Zener: Saturday, November 12, 1985," by Frederick Seitz in *J. Appl. Phys.*, Vol. 60, **1986**, pp. 1865–1867. © American Institute of Physics. Reprinted with permission.)

7.4.2 The Half-Metallic Ferromagnet Model

More sophisticated theories have attempted to addressed certain aspects of CMR that are not explained by Zener's simple double-exchange mechanism, such as the high insulating-like resistivity above the transition temperature in the perovskite phases. The electronic structures of many solids exhibiting CMR have been calculated using the local spin-density approximation (LSDA) method, in order to investigate the exchange interactions. One technique commonly utilized was not discussed in earlier chapters—*spin-polarized (SP) calculations*, in which separate band structures are calculated for spin-up and spin-down electrons. Such calculations are useful for studying itinerant ferromagnetism (ferromagnetism due to itinerant electrons) or any other SP configuration where the numbers of spin-up and spin-down electrons are not equal.

According to the Stoner criterion for itinerant ferromagnetism (Eq. 7.42), when the inequality $I \cdot N(E_F) > 1$ is satisfied, a system is predicted to be unstable in the nonmagnetic state. Stabilization can be achieved if the states split by intraband polarization. SP calculations on the perovskite $La_{1-x}Ca_xMnO_3$ exhibiting CMR strongly indicated such a half-metallic character, that is, the existence of a *metallic* majority spin band and a *nonmetallic* minority spin band in the ferromagnetic phase (Pickett and Singh, 1996).

Hybridization between the Mn d states and the oxygen p states was found to be strongly spin dependent—the *majority* Mn d band overlaps the O p band, while the *minority* Mn d band and O p band are nonoverlapping and hybridize much more weakly. Furthermore, it has been suggested that local lattice distortions arising from disorder on the Ca^{2+}/La^{3+} sites tend to localize the minority states near the Fermi level (E_F lies below a mobility edge). In the CMR regime, this results in nonconducting minority states, but has little effect on the strongly hybridized metallic majority bands. Separate spin-resolved photoemission measurements (Park et al., 1998) and scanning tunneling spectroscopy (Wei et al., 1997) on $La_{0.7}Sr_{0.3}MnO_3$ have provided direct experimental evidence that the mixed-valence manganite perovskite oxides exhibiting CMR have a spin polarization at E_F of 100% and are indeed half-metallic ferromagnets below the transition temperature.

With the possibility of spin fluctuation, the half-metallic ferromagnetic model is able to account for the high resistivity above the transition temperature in zero applied field, which the double-exchange model cannot explain. More importantly, this model does not explicitly depend on the presence of mixed valency to rationalize the CMR effect, even though mixed valency is implied in nonstoichiometric compositions. In fact, the half-metallic model can also explain the CMR observed in the stoichiometric pyrochlore $Tl_2Mn_2O_7$, which is metallic at high temperatures in the absence of a magnetic field (Matar et al., 1997).

It is possible to gain insight into magnetotransport properties from simpler nonspin polarized electronic structure calculations. For example, in $Tl_2Mn_2O_7$ the Mn^{4+}–O–Mn^{4+} bond angle of 133° is in the range for which a crossover fromanti-ferromagnetic-to-ferromagnetic interaction is expected (Shimikawa et al., 1997). The magnetic moments are disordered above the ferromagnetic transition temperature, and thus they act as scattering centers for the conduction electrons. When a sufficiently strong magnetic field is applied, the spins become ordered, which decreases the scattering and resistivity. An early extended Hückel tight-binding calculation (Seo et al., 1997) revealed a partially filled Tl $6s$ band resulting from small overlap with the Mn t_{2g} block band. The carrier density at the Fermi level was shown to be very low. Hence, transport should be strongly affected by spin ordering, which supports the simple spin scattering/superexchange mechanism for CMR in this compound.

Recent investigations have also studied the dependence of CMR on microstructural features. In manganese perovskite thin films, for example, film strain, oxygen deficiency, and chemical disorder have all been found to drastically affect the magnetotransport behavior (Panagiotopoulos et al., 1998). Likewise, in powder samples of $La_{0.66}Ca_{0.33}MnO_3$ ($T_C \sim 352°C$), the magnetoresistance appears to be

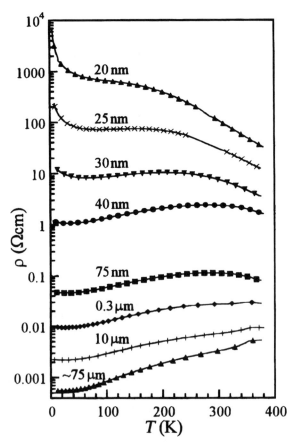

Figure 7.8 The zero-field resistivity of powder samples of $La_{0.66}Ca_{0.33}MnO_3$ are highly dependent on the grain size. (After Fontcuberta et al., 1999, Nedkov and Ausloos, editors, *Nano-Crystalline and Thin Film Magnetic Oxides*. Copyright © Kluwer Academic Publishers. Reproduced with permission.)

dominated by the tunneling of spin-polarized electrons *between* grains. Carriers can only cross from one grain to another if the corresponding majority bands in both sides are energetically coincident, that is, if there is no change in spin of the itineant electron (Fontcuberta et al., 1998). As expected, an intrinsic grain-size dependence on the magnetoresistance is observed. This is illustrated in the zero-field resistivity plot in Figure 7.8. It can be seen that coarse-grained samples have a metallic-like resistivity below T_C, but the M–NM transition becomes broader with decreasing grain size. In addition, as the grain size is lowered, the zero-field resistivity increases (due to a larger number of interfaces), spanning seven orders of magnitude, and eventually washes out the M–NM transition. However, it can be recovered in a field of a few Oe.

7.5 MAGNETOSTRICTION

Some magnetic materials can convert magnetic energy into mechanical energy and vice versa. This is called the *magnetostrictive effect*. Upon application of a magnetic field, magnetostrictive materials expand (positive magnetostriction) or contract (negative magnetostriction) along the direction of magnetization. This transduction capability allows magnetostrictive materials to be used as actuators and sensors. The first materials found to exhibit magnetostriction were nickel, cobalt, and iron, as well as alloys of these metals. Subsequently it was discovered that terbium and dysprosium displayed much larger magnetostrictive strains, but only at low temperatures. Alloying with iron to form the intermetallic compounds $TbFe_2$ and $DyFe_2$ bring the magnetostrictive properties to room temperature. However, due to the magnetic anisotropy of these materials, large strains are only produced in the presence of very large magnetic fields. Alloying these two compounds together to produce $Tb_{0.27}Dy_{0.73}Fe_{1.95}$ (Terfenol-D) greatly reduces the required magnetic-field strength.

The atomic mechanism of magnetostriction is complex. However, two processes can be distinguished at the domain level, which depend on the response of a material to an external magnetic field. The first process involves the migration of domain walls and the second involves the reorientation of domains. The rotation and movement of domains cause a physical length (dimension) change in the material. Magnetostriction is actually exhibited by most magnetic materials, but the effect is usually very small. By contrast, Terfenol-D is capable of strains as high as 1500 ppm (compared to 50 ppm for cobalt and iron alloys) and is currently the most widely used magnetostrictive material.

7.6 DIELECTRIC PROPERTIES

In an electrical conductor, an applied electric field, E, produces an electric current—ions in the case of an ionic conductor or electrons in the case of an electronic conductor. In insulating solids, the only response to an applied electric field is a static spatial displacement of the bound ions or electrons, resulting in an electrical polarization, P, or net dipole moment per unit volume:

$$P = \varepsilon_0 \chi E \qquad (7.43)$$

where ε_0 is the vacuum permittivity (8.8542×10^{-12} F/m), and χ is the *dielectric susceptibility*. Dielectric susceptibility, like magnetic susceptibility, is a second-rank tensor. The *dielectric constant* (also known as the *relative permittivity*), k, is defined as:

$$k = \varepsilon/\varepsilon_0 \qquad (7.44)$$

The dielectric constant defines the ability of a material to store an electric charge (capacitance). The capacitance of a parallel-plate capacitor, for example, is given by:

$$C = kA/d \qquad (7.45)$$

where A is the area of the plates, d is the distance separating them, and k is the dielectric constant of the material between the plates. Because C also equals Q/V (charge/voltage), the stored charge increases with materials of a higher dielectric constant. One intensive area of research in today's microelectronics industry concerns the search for new and suitable *low-k* dielectrics, materials that have a dielectric constant lower than that of SiO_2, or $k < 3.9$. Low-k dielectrics reduce the amount of capacitve coupling, or "cross talk," between the metal interconnects (analogous to the plates of a capacitor) of ever-decreasing line spacing in very high packing density integrated circuits.

As in the case of magnetism, the *internal* electric field inside a dielectric is not the same as the externally applied field, due to the local electric fields produced by the individual electric dipoles. For an isolated atom, the electric dipole moment, p, is analogous to Eq. 7.43, namely:

$$p = \alpha E_{ex} \qquad (7.46)$$

where α is the polarizability (a tensor). For solids, the bulk polarization, P, can be obtained by summation of the individual dipole moments:

$$P = \sum_i n_i p_i \qquad (7.47)$$

The polarizability in Eq. 7.46 may have several origins. First, the polarization may be *induced* by the external field. This induced polarization may originate from the polarizability associated with the local electronic charge cloud around a nucleus. It is given by $4\pi\varepsilon_0 r^3$, with r being the atomic radius. Other origins include bond polarizability in covalent solids and the ionic polarizability due to an electric field displacing cations and anions in opposite directions. A second type of contribution to the electric polarization is the field-induced reorientation of *permanent* electric dipoles in a material. This type is much more common in isolated molecules than in crystals. However, in certain solids (e.g., silicas and polymers), the presence of polar side-groups may create permanent dipoles. These are termed *paraelectric* solids. Not all the aforementioned mechanisms are active in all materials. For example, in ionic solids, the ionic polarizability is dominant at low frequencies. At high frequencies, however, the atomic polarizability becomes the dominant contribution to the dielectric constant even in ionic solids.

The applied electric field, E, and the polarization, P, are related through a second-rank tensor called the dielectric susceptibility, χ. Three equations, each containing three terms on the right-hand side, are needed to describe the susceptibility exactly:

$$\begin{aligned}
P_1 &= \varepsilon_0\chi_{11}E_1 + \varepsilon_0\chi_{12}E_2 + \varepsilon_0\chi_{13}E_3 \\
P_2 &= \varepsilon_0\chi_{21}E_1 + \varepsilon_0\chi_{22}E_2 + \varepsilon_0\chi_{23}E_3 \\
P_3 &= \varepsilon_0\chi_{31}E_1 + \varepsilon_0\chi_{32}E_2 + \varepsilon_0\chi_{33}E_3
\end{aligned} \tag{7.48}$$

Hence, the nine components of the susceptibility can be written in a square array as:

$$\chi = \begin{pmatrix} \chi_{11} & \chi_{12} & \chi_{13} \\ \chi_{21} & \chi_{22} & \chi_{23} \\ \chi_{31} & \chi_{32} & \chi_{33} \end{pmatrix} \tag{7.49}$$

The first suffix of each tensor component gives the row and the second the column in which the component appears. The χ_{23} term, for example, measures the component of the polarization parallel to x_2 (usually the y direction in a Cartesian coordinate system) when a field is applied parallel to x_3 (the z direction). The susceptibility tensor must conform to any restrictions imposed by crystal symmetry, that is, Eqs. 5.5–5.9.

The temperature dependence to the electric susceptibility depends on the polarization mechanism. For nonpolar molecules, the susceptibility is independent of temperature. For polar molecules or paraelectric solids, the electric susceptibility follows an equation equivalent to the Langevin function for paramagnetic susceptibility (Eq. 7.6). When applied to paraelectrics, however, this equation is called the Langevin–Debye or Debye formula (Debye, 1912).

7.6.1 Piezoelectricity

Piezoelectric crystals are transducers that generate an oscillating electrical polarization when subjected to an external oscillating mechanical stress, and vice versa. The brothers Paul-Jacques and Pierre Curie discovered the piezoelectric effect in 1880 when they compressed certain crystals along certain axes (Curie and Curie, 1880). The reciprocal behavior was deduced from thermodynamic principles a year later by Gabriel Lippman (1845–1921) and quickly verified by the Curies. The natural frequency of vibration (the *resonant frequency*) of a piezoelectric crystal, or one of its overtone frequencies, may be used to stabilize the frequency of a radio transmitter or other electronic oscillator circuit, or to provide a clock for digital circuits. That is because if the frequency applied to a crystal is *not* its natural frequency, only low-amplitude vibrations are induced. This type of opposition to an ac or volatage is known as *reactance*. When the applied frequency gets closer to the natural frequency of the crystal, the amplitude of its vibration increases, which decreases the reactance felt by the external circuit. Quartz is the most commonly used piezoelectric crystal because of its very low coefficient of thermal expansion. This allows a quartz clock, for example, to maintain its accuracy as the temperature changes.

The origin of the piezoelectric effect in a material is the vector displacement of ionic charges. The phenomenon can only be observed in noncentrosymmetric crystals, that is, those not containing an inversion center. Many crystals containing tetrahedral units, in addition to quartz, such as ZnO and ZnS, are piezoelectric. Application of a shearing stress along one of the polar axes of a tetrahedron results in a displacement of the cation charge relative to the center of the anion charge. It follows that the formation of such an electric dipole is not possible in crystals with inversion centers. Some centrosymmetric crystals exhibit a slight distortion from their ideal structure, which removes the inversion center and allows observation of the piezoelectric effect. For example, the ideal cubic perovskite structure possesses a center of inversion, which precludes piezoelectric behavior. However, slight tetragonal and rhombohedral distortion from cubic symmetry is responsible for the piezoelectric effect in the perovskite oxides $PbTiO_3$ and $PbZrO_3$, and for enhancing piezoelectricity in the solid solution $Pb(Zr,Ti)O_3$ (PZT).

The induced polarization in a piezoelectric, P_i, is a first-rank tensor (vector), and mechanical stress, σ_{jk}, is a second-rank tensor (nine components), which is represented in a Cartesian coordinate system with axes x, y, and z, as:

$$\sigma = \begin{pmatrix} \sigma_{xx} & \sigma_{xy} & \sigma_{xz} \\ \sigma_{yx} & \sigma_{yy} & \sigma_{yz} \\ \sigma_{zx} & \sigma_{zy} & \sigma_{zz} \end{pmatrix} \qquad (7.50)$$

The general form of the relation between P_i and σ_{jk} can be written as:

$$P_i = \sum_{j,k} d_{ijk} \sigma_{jk} \qquad (7.51)$$

From tensor algebra, the tensor property relating two associated tensor quantities, of rank f and rank g, is of rank $(f + g)$. Hence, the physical property connecting P_i and σ_{jk} is the third-rank tensor known as the piezoelectric effect, and it contains $3^3 = 27$ *piezoelectric strain coefficients, d_{ijk}*:

The size of this matrix of coefficients, however, is reduced by the fact that nonzero stresses require $\sigma_{jk} = \sigma_{kj}$. Hence, of the nine components to the stress tensor, only six are independent. Consequently, pairs of the piezoelectric strain coefficients like d_{ijk} and d_{ikj}, for example, cannot be separated and it is assumed that $d_{ijk} = d_{ikj}$. This reduces the number of independent piezoelectric strain coefficients from 27 to 18. To simplify notation, the subscripts of ij are normally replaced by a single subscript, n, following the convention:

$$1 \rightarrow xx, \qquad 2 \rightarrow yy, \qquad 3 \rightarrow zz, \qquad 4 \rightarrow yz, \qquad 5 \rightarrow xz, \qquad 6 \rightarrow xy$$

The piezoelectric strain coefficiensts now take the form d_{in} and the relation between the applied stress, σ_n, and induced polarization, P_i, can be expressed in final matrix-like form as:

$$\begin{pmatrix} P \\ P_2 \\ P_3 \end{pmatrix} = \begin{pmatrix} d_{11} & d_{12} & d_{13} & d_{14} & d_{15} & d_{16} \\ d_{21} & d_{22} & d_{23} & d_{24} & d_{25} & d_{26} \\ d_{31} & d_{32} & d_{33} & d_{34} & d_{35} & d_{36} \end{pmatrix} \begin{pmatrix} \sigma_1 \\ \sigma_2 \\ \sigma_3 \\ \sigma_4 \\ \sigma_5 \\ \sigma_6 \end{pmatrix} \qquad (7.52)$$

From Eq. 7.51, each component of the electric polarization is given in terms of all six components of the applied stress. For example, P_1 is:

$$P_{xx} = P_1 = \begin{pmatrix} d_{11}\sigma_1 + d_{16}\sigma_6 + d_{15}\sigma_5 \\ d_{16}\sigma_6 + d_{12}\sigma_2 + d_{14}\sigma_4 \\ d_{15}\sigma_5 + d_{14}\sigma_4 + d_{13}\sigma_3 \end{pmatrix}$$

Alternatively, in piezoelectric crystals, applied electric fields, E_i, generate mechanical strains, ε_n, that are also related by the piezoelectric coefficients:

$$\begin{pmatrix} \varepsilon_1 \\ \varepsilon_2 \\ \varepsilon_3 \\ \varepsilon_4 \\ \varepsilon_5 \\ \varepsilon_6 \end{pmatrix} = \begin{pmatrix} d_{11} & d_{12} & d_{13} & d_{14} & d_{15} & d_{16} \\ d_{21} & d_{22} & d_{23} & d_{24} & d_{25} & d_{26} \\ d_{31} & d_{32} & d_{33} & d_{34} & d_{35} & d_{36} \end{pmatrix} \begin{pmatrix} E_1 \\ E_2 \\ E_3 \end{pmatrix} \qquad (7.53)$$

A centrosymmetric stress cannot produce a noncentrosymmetric polarization in a centrosymmetric crystal. Electric dipoles cannot form in crystals with an inversion center. Hence, only the 20 noncentrosymmetric point groups (Table 2.2) are associated with piezoelectricity (the non-centrosymmetric cubic class *432* has a combination of other symmetry elements that preclude piezoelectricity). The piezoelectric strain coefficients, d_{in} for these point groups are given in Table 7.6, where, as expected, crystal symmetry dictates the number of independent coefficients. For example, triclinic crystals require the full set of 18 coefficients to describe their piezoelectric properties, but monoclinic crystals require only 8 or 10, depending on the point group.

TABLE 7.6 Piezoelectric Strain Coefficients in the Noncentrosymmetric Point Groups

Class	Point Group	Piezoelectric Strain Coefficients
Triclinic	*1*	$d_{11}, d_{12}, d_{13}, d_{14}, d_{15}, d_{16}, d_{21}, d_{22}, d_{23}, d_{24}, d_{25}, d_{26}, d_{31}, d_{32},$ $d_{33}, d_{34}, d_{35}, d_{36}$
Monoclinic	*2*	$d_{14}, d_{16}, d_{21}, d_{22}, d_{23}, d_{34}, d_{36}$
	m	$d_{11}, d_{12}, d_{13}, d_{15}, d_{24}, d_{26}, d_{31}, d_{32}, d_{33}, d_{35}$
Orthorhombic	*222*	d_{14}, d_{25}, d_{36}
	mm2	$d_{15}, d_{24}, d_{31}, d_{32}, d_{33}$
Tetragonal	*4*	$d_{14} = -d_{25}, d_{15} = d_{24}, d_{31} = d_{32}, d_{33}$
	4	$d_{14} = d_{25}, d_{15} = -d_{24}, d_{31} = -d_{32}, d_{36}$
	422	$d_{14} = -d_{25}$
	4mm	$d_{15} = d_{24}, d_{31} = d_{32}, d_{33}$
	42m	$d_{14} = d_{25}, d_{36}$
Cubic	*43m; 23*	$d_{14} = d_{25} = d_{36}$
Trigonal	*3*	$d_{11} = -d_{12}, d_{14} = -d_{25}, d_{15} = d_{24}, d_{16} = -2d_{22},$ $d_{21} = -d_{22}, 2d_{26}, d_{31} = d_{32}, d_{33}$
	32	$d_{11} = -d_{12}, d_{14} = -d_{25}, d_{26} = -2d_{12}$
	3m	$d_{15} = d_{24}, d_{16} = -d_{22}, d_{21} = -d_{22}, d_{31} = d_{32}, d_{33}$
Hexagonal	*6*	$d_{14} = -d_{25}, d_{15} = d_{24}, d_{31} = d_{32}, d_{33}$
	6mm	$d_{15} = d_{24}, d_{31} = d_{32}, d_{33}$
	622	$d_{14} = -d_{25}$
	6	$d_{11} = -d_{12}, d_{21} = -d_{22}, d_{16} = -2d_{22}, d_{26} = -d_{12}$
	6m2	$d_{21} = -d_{22}, d_{16} = -2d_{22}$

Source: Nye, J. F. 1957, *Physical Properties of Crystals: Their Representation by Tensors and Matrices*, Oxford University Press, Oxford.

7.6.2 Pyroelectricity

In pyroelectric materials (e.g., ZnO, LiNbO$_3$, and K(Nb,Ta)O$_3$), temperature changes induce changes in the electric polarization. While the temperature changes, the ions of the lattice shift, setting up a polarization current and thereby generating an electric field. When the temperature stops changing, the polarization current disappears. This effect has been known since ancient times and was given the name pyroelectricity in 1824 by Sir David Brewster. The theoretical principles behind the effect were studied by William Thomson and Woldemar Voigt in the late 1800s. Today, pyroelectric crystals are used as infrared detectors and minature X-ray generators. As with piezoelectricity, the pyroelectric effect requires the presence of a permanent electric dipole, arising from an electrical dipole moment in each primitive unit cell, that is, a noncentrosymmetric crystal. Actually, all pyroelectric crystals are also piezoelectric. However, because of Neumann's principle, the converse is not true. That is, since the electric polarization must have the point group symmetry, there can be no component of a polarization vector perpendicular to a mirror plane or twofold rotation axis, since it would transform into the negative of itself. Thus, only half of the 20 point groups associated with piezoelectricity also give rise to pyroelectricity. These 10 groups are one and the same with the 2D point groups: *1*, *2*, *m*, *2mm*, *4*, *4mm*, *3*, *3m*, *6*, and *6mm*.

7.6.3 Ferroelectricity

Ferroelectricity is a subclass of pyroelectricty discovered in "Rochelle salt" (potassium sodium tartrate) in 1921 by Joseph Valasek (1899–1982) at the University of Minnesota (Valasek, 1921). The symmetry constraints of ferroelectric crystals are identical to those of pyroelectric ones, which also make all ferroelectric crystals piezoelectric. Both ferroelectric crystals and pyroelectric crystals spontaneously generate an electric polarization, in the absence of an external electric field. However, unlike pyroelectric materials, ferroelectric materials exhibit a remant (residual) polarization that is *reversible* upon application of a field with opposite polarity greater than the coercive field. Ferroelectric crystals also possess a domain structure, similar to ferromagnets and, hence, exhibit hysteresis in their *P–E* (polarization–field) curves. The electric dipole moments in ferroelectrics spontaneously align below the *ferroelectric Curie temperature*, T_{Cf}. Above this temperature, a high dielectric constant is still obtained, but no residual polarization is retained in the absence of an applied field. *Ferrielectric crystals* are a subset in which the net polarization vectors are oriented in different directions in neighboring domains.

REFERENCES

Anderson, P. W. 1950, *Phys. Rev.*, *79*, 350.

Bonesteel, N. E. 1993, *Phys. Rev. B*, *47*, 11302.

Brillouin, L. 1927, *J. Phys.*, *8*, 74.

Chan, J. Y.; Wang, M. E.; Rehr, A.; Kauzlarich S. M.; Webb, D. J. 1997a, *Chem. Mater.*, *9*, 2131.

Chan, J. Y.; Kauzlarich S. M.; Klavins, P.; Shelton R. N.; Webb, D. J. 1997b, *Chem. Mater.*, *9*, 3132.

Chan, J. Y.; Kauzlarich S. M.; Klavins, P.; Shelton R. N.; Webb, D. J. 1998, *Phys. Rev. B*, *57*, R8103.

Clausius, R. 1879, *Die mechanische Behandlung der Electricität*, Vieweg, Braunschweig, Germany.

Curie. J. P.; Curie, P. 1880, *C. R. Acad. Sci.*, *91*, 294.

Curie, P. 1895, *Ann. Chim. Phys.*, *5*, 289.

Debye, P. 1912, *Phys. Z.*, *13*, 97.

Dietl, T; Haury, A.; Merle'dAubigné, Y. 1997, *Phys. Rev. B*, *55*, R3347.

Dietl, T.; Matsukura, F.; Ohno, H. 2002, *Phys. Rev. B*, *66*, 33203.

Drago, R. S. 1992, *Physical Methods for Chemists*, Saunders College Publishing, Ft. Worth, TX.

Dzialoshinski, I. 1958, *J. Chem. Phys. Solids*, *4*, 241.

Fontcuberta, J.; Balcells, L. L.; Martínez, B.; Obradors, X. 1998, in Nedkov, I, Ausloos, M., editors, *Nano-Crystalline and Thin Film Magnetic Oxides*, Kluwer Academic Publishers, Dordrecht, The Netherlands.

Gardner, J. S.; Gaulin, B. D.; Berlinsky, A. J.; Waldron, P.; Dunsiger, S. R.; Raju, N. P.; Greedan, J. E. 2001, *Phys. Rev. B*, *64*, 224416.

Gomonay, H.; Loktev, V. M. 2002, *J. Phys: Condens. Matter*, *14*, 3959.

Goodenough, J. B. 1955, *Phys. Rev.*, *79*, 564.

Goodenough, J. B. 1958, *J. Phys. Chem. Solids*, *6*, 287.

Goodenough, J. B. 1966, *Magnetism and the Chemical Bond*, 2nd print., Interscience Publishers, New York.

Goodenough, J. B. 1971, *in Progress in Solid State Chemistry*, Vol. 5, Reiss, H, editors, Pergamon Press, Oxford.

Heisenberg, W. 1928, *Z. Phys.*, *49*, 619.

Jang, Y. I.; Chou, F. C.; Chiang, Y.-M. 1999, *Appl. Phys. Lett.*, *74*, 2504.

Kanamori, J. 1959, *J. Phys. Chem. Solids*, *10*, 87.

Kasuya, T. 1956, *Prog. Theor. Phys.*, *16*, 45.

Kramers, H. 1934, *Physica*, *1*, 182.

Landé, A. 1923, *Z. Phys.*, *15*, 189.

Langevin, P. 1905, *J. Phys.*, *4*, 678.

Li, Y. Y. 1956, *Phys. Rev.*, *101*, 1450.

Liu, J. Z.; Chang, I. C.; Irons, S.; Klavins, P.; Shelton, R. N.; Song, K.; Wasserman, S. R. 1995, *Appl. Phys. Lett.*, *66*, 3218.

Lorentz, H. A. 1906, *The Theory of Electrons*, Teubner, Leipzig.

Maignan, A. Simon, C.; Caignaert, V.; Raceau, B. 1996, *J. Magn. Magn. Mat.*, *L5*, 152.

Matar, S. F.; Subramanian, M. A.; Etourneau, J. 1997, *J. Mater. Chem.*, *7(8)*, 1457.

Moriya, T. 1960, *Phys. Rev.*, *120*, 91.

Mossottii, O. F. 1836, *Sur les forces qui régissent laconstitution intime des corps*, Turin, Italy.

Néel, L. 1954, *Proc. Int. Conf. on Theoretical Physics*, Science Council of Japan, Tokyo, 701.

Nye, J. F. 1957, *Physical Properties of Crystals*, Oxford University Press, London.

O'Connor, C. J. 1982, *Prog. Inorg. Chem.*, *29*, 203.

Panagiotopoulos, I.; Pissas, M.; Christides, C.; Kallias, G.; Psycharis, V.; Moutis, N.; Niarchos, D. 1998, in Nedkov, I., Ausloos, M., editors, *Nano-Crystalline and Thin Film Magnetic Oxides*, Kluwer Academic Publishers, Dordrecht, The Netherlands, and references 32–38 therein.

Park, J-H.; Vescovo, E.; Kim, H.-J.; Kwon, C.; Ramesh, R.; Venkatesan, T. 1998, *Nature*, *392*, 794.

Pickett, W. E.; Singh, D. J. 1996, *Phys. Rev. B*, *53*, 1146.

Raju, N. P.; Greedan, J. E.; Ferguson, M. J.; Mar, A. 1998, *Chem. Mater.*, *10*, 3630.

Ramirez, A. P.; Cava, R. J. Krajewski, J. 1997a, *Nature*, *386*, 156.

Ramirez, A. P. 1997b, *J. Phys: Condens. Matter.*, *9*, 8171.

Ruderman, M. A.; Kittel, C. 1954, *Phys. Rev.*, *96*, 99.

Russell, H. N.; Saunders, N. 1925, *Astrophys. J.*, *61*, 38.

Seo, D. K.; Whangbo, M.-H.; Subramanian, M. A. 1997, *Solid State Commun.*, *101*, 417.

Serpil Gönen, Z.; Fournier, P.; Smolyaninova, V.; Greene, R.; Araujo-Moreija, F. M.; Eichhorn, B. 2000, *Chem. Mater.*, *12*, 3331.

Shimakawa, Y.; Kubo, Y.; Manako, T. 1996, *Nature*, *379*, 53.

Shimakawa, Y.; Kubo, Y.; Manako, T.; Susho, Y. V.; Argyriou, D. N.; Jorgensen, J. D. 1997, *Phys. Rev. B*, *55*, 6399.

Simpson, A. W. 1970, *Phys. Stat. Sol.*, *40*, 207.

Slater, J. C. 1953, *Q. Progr. Rep. M.I.T.*, July 15,1; Oct. 15, 1.

Sorenson, C. M. 2001, "Magnetism," in *Nanoscale Materials in Chemistry*, Wiley-Interscience, New York.

Stoner, E. C. 1938, *Proc. R. Soc. London*, *165*, 372.

Urushibara, A.; Moritomo, Y.; Arima, Y.; Asamitsu, A.; Kido, G.; Tokura, Y. 1995, *Phys. Rev. B*, *51*, 14103.

Valasek, J. 1921, *Phys. Rev.*, *17*, 475.

Van Vleck, J. H. 1932, *The Theory of Electric and Magnetic Susceptibilities*, Oxford University Press, Oxford.

Van Vleck, J. H. 1937, *J. Chem. Phys.*, *5*, 320.

Walstedt, R. E.; Kummer, R. B.; Geschwind, S.; Narayanamurti, V.; Devlin, G. E. 1979, *J. Appl. Phys.*, *50(3)* 1700.

Wei, J. Y. T.; Yeh, N.-C.; Vasquez, R. P. 1997, *Phys. Rev. Lett.*, *79*, 5153.

Weihe, H.; Güdel, H. U. 1997, *Inorg. Chem.*, *36*, 3632.

Weiss, P. 1907, *J. Phsique*, *6*, 667.

Yosida, K. 1957, *Phys. Rev.*, *106*, 893.

Zener, C. 1951a, *Phys. Rev.*, *81*, 440.

Zener, C. 1951b, *Phys. Rev.*, *83*, 299.

Zener, C. 1951c, *Phys. Rev.*, *82*, 403.

Zener, C. 1955, *Trans AIME*, *203*, 619.

Zuotao, Z.; Yufang, R. 1996, *J. Solid State Chem.*, *121*, 138.

CHAPTER 8

Optical Properties of Materials

The interaction of light with matter has fascinated people since ancient times. The color of an object is the result of this interaction. In modern terms, we describe this interaction as *spectroscopy*. In this chapter, we examine how the optical properties of a material are the result of its chemical composition and structure. We then present several examples of technologically relevant applications of the manipulation of the optical properties to achieve a desired performance.

8.1 MAXWELL'S EQUATIONS

In one sense, this is an easy topic. All of the interactions of light with matter can be described with Maxwell's equations (Griffiths, 1981). However, for the materials chemist faced with the problem of designing a glass lens that does not reflect visible light, Maxwell's equations in their native state do not appear to offer a straightforward solution. Fortunately, Maxwell's equations have been solved for most of the problems encountered in materials design. Let us examine one such case.

Imagine a material that is diamagnetic, transparent, and is an insulator: a piece of glass! When solving Maxwell's equations for the interaction of light with this system, several important and unexpected results are realized. First, a plane-polarized light beam entering this material is split into two plane-polarized beams. The beams are orthogonal to each other with neither necessarily parallel to the incoming beam. Second, the two beams do not necessarily travel at the same speed through the material. Upon exiting the material, the two beams do not necessarily recombine. It is possible to observe two exit beams for one entrance beam.

So Maxwell's equations describe phenomena that are difficult to anticipate intuitively. Historically, physicists have manipulated these laws and recast them in forms specific to particular systems. For example, to calculate the magnetic field a distance r from a straight wire carrying current i, the relevant equation is

$$B = \frac{\mu_0 i}{2\pi r} \tag{8.1}$$

Principles of Inorganic Materials Design   By John N. Lalena and David A. Cleary
ISBN 0-471-43418-3   Copyright © 2005 John Wiley & Sons, Inc.

where μ_0 is the permeability of free space. However the Maxwell's equation responsible for this equation is

$$\bar{V} \times \boldsymbol{B} = \mu_0 \boldsymbol{J} \tag{8.2}$$

known as Ampère's law, where boldface quantities are vectors. The current density is $\boldsymbol{J}$.

In the same spirit, we present the results of the applications of Maxwell's equations to a number of systems and exploit the resulting simplified expressions.

As just mentioned, we begin by considering the diamagnetic, transparent, insulator from above. When such a sample is exposed to an electric field, it responds. The electrons around each atom move in such a way as to reduce the potential energy (or minimize the Gibbs free energy if the system is at a constant temperature and pressure) of the system. Electrons have a low mass compared to the nucleus that they surround and can move in response to the electric field. If the field is oscillating, the electrons will oscillate as well with the same frequency. The electric field, $\boldsymbol{E}$, induces in the material a polarization,

$$P = \varepsilon_0 \chi_e \boldsymbol{E} \tag{8.3}$$

where initially we assume that the direction of $\boldsymbol{E}$ and $\boldsymbol{P}$ is the same. The proportionality constant is a product of two terms. The first is ε_0, the permittivity of free space, the second is χ_e, the electric susceptibility. The relation holds as long as the electric field $\boldsymbol{E}$ is not too large. Later in this chapter we consider what happens to $\boldsymbol{P}$ when $\mathbf{E}$ becomes large. The induced polarization $\mathbf{P}$ creates its own electric field, which in turn affects $\boldsymbol{P}$. The net electric field present inside our diamagnetic, transparent, insulator is $\boldsymbol{E}$. We add to $\boldsymbol{E}$ the polarization $\boldsymbol{P}$ to create the electric displacement, $\boldsymbol{D}$

$$\boldsymbol{D} = \varepsilon_0 \boldsymbol{E} + \boldsymbol{P} \tag{8.4}$$

$$\boldsymbol{D} = \varepsilon_0 (1 + \chi_e) \boldsymbol{E} \tag{8.5}$$

$$\boldsymbol{D} = \varepsilon \boldsymbol{E} \tag{8.6}$$

where $\epsilon = \epsilon_0 (1 + \chi_e)$. In this equation, ϵ is known as the permittivity of the material. It is this relationship, the electric displacement vector to the electric field, that forms the basis of our entire discussion concerning the optical properties of materials.

Solving Maxwell's equations for a plane wave traveling through a diamagnetic, transparent, insulator is somewhat involved (Nye, 1985). The equations are simplified somewhat by the these conditions: diamagnetic sets $\boldsymbol{B} = -\mu_0 \boldsymbol{H}$ and insulator sets $\boldsymbol{J} = 0$. The incoming light $\boldsymbol{E}$ is modeled as a plane wave. The derivation begins with

$$\bar{V} \times \boldsymbol{E} = -\partial \mu_0 \boldsymbol{H} / \partial t \tag{8.7}$$

where the equation for E is

$$E = E_0 \exp\left\{ i\omega\left(t - \frac{\mathbf{r} \bullet \mathbf{1}}{v} \right) \right\} \tag{8.8}$$

In the end, a relationship between D and E is realized:

$$\mu_0 v^2 D - E + \mathbf{1}(\mathbf{1} \cdot E) = 0 \tag{8.9}$$

where $\mathbf{1}$ is a unit vector perpendicular to the plane wave, and v is the velocity of the wave. Before applying this equation, we consider the relationship between D and E further. If these two vectors were collinear, the relationship between them would be a simple constant of proportionality

$$D = \varepsilon E \tag{8.10}$$

However, in the general case, D and E will not be collinear. Therefore a tensor is necessary to describe the relationship between D and E

$$D_i = \sum_j \varepsilon_{ij} E_j \tag{8.11}$$

For example, for an arbitrary orientation of E,

$$D_x = \varepsilon_{xx} E_x + \varepsilon_{xy} E_y + \varepsilon_{xz} E_z \tag{8.12}$$

meaning the a component of the E vector along the y-axis contributes to the component of D along the x-axis. If the E vector is along the x-axis, this equation simplifies to

$$D_x = \varepsilon_{xx} E \tag{8.13}$$

We see that D and E, in the general case, are connected by a second-rank tensor. This tensor can be either the permitivitty tensor or the dielectric-constant tensor, depending on how the individual components are expressed. In the preceding equation, the tensor is a permitivitty tensor. To convert it to a dielectric-constant tensor, each element, ϵ_{ij} is divided by the permitivitty of free space, ϵ_0,

$$D_x = \varepsilon_0 K_{xx} E \tag{8.14}$$

Now we have the two things we need to progress in our understanding of the interaction of light with matter:

1. The wave relation between E and D, $\mu_0 v^2 D - E + \mathbf{1}(\mathbf{1} \cdot E) = 0$;
2. The tensor relation between E and D, $D_i = \sum_j \varepsilon_{ij} E_j$.

With these two relationships, we can examine what happens when light enters a diamagnetic, transparent insulator. To simplify the relationship between D and E, we examine the permittivity tensor ϵ_{ij} in detail. This is a square symmetric tensor. Hence, it is possible to transform it into a diagonal tensor:

$$
\begin{vmatrix} \varepsilon_{xx} & 0 & 0 \\ 0 & \varepsilon_{yy} & 0 \\ 0 & 0 & \varepsilon_{zz} \end{vmatrix} = T^{-1} \begin{vmatrix} \varepsilon_{11} & \varepsilon_{12} & \varepsilon_{13} \\ \varepsilon_{12} & \varepsilon_{22} & \varepsilon_{23} \\ \varepsilon_{13} & \varepsilon_{23} & \varepsilon_{33} \end{vmatrix} T \tag{8.15}
$$

where T is a rotation matrix and T^{-1} is its inverse. When D and E are referred to the principal axes of the permitivitty tensor, the second equation simplifies in component form to

$$
D_x = \varepsilon_{xx} E_x \qquad D_y = \varepsilon_{yy} E_y \qquad D_z = \varepsilon_{zz} E_z \tag{8.16}
$$

which allows the components of the first equation to be written

$$
D_x = \frac{1_x(\mathbf{1} \cdot \mathbf{E})}{\frac{1}{\varepsilon_{xx}} - \mu_0 v^2} \tag{8.17}
$$

The electric displacement vector D is perpendicular to the unit vector $\mathbf{1}$, so

$$
\mathbf{D} \cdot \mathbf{1} = 0 \tag{8.18}
$$

or in component form

$$
D_x 1_x + D_y 1_y + D_z 1_z = 0 \tag{8.19}
$$

which produces

$$
\frac{1_x^2}{\frac{1}{\varepsilon_{xx}} - \mu_0 v^2} + \frac{1_y^2}{\frac{1}{\varepsilon_{yy}} - \mu_0 v^2} + \frac{1_z^2}{\frac{1}{\varepsilon_{zz}} - \mu_0 v^2} = 0 \tag{8.20}
$$

The l_x, l_y, and l_z are the direction cosines for the unit vector $\mathbf{1}$. For a light wave traveling along the x-axis, $l_x = 1$, $l_y = 0$, and $l_z = 0$, and the resulting equation is

$$
\left(\frac{1}{\varepsilon_{yy}} - \mu_0 v^2 \right) \left(\frac{1}{\varepsilon_{zz}} - \mu_0 v^2 \right) = 0 \tag{8.21}
$$

This equation has two positive roots for v

$$
v_1 = \frac{1}{\sqrt{\varepsilon_{yy} \mu_0}} \qquad v_2 = \frac{1}{\sqrt{\varepsilon_{zz} \mu_0}} \tag{8.22}
$$

This is a remarkable result. It states that the incoming light will be split into two waves, the waves will be polarized along y and z, and will travel at different speeds.

It is common to express this speed as a fraction of the speed of light in a vacuum, c, using the refractive index, n. Hence

$$v_1 = \frac{c}{n_1} \qquad v_2 = \frac{c}{n_2} \qquad (8.23)$$

This is our first, and most important, optical property of a material: *refractive index*. Developing the refractive index using Maxwell's equations is the most comprehensive and general approach, although the mathematics can become difficult and at times obscure the physical manifestations. It is our objective here to show how the optical properties of materials can be engineered by exploiting a parameter such as the refractive index. Hence, we shift our attention away from the fundamental origin of the refractive index and move toward a more phenomenological approach.

8.2 REFRACTIVE INDEX

Earlier we outlined that a plane-polarized light wave upon entering a material will split into two waves traveling at different speeds. This result is based upon the three components of the permitivitty tensor ϵ_{ij} being unequal. This is the most general result. If we focus for a moment on crystalline materials, specifically, single crystals, we can begin to simplify this general result. The symmetry of a crystal will determine whether the three components of the permitivitty tensor are all different, whether two are equal, or whether all three are the same. Working backwards, all three components of the permitivitty tensor will be equal for cubic space groups (numbers 195–230) (Burns and Glazer, 1978). In these crystals, when light polarized along the z-axis propogates through the crystal along the x-axis, the two resulting waves in the material propogate at the same velocity

$$v_1 = \frac{c}{n_1} \qquad v_2 = \frac{c}{n_1} \qquad (8.24)$$

Similarly, light propogating along the y or z axes is also split into two waves where both travel at c/n_1. Regardless of the orientation of the incoming light with respect to the crystal axes, the two resulting waves inside the crystal travel at c/n_1. Such a material is referred to as optically *isotropic*. For the tetragonal space groups (# 75–142), hexagonal space groups (#168–194), and trigonal space groups (#143–167), two of the principal values of the permitivitty tensor are equal. These space groups have a unique axis: the high symmetry rotation axis, fourfold for tetragonal, sixfold for hexagonal, and threefold for trigonal. Typically this axis is referred to as the z-axis of the crystal. For light propogating along this axis, the two resulting waves have equal speeds of c/n_1. For light traveling along either x or y, however, the resulting two waves travel at c/n_1 and c/n_2. Moreover, since these orthognal waves began in phase, and one is traveling faster than the other, when they exit they are no longer necessity in phase. The exact phase difference will depend on the thickness of the crystal. What happens when these two waves exit the crystal?

Before answering this question, we need to return to the isotropic case and look more closely at what happens when light crosses a refractive index boundary. We have already established that two waves, orthogonal in their polarization, will develop and travel at equal speeds. What we have not addressed is the direction of propogation. Because the speed of the light must slow down in the medium with higher refractive index, the direction of propogation must change. This change in the direction of propogation is called *refraction*. There are a number of ways to derive a quantitative description of refraction, the most complicated being the use of Maxwell's equations at the air/crystal boundary. A simpler argument relies on the fact that while the speed and wavelength of the light may change as the light moves from one refractive index to another, the *frequency* cannot.

The frequency is established by the source of the light. The resulting electromagnetic wave is produced as a result of the driven oscillations of electrons. As the light propogates through any medium, the incoming light drives the electrons at the frequency of the incoming light. These driven electrons, in turn, radiate light at exactly the same frequency as the incoming light. A little latter in this chapter, we consider the fascinating phenomenon where the driven electrons radiate not only the incoming frequency, but integral multiples of it as well.

In order for the electromagnetic waves to remain continuous in a region of longer wavelength and shorter wavelength with equal frequencies, the direction of propogation must change. The equation governing this change in propogation direction is called Snell's law, after the Dutch mathematician Willebrord van Roijen Snell (1580–1626) (Halliday et al., 2001)

$$n_1 \sin \theta_1 = n_2 \sin \theta_2 \tag{8.25}$$

where n_1 is the refractive index of air and n_2 is the refractive index of the isotropic crystal. The angle, θ, is measured from the normal to the surface of the crystal. If we take a typical isotropic material such as sodium chloride (cubic space group # 225) with $n = 1.541$, light hitting the surface at 45° would bend toward the normal such that

$$\sin^{-1} \theta_2 = \frac{n_{air} \sin(45°)}{1.541} \tag{8.26}$$

and θ_2 would equal 27.3°, as shown in Figure 8.1. If the light strikes the crystal along the normal such that $\theta_1 = 0$, then it will continue through the crystal along the normal and not undergo refraction, $\theta_2 = 0$.

A simple consequence of refraction is that it makes in possible to quickly distinguish two clear colorless materials. For example, in glassblowing, it is critical that the glassblower keep Pyrex glass and quartz glass separate. If a piece of quartz is blown directly onto a piece of Pyrex, the joint will break when the combined

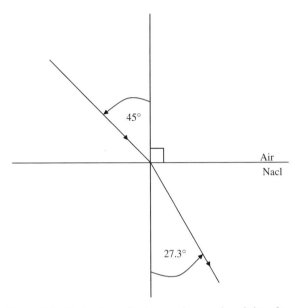

Figure 8.1 Refraction of light crossing an air–salt interface.

pieces cool. The reason is that the Pyrex shrinks upon cooling and the quartz does not. If Pyrex and quartz were different colors, distinguishing them would be trivial. Unfortunately, they are both clear, colorless glasses. Fortunately, they have different refractive indices. The organic liquid 2-methyl-1,3-cyclohexadiene has a refractive index of 1.4662 at 18°C for the sodium D wavelength of light ($\lambda = 589$ nm), which matches the refractive index of quartz pretty well at 1.46008 at 546.1 nm. Therefore, when you insert a piece of Pyrex glass into this liquid, it essentially vanishes. A quartz tube, on the other hand is clearly visible due to the refraction of the light at the glass–liquid interface. The same clever trick can be accomplished by matching the index of refraction of the Pyrex with an organic liquid, such as 2-allylphenol. In addition to refraction, reflection also occurs when light crosses a refractive index boundary. We treat this later in this chapter, but it, too, contributes to the disappearance of a glass rod in a suitable liquid.

Having established that refraction occurs when light travels from one refractive index to another, we return to the question: *What happens when these two waves exit the crystal?* Let us consider the simplest case first. Assume the incoming light is plane polarized and that is strikes the surface of a crystal at 90° and therefore does not refract. Assume also that the crystal is of the cubic class so that all three refractive indices are equal. The plane-polarized light breaks into two plane-polarized beams. The beams travel at the same speed, since all three refractive indices are equal. The beams exit the crystal at 90°, again avoid refracting, and the orthogonal in-phase beams add to reproduce the original plane-polarized beam traveling at the original speed in air.

Now consider the same nonrefracting 90° geometry, but this time let the crystal be from the hexagonal, tetragonal, or rhombic class. Assume the unique axis

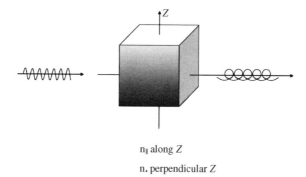

$n_{\parallel}$ along Z

$n_{\perp}$ perpendicular Z

Figure 8.2   Conversion of linearly polarized light to elliptically polarized light on passing through a birefringent crystal.

coincides with the plane of polarization for the incoming light. Again, the incoming light, avoiding refraction, splits into two orthogonal beams, but this time the one beam travels slower than the other. When they emerge, they again add vectorially, but this time the phase relationship depends on the thickness of the crystal. For an arbitrary thickness, the two plane-polarized beams, phase shifted, will add to produce elliptically polarized light, as shown in Figure 8.2. Under special thickness conditions, the outcoming light could be circularly or even linearly polarized (Hecht and Zajac, 1974). Circular polarization will result when the phase difference between the two emerging beams is one quarter wavelength, and linear polarization will result when the phase difference is one half wavelength. In the case of a calcite crystal with n along the unique axis equal to 1.486 and n perpendicular to it equal to 1.658, a 100-micron-thick crystal will produce a phase shift of 62° for 532-nm light.

The importance of considering the normal incidence first now becomes apparent. For light moving from one refractive index to another, uniaxial material at an angle less than 90° refraction occurs. Moreover, because the medium accepting the light has two refractive indices, the incoming light will undergo two refractions. As we have already discussed, the incoming plane-polarized light will be split into two orthogonal beams, each traveling at a speed dictated by the refractive index for that orientation. Hence the incoming beam is split into two physically separate beams. In a case of a large difference between the two refractive indices, the two beams exit the crystal displaced from one another. Returning the calcite example, when light enters a calcite crystal, as shown in Figure 8.3, the two beams separate to such an extent that two images are observed in transmission.

One of the main goals of the optical materials engineer is to control the direction of light flow. For the electrical engineer, the analogous task of controlling the flow of electrons is a straightforward task accomplished by using wire, resistors, diodes, and the like. For the optical engineer, light flow is controlled by taking advantage of the refractive index. We have already seen how light bends, or refracts, as it passes from one refractive index to another. Returning to Eq. 8.25, Snell's law, we consider

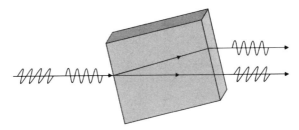

Figure 8.3 Physical separation of a light beam into its two polarizations.

the case where the angle of the incoming beam and the two refractive indices combine to produce a solution to Snell's law that predicts an angle of refraction with a sine value greater than 1. For example, in Figure 8.4a, consider light traveling from a glass with refractive index 1.5 to air with refractive index 1.0. If the light strikes the glass at an angle of 20° with respect to the normal, the refracted beam will leave the glass at an angle of 30° with respect to the normal. As expected, the light bends away from the normal as the light travels from a higher refractive index to a lower refractive index. If the angle of the incoming beam is increased beyond 41°, the solution to Snell's law requires an angle with a sine value greater than 1. Under this circumstance, the light will not pass from the glass to the air, but instead it will reflect off the glass and remain within the glass as shown in Figure 8.4b. This phenomenon, termed *total internal reflection*, is a possibility anytime light travels from a higher refractive index to a lower one. Determining the minimum angle required to observe total internal reflection, the critical angle, involves solving Snell's law with $\theta_2 = 90°$.

An important area where this effect has been exploited is fiber-optical cables (Palais, 1998). The goal of the optical engineer in this case is to keep the light in the cable and prevent it from escaping. The goal is achieved by encasing a transparent fiber with a material of higher refractive index, as shown in Figure 8.5. This encasement, known as the cladding, is typically pure SiO_2, with a refractive index of 1.46 at the wavelength of operation. The core, where the light actually travels, has a higher refractive index, typically 1.48. Hence, light in the core is confined there by total internal reflection, since most reflections occur at an angle of incidence much greater than the critical angle, 40.3° in this case.

This production of a core with refractive index higher than the cladding raises the important question of how the refractive index can be manipulated. To answer this question, we examine in detail what factors contribute to the numerical value of the refractive index. First, consider what is responsible for the transmission of light through a material. The material is viewed as a collection of static cations with electrons elastically attached. The incoming light sets the electrons in motion at the frequency of the light. These electrons constitute oscillating charge at frequency ν. This produces electromagnetic radiation at frequency ν and the light is reconstituted and continues on its way. This oscillating field combines with the incoming field to produce a net field at a point removed from the material that is

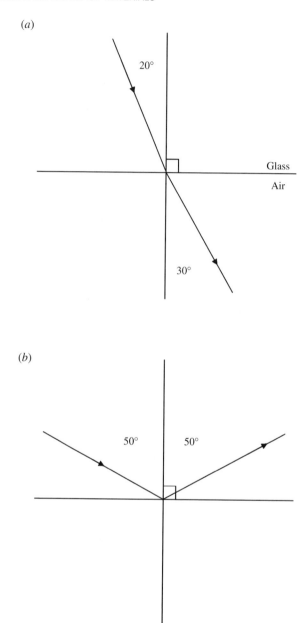

Figure 8.4   (a) Refraction versus (b) total internal reflection.

phase shifted relative to light that did not experience the material as shown in Figure 8.6. This phase shift can be viewed phenomonologically as resulting from a slowing down of the wave inside the medium where the effective velocity is c/n, as before.

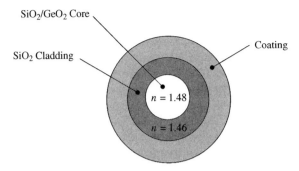

Figure 8.5 Cross section of a fiber-optic cable.

A detailed treatment of this model results in an expression for the refractive index shown in the following equation: (Feynman et al., 1963)

$$n = 1 + \frac{Nq_e^2}{2\varepsilon_0 m(\omega_0^2 - \omega^2)} \qquad (8.27)$$

where N is the number of charges, q_e is the charge per electron, m is the mass of the electron, ω_0 is the natural frequency of oscillation, and ω is the driving frequency of the incoming light. From this expression, we gain an insight into what factors contribute to the numerical value of the refractive index.

First, the density of charges affect the refractive index. All other factors being equal, as the atomic number of the constituent elements increases, the refractive index of a material will also increase. This can be seen with a simple listing of a few minerals in Table 8.1.

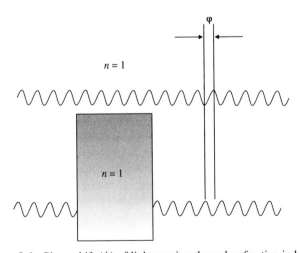

Figure 8.6 Phase shift (ϕ) of light passing through refractive index, n.

TABLE 8.1   Refractive Index as a Function of Electron Density

Compound	n (at 0.5 µm)
KCl	1.496
KBr	1.568
KI	1.673

TABLE 8.2   Refractive Indices of SIO$_2$ Phases

SiO$_2$ Phase	Density	Refractive Index
Tridymite	2.3	1.47
Cristrobalite	2.4	1.49
Keatite	2.5	1.5
Quartz	2.65	1.52
Ocesite	2.9	1.57
Stishovite	4	1.8

Electron density can also be changed by compression of the sample. This is illustrated by the many forms of silicon dioxide listed in Table 8.2. The fundamental frequency, ω_0, of the electron–nucleus harmonic oscillator model is determined by the type of bond in which the electron resides. Single bonds have a lower fundamental frequency than double bonds, which in turn will be lower than triple bonds. Hence the refractive index of benzene and cyclohexane (Figures 8.7 and 8.8) are different, as shown in Table 8.3.

This simple correlation holds when one considers the optical fiber again. In this case, the cladding is pure SiO$_2$ with a refractive index of 1.46 at the operating wavelength. The core is a mixture of SiO$_2$ and GeO$_2$. The specific mechanism by which

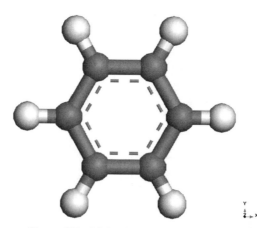

Figure 8.7   Molecular structure of benzene.

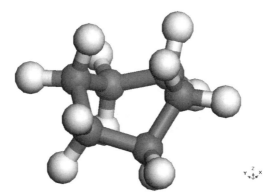

Figure 8.8   Molecular structure of cyclohexane.

the dopant, germanium, raises the refractive index is complex and the subject of investigation; however, it follows the qualitative trend that a higher atomic number will produce a higher refractive index.

Beyond using the refractive index to keep light in a fiber-optical cable with total internal reflection, refractive index manipulation can be used to prevent reflection (Halliday et. al, 2001). The simplest antireflection strategy involves coating an optical element such as a lens or a window with a thin layer of transparent material with a refractive index between that of the air and the optical material, for example, glass. The exact relationship, assuming normal incidence, is

$$\frac{n_{air}}{n_{layer}} = \frac{n_{layer}}{n_{glass}} \tag{8.28}$$

Hence with $n_{air} = 1.0$ and $n_{glass} = 1.5$, the layer should have a refractive index of $\sqrt{1.5}$, or 1.225. Another important parameter for the film is thickness. The film thickness must be an odd integral multiple of $\lambda/4$ of the wavelength of light to be affected. To see how a film of this thickness and refractive index accomplishes attenuation or extinction of reflected light, we examine Figure 8.9.

In Figure 8.9, the light reflected from the front surface of the film is phase shifted by 180°, because the light is traveling from a low refractive index to a high refractive index. The light reflecting off of the film–glass interface is also phase shifted by 180° for the same reason. Hence, the two reflected beams suffer no relative phase shift due to reflection. However, the beam that traverses the film does undergo a phase shift relative to the beam that does not, similar to what was depicted in

TABLE 8.3   Refractive Index Versus Bond Type

Compound	Refractive Index
Benzene	1.5010
Cyclohexane	1.4260

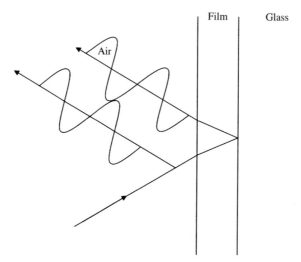

Figure 8.9   Reflections from the front and back surfaces of thin film.

Figure 8.6. If the two beams can be set to a phase shift of 180° relative to each other, they will destructively interfere and the reflection will be eliminated. To achieve this retardation, the beam traversing the film must do so through a minimum thickness, d, such that

$$d = \frac{\lambda}{2 \times n} \tag{8.29}$$

where n is the refractive index of the film and λ is the wavelength of light. The assumption here is that the light is hitting the film at normal or near normal incidence. More sophisticated antireflection coatings use multiple layers of varying refractive indices, resulting in antireflection for a wider range of incident angle and a wider range of wavelengths.

8.3   ABSORPTION

In these examples of using the refractive index to control and manipulate light, we have assumed that the materials are all transparent to the incoming light. Now we consider the case where the material absorbs the light. Interestingly, however, we can introduce this concept without leaving our discussion of the refractive index. In Eq. 8.27 we showed, the result of deriving the refractive index from a model of elastically bound electrons. Incoming light sets these electrons in oscillatory motion at the frequency of the incoming light and the light propogates through the material unattenutated. However, if this model for the oscillating electrons is modified to include a damping coefficient, a reasonable modification given that the oscillations will not continue indefinitely in the absence of the driving frequency, we arrive an equation for n (Feynman et al., 1963):

$$n = 1 + \frac{Nq_e^2}{2\varepsilon_0 m(\omega_0^2 - \omega^2 + i\gamma\omega)} \tag{8.30}$$

where a new term γ has been added to account for the damping. With this view of the refractive index, we can recast our expression for n as

$$n = n_r - in_i \tag{8.31}$$

where we express the refractive index as a complex number. In conventional notation, this complex refractive index is written

$$n = n - ik \tag{8.32}$$

retaining the variable n for the real part of the complex refractive index, so as to be consistent with common notation. Returning to our description of a plane wave traveling inside a material with dielectric constant, n,

$$E(r,t) = E_0 \exp\left\{ i\omega\left(t - \frac{r \bullet 1}{v}\right)\right\} \tag{8.33}$$

where $v = c/n$, this equation can be rearranged after substituting in the complex form of n to yield

$$E(r,t) = E_0 \exp\left\{\frac{-\omega k r \bullet 1}{c}\right\} \exp\left\{i\omega\left(t - \frac{nr \bullet 1}{c}\right)\right\} \tag{8.34}$$

The net effect of inserting the complex refractive index, derived from the damped oscillation of the electrons, is that the plane wave traveling through a material with refractive index $n - ik$ undergoes attenuation, as shown in Figure 8.10. Hence the

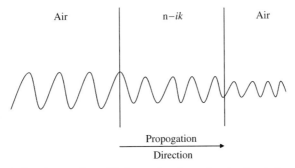

Figure 8.10   Attenuation of light passing through refractive index $n - ik$.

material is absorbing energy from the light. In a material that does not absorb light, the refractive index is essentially real, and for those strongly absorbing light, the refractive index is predominantly imaginary.

When considering absorption of light by matter, it is important to distinguish *extended materials* from *molecular materials*. Molecular materials consist of discreet molecular units bound together through van der Waals interactions, hydrogen-bonding, or dipolar interactions. Extended materials do not have a discreet molecular unit and are bonded together through ionic bonding, covalent bonding, or metallic bonding. Our picture of absorption is appropriate for either class of materials. However, each class has its own highly developed model to further describe the absorption of light.

For example, in the case of extended materials, where the electronic structure of the materials is described using the band-structure approach outlined in Chapter 4, the absorption of light is viewed as occurring as the result of an electron absorbing the energy of a photon and being promoted from one band to another. Hence, we observe a connection between the band gap of a material and its color. Silicon has a band gap at room temperature of 1.12 eV, and therefore absorbs radiation with this energy or greater, as shown Figure 8.11 (Ashcroft and Mermin, 1976). Alternatively, we could say that for wavelengths of light less than 1100 nm, the imaginary component of the refractive index of silicon is large. The imaginary component of the refractive index of silicon versus wavelength is plotted in Figure 8.12, showing its increasing magnitude as the band-gap energy is approached. Often, experimentalists

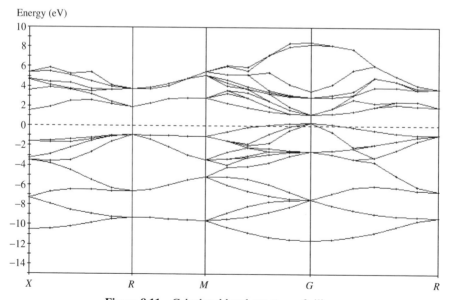

CASTEP Band Structure

Figure 8.11 Calculated band structure of silicon.

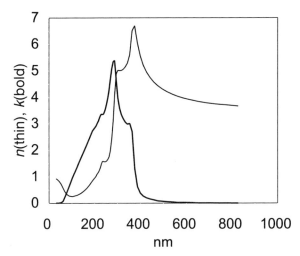

Figure 8.12   Real and imaginary refractive index components of silicon versus wavelength.

are left with measuring the refractive index of highly absorbing materials because the absorption is too high. With a large imaginary component in the refractive index, the material is highly absorbing, and, as discussed next, at the same time highly reflective. Hence, the reflectivity becomes the accessible experimental parameter.

The reflectivity, R, from a surface under normal incident illumination is (Fox, 2001):

$$R = \frac{(n-1)^2 + k^2}{(n+1)^2 + k^2} \tag{8.35}$$

where the real and imaginary components are as before. With a highly absorbing material, $k \gg n$ and $R \approx 1$. This same equation also explains why you can see your reflection is a glass window pane: with $n = 1.5$ and $k \approx 0$, $R = 0.04$, which is the 4% reflection we expect from a glass–air interface. This phenomenon, that highly absorbing (large k) materials are also highly reflecting, accounts for the difference in color of a highly absorbing material when viewed in transmission versus reflection. A sample of ink will appear one color in transmission, but a different color in reflection.

The process of absorption in chemistry is often treated with Beer's law (after the German scientist August Beer (1825–1863)), rather than considering the imaginary component of the refractive index. The two approaches are equivalent. According to Beer's law, the absorption, A, of a sample is related to the concentration of the absorbing species, the optical path length, and the molar absorbtivity, ϵ:

$$A = \varepsilon bc \tag{8.36}$$

The imaginary component of the refractive index, k, and ϵ are related by

$$k = \frac{\epsilon\lambda}{4\pi} \tag{8.37}$$

Just as k varies with wavelength, so too does ϵ. The absorbance, A, is equal to the log of the ratio of the light intensity (power) versus the attenuated light intensity:

$$A = -\log_{10}\frac{I}{I_0} \tag{8.38}$$

Substituted into Eq. 8.36, with both sides raised to the power of 10, we get

$$I = I_0 10^{-\epsilon bc} \tag{8.39}$$

We began this section discussing the absorption of light by a material. We specifically mentioned an extended material where we could envision the absorbed light transferring its energy to an electron in the conduction band and promoting it to the valence band. In addition to extended materials where a band-structure picture is appropriate for modeling the electronic structure, we should also consider molecular materials, where the band structure is not appropriate. We term such materials *molecular materials*. In these materials, the electronic structure is determined by a single molecular unit, and the effects on intermolecular interactions is often ignored. The optical properties of molecular materials is developed using quantum mechanics and solving the appropriate Schrodinger equation (Levine, 2003):

$$\hat{H}\Psi_i = E\Psi_i \tag{8.40}$$

Typically the time-independent Hamiltonian is used and the stationary-state energy levels are determined.

In order to determine where in the electromagnetic radiation spectrum the sample is going to absorb, two conditions must be met:

1. The energy of the radiation, expressed as the energy of a photon of that radiation, $E = h\nu$, must equal the energy difference between two quantum mechanical states, Ψ_i and Ψ_f, where the subscripts i and f stand for initial and final, indicating the two quantum states whose energy difference matches the photon's energy. This is called the *resonance condition*.

2. After meeting the first condition, a second condition must be met. Expressed mathematically it is

$$\mu_{fi} = \left[\int \psi_f^* \hat{\mu}\psi_i d\tau\right]^2 \tag{8.41}$$

which is called the transition moment. The magnitude of μ_{fi} determines the intensity of the absorption and can range from zero (a forbidden transition) to a large number (an intense transition). This second condition is called the *selection rules*.

Our purpose here is to comment on a particular feature of the optical properties of molecular materials as it relates to Beer's law. The magnitude of the absorptivity, ϵ, in Beer's law (or μ_{fi} in quantum mechanics) can be related to particular types of molecular entities responsible for the transition. For example, molar absorptivity values of ≈ 10 are typical for optical transitions occurring within transition metal ions (Dunn et al., 1965). Molar absorptivity values can reach values of 10,000 to 15,000 for $\pi \rightarrow \pi*$ occurring within an organic species. The integral in Eq. 8.41 becomes very large, and hence the molar absorptivity becomes large, when the electron-charge distribution in the two states Ψ_i and Ψ_f are markedly dissimilar. This results in a large oscillating dipole moment as the species oscillates between these two states. Such a large oscillating dipole moment couples strongly to electromagnetic radiation if the resonance condition is met. Hence, for example, in the design of dyes and inks, a large number of delocalized electrons can result in in tensely colored materials. Rhodamine 6G, a common laser dye, shown in Figure 8.13, has a molar absorptivity of $\approx 100,000$ at its peak absorption, 529.75 nm (in ethanol) (**).

While molecular materials lack the robustness of extended materials, this correlation between molecular entities and absorptivity values makes molecular materials an important area for those attempting to design materials with specific optical properties. Not only does the chemist of synthetic products have control over the optical properties, the synthesis of molecular materials is also more amenable to specific designs. We discuss synthetic strategies more in Chapter 12.

We have developed the idea of the central role that the refractive index plays in describing all optical properties of materials. As already stated, it is not so much the central role of the refractive index, but the central role of Maxwell's equation that produces the phenomenological quantity refractive index. From the perspective of the optical materials engineer or producer, the refractive index clearly holds the key to producing desired materials. We have already addressed those chemical factors

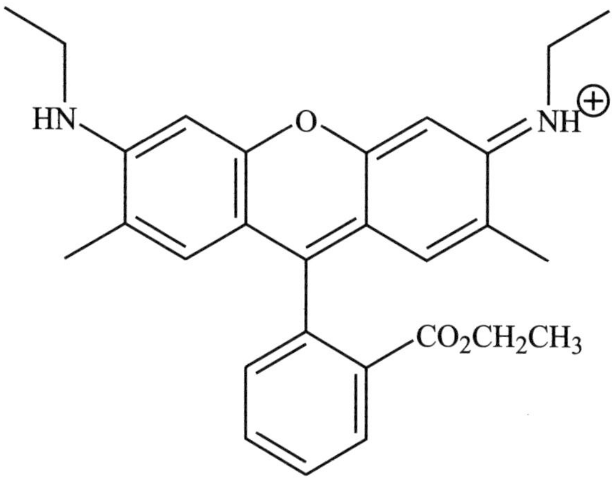

Figure 8.13 Molecular structure of Rhodamine 6G.

(electron density, bond type) that influence the magnitude of the refractive index. We now examine how physical manipulation of a material influences its refractive index.

8.4 NONLINEAR EFFECTS

An isotropic material such as sodium chloride, with all three values of the refractive index equal, $n_x = n_y = n_z$, can be made optically anisotropic by the application of an electric field or a magnetic field along one axis. When a magnetic field is used, the resulting birefringence is called the Faraday effect or the magnetooptic effect. When an electric field is used, the effect is called the Kerr effect or the Pockels effect. This ability of an externally applied field to alter the optical properties of a material, specifically inducing birefringence, would remain at the level of curiosity, except when applied to optially nonlinear materials. That is the next subject we discuss, and after doing so, we return to induced birefringence as an example of a technologically important process that relies on the design and manipulation of materials with specific refractive index features.

We have used the driven oscillator model to gain an insight into what factors influence the refractive index. As we begin to explore impressive applications of manipulations of the refractive index, we revisit the assumptions in the simple driven oscillator model. One assumption was that the electrons are not excessively displaced as a result of their interaction with the varying electric field from the light. This is an important assumption. Not only does it make the mathematics manageable, it results in a solution to the problem where the frequency of the resulting light is equal to the frequency of the driving field. Hence, no change in color. However, let us relax that assumption, and assume the electron displacement is large. We return to the simple Hooke's law spring

$$F = kx \tag{8.42}$$

and add additional terms to the force to account for the anharmonic character of the driven oscillation

$$F = kx + ax^2 + bx^3 + cx^4 + \cdots \tag{8.43}$$

The solution to the damped forced-oscillator differential equation proceeds as follows. Begin with (Braun, 1978)

$$m\frac{d^2x}{dt^2} + \Gamma\frac{dx}{dt} + \omega_0^2 x + ax^2 = F \tag{8.44}$$

where the driving force, F, is expressed as a sum of two frequencies:

$$F = \frac{q}{m}[E_1 \cos(\omega_i t) + E_2 \cos(\omega_2 t)] \tag{8.45}$$

Our goal is to solve Eq. 8.44 for x, since we can express the polarization P, as

$$P = Nqx \qquad (8.46)$$

Ignoring the anharmonic ax^2 term and higher terms, the solution for $x(t)$ is

$$x(t) = \frac{(q/m)E_1}{(\omega_0^2 - \omega_1^2 - i\omega_1\Gamma)} \exp(-i\omega_1 t) \qquad (8.47)$$

with a similar expression for the second frequency, ω_2. The frequency of the oscillation remains at the driving frequency. However, if the first anharmonic term is included, and assumed to be small, the frequency of the oscillation contains the driving frequency as well as a dc component and a frequency at twice the driving frequency.

The Hooke's law treatment resulted in a polarization which was linear in the electric field

$$P = a_{ij}E \qquad (8.48)$$

Now we anticipate contributions to P from terms higher than one in E (Boyd, 1992)

$$P = a_{ij}E + \beta_{ijk}E^2 + a_{ijkl}E^3 + \cdots \qquad (8.49)$$

where each contribution following the linear term is smaller than the previous one. If we stop at the second-order term in E and express E as before

$$E(r,t) = E_0 \exp\left\{ i\omega\left(t - \frac{r \bullet 1}{v} \right) \right\} \qquad (8.50)$$

then upon squaring E, we see again the doubled frequency. The nonlinear relationship between P and E can be represented in a simplified plot shown in Figure 8.14. The consequences of this nonlinearity become apparent when one considers the P resulting from an oscillating E. A comparison of the P vs. E for the linear and nonlinear regions is shown in Figure 8.15. When P is linearly proportional to E, the oscillation in E is reproduced by P multiplied by a scaling factor, the polarizability. However, when P is quadratically proportional to E, the response to a sinusoidal oscillation in E becomes more interesting. As the plot shows, the peaks are sharper, and the troughs are broader. A Fourier analysis of this output reveals three features:

1. The main frequency component remains equal to the driving frequency;
2. A frequency of 2 times the driving frequency is present;
3. A dc offset is present.

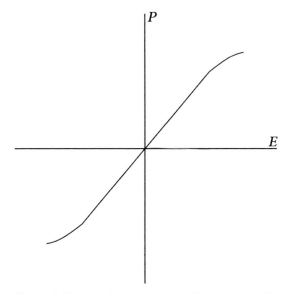

Figure 8.14   Polarization versus applied electric field.

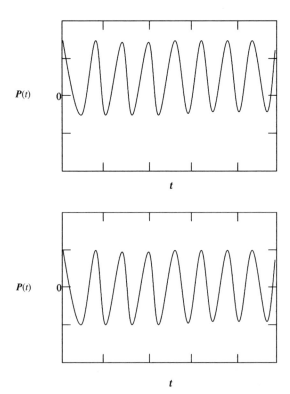

Figure 8.15   Polarization response for a (upper) large and (lower) small oscillating electric field.

Points 1 and 3 can be identified by inspection, point 2 is more subtle, but it comes about from the trigonometric identity

$$\cos^2(\theta) = \frac{1}{2}\cos(2\theta) + \frac{1}{2} \tag{8.51}$$

Whether through the mathematical analysis or graphical analysis, the conclusion remains the same: in the presence of high-intensity light, materials can produced light at harmonics of the driving frequency. As a general phenomenon, this is not a dramatic revelation. Harmonic distortion is a common feature of electronic amplifiers. When the input to an audio amplifier is too high, the output to the speaker will contain harmonics of the input frequency. In audio electronics, this is referred to an *harmonic distortion*. In optical properties of materials, it is called *harmonic generation*. The mathematics describing both are the same, because in both cases, the system is a driven damped oscillator.

An important remaining question is: Will all materials produce these multiple harmonics? The answer to that is straightforward, but the explanation may not be. The answer is yes, all materials produce the third, fifth, seventh, and so forth, odd harmonics. In order to observe the even harmonics, a material must lack a center of inversion in the crystallographic space group. Such space groups are called noncentrosymmetric. This is an important consideration for the materials chemist designing new materials for use as even harmonic generators, the only commercially important ones being the second harmonic generation materials.

Both second-order and third-order materials have technological applications because of their ability to convert low-frequency light to high-frequency light. However, the efficiency with which they are able to accomplish this feat decreases dramatically from second order to third order. Even the second-order process is small compared to the first-order process. From the design point of view, one faces a dilemma: avoid the symmetry constraint and live with the low efficiency of third-order materials or adhere to the symmetry constraints and reap the benefit of better conversion. In Chapter 12, we will say more about synthetic strategies as they relate to producing nonlinear optical materials.

As we said earlier, birefringence can be induced in a material with the application of an electric field. This effect finds an application in the important field of electrooptic modulators. The effect is outlined in Figure 8.16. When the refractive

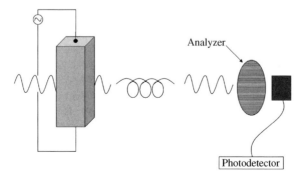

Figure 8.16   Electrooptic modulator.

index of the material is altered as a result of the application of the electric field, the change in the linear polarization of incident beam to elliptical polarization of the outgoing beam allows some component of the outgoing beam to pass through the analyzer. Hence, the light level at the detector can be switched from high to low with the application of the electric field. In order to make this a practical device, the field must be switchable at a high rate of speed (MHz). This high switching speed requires that the applied voltage be small. In order to get enough change in the polarization for the device to work, the material will require large second-order coefficients, and therefore the material must be noncentrosymmetric.

As a final example of manipulating the optical properties of a material to produce a desired effect, we consider upconversion (Risk et al., 2003). In one sense this is similar to nonlinear optical processes, since the net result is the production of high-energy light from low-energy light. However, the details are different. One immediate advantage of upconversion is that it does not impose the same symmetry requirements as the second-order nonlinear effects. However, it does have its own limitations. Upconversion requires that the excited state of a chromophore have a long lifetime. The tripositive lanthanides meet the condition of long-lived excited states, and therefore represent the bulk of the work in this area.

An example of upconversion is as follows

$$Er^{+3}\left({}^{4}I_{15/2}\right) + h\nu_1 \rightarrow Er^{+3}\left({}^{4}I_{11/2}\right)$$
$$Er^{+3}\left({}^{4}I_{11/2}\right) + h\nu_1 \rightarrow Er^{+3}\left({}^{4}F_{7/2}\right)$$
$$Er^{+3}\left({}^{4}S_{3/2}\right) \rightarrow Er^{+3}\left({}^{4}I_{15/2}\right) + h\nu_2$$

where the emitted light, $h\nu_2$, is of higher energy than the absorbed light. A radiationless transition connects the ${}^{4}F_{7/2}$ state and the ${}^{4}S_{3/2}$ state. In the preceding example, the absorbed light is 970 nm and the emitted light is 540 nm and 525 nm, hence the term upconversion. In order for this process to work, it is important that the Er^{+3} not be too close together. If the Er^{+3}–Er^{+3} distance is too short, the lifetime of the excited state is decreased. Hence, this material is prepared with the Er^{+3} doped into a lattice.

We include upconversion because it is a case, unlike all the others presented here, where the refractive index is not a concern in the design of the material. For upconversion, the important parameters for the designer include absorption bands and excited-state lifetimes. Finding the blank host to hold the emitters and keep them separated from each other is also part of the design task. In addition, this host often serves as the primary absorber of the light to be upconverted. In the case of Er^{+3}, the host contains Yb^{+3} for this purpose.

8.5 SUMMARY

In this chapter we have considered the optical properties of materials from a design point of view. The emphasis is on synthetic design. The overiding optical parameter is the refractive index. This comes as no surprise given that it owes its origin to Maxwell's equations. Moreover, we have examined the consequences of how light interacts with a material, given the material's refractive index or indices. Finally, we offered some specific examples where the optical properties of a material were deliberately imposed by controlling the refractive index. We complemented those examples with a case where the refractive index was not critical.

REFERENCES

Ashcroft, N. W., Mermin, N. D. 1976, *Solid State Physics, Saunders College, Philadelphia, PA.*

Braun, M. 1978, *Differential Equations and Their Applications, Springer-Verlag, New York.*

Boyd, R. W. 1992, *Nonlinear Optics*, Academic Press, San Diego, CA.

Burns G., Glazer, A. M. 1978. *Space Groups for Solid State Scientists*, Academic Press, New York.

Dunn, T. M., McClure, D. S., Pearson, R. G. 1965, *Some Aspects of Crystal Field Theory*, Harper & Row, New York.

Feynman, R. P., Leighton, R. B., and Sands, M. *The Feynman Lectures on Physics*, Addison-Wesley, Reading, MA.

Fox, M. 2001, *Optical Properties of Solids*, Oxford University Press, Oxford.

Griffiths, D. J. 1981, *Introduction to Electrodynamics*, 3rd ed., Prentice Hall, Saddle River.

Halliday, D., Resnick, R., Walker, J. 2001, *Fundamentals of Physics*, 6th ed., John Wiley & Sons, New York.

Hecht, E., Zajac, A., 1974, *Optics*, Addison-Wesley, Reading, MA.

Levine, I. N. 2001, *Physical Chemistry*, 5th ed., McGraw-Hill Book Company, New York.

Nye, J. F. *Physical Properties of Crystals*, 1985, Oxford University Press: Oxford.

Palais, J. C. 1998, *Fiber Optic Communications*, 4th. ed., Prentice Hall, New York.

Risk, W. R., Gosnell, T. R., Nurmikko, A. V., 2003, *Compact Blue-Green Lasers, Cambridge University Press, Cambridge.*

CHAPTER 9

Mechanical Properties

The fact that many primarily nonmechanical engineering applications require materials with specific mechanical properties is sometimes underappreciated by those who are not materials scientists or mechanical engineers. At first glance, it might seem as though a given material holds great promise in a particular application due to some potentially exploitable property (e.g., a transport, magnetic, or optical property). In actuality, applicability can be significantly reduced for a variety of reasons, frequently because of inadequate mechanical behavior.

Many solids, particularly ceramics, exhibit very limited plasticity, in which case, fabrication into the desired shape, or form, is difficult or not cost effective. Furthermore, successful fabrication does not ensure the material will have an adequate lifetime once in use. Metals subjected to relatively low stresses at elevated temperatures, for instance, may undergo *creep*, defined as any time-dependent plastic deformation at constant stress and temperature. Ceramics are normally very resistant to creep, until very high temperatures are reached. Alternatively, *fatigue* arises from repeated, or cyclic, stresses such as mechanical loading, thermal fluctuations, or both (thermomechanical fatigue). Fatigue is of concern in both metals and ceramics.

9.1 BASIC DEFINITIONS

When mechanical forces act on a body, the body is said to be in a state of *stress*, defined as the force per unit area over which the force is distributed. Stress is usually denoted by the symbol σ. Consider the cube shown in Figure 9.1, where the force acting on each face has been resolved into three components, which point in the positive direction in this diagram. The convention normally used is that the component of the force exerted in the $+i$ direction, and transmitted across the cube face that is perpendicular to the j direction, is denoted by σ_{ij}. There are thus nine components to stress; it is a second-rank tensor, which was first formalized by the French mathematician Augustin-Louis Cauchy (1789–1857) in 1822. A body is under *homogeneous* stress when the forces acting on each pair of opposite faces are

Principles of Inorganic Materials Design By John N. Lalena and David A. Cleary
ISBN 0-471-43418-3 Copyright © 2005 John Wiley & Sons, Inc.

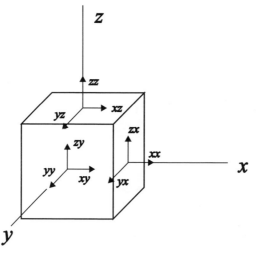

Figure 9.1 The forces on the faces of a unit cube in a body under homogeneous stress.

equal in magnitude but opposite in sign. In other words, if the stress is homogenous, there is no *net* force acting on the cube.

Stress causes an object to *strain*, or deform. Strain is denoted as ε and is equal to the change in length divided by the initial length, $\Delta L/L_0$. Like stress, strain too is a second-rank tensor. With homogeneous strain, the deformation is proportionately identical for each volume element of the body and for the body as a whole. If the strain is the same in all directions, it is *isotropic*. Deformation can be categorized into two regimes, as depicted in Figure 9.2, which shows the stress–strain curve for

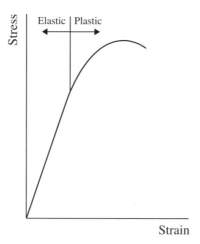

Figure 9.2 Stress–strain behavior. With elastic (reversible) deformation, stress and strain are linearly proportional in most materials (exceptions include polymers and concrete). With plastic (permanent) deformation, the stress–strain relationship is nonlinear.

three "typical" classes of materials. The first regime is *elastic deformation*, which occurs when a solid object is subjected to an applied stress that is small enough that the object returns to its original dimensions once the stress is released. That is, the deformation exists only while the stress is applied. The response of the material in this regime is linear (it obeys Hook's law). In 1727 the Swiss mathematician Leonhard Euler (1707–1783) first expressed this linear relationship in terms of strain and stress: $\sigma = E\varepsilon$, but where the proportionality constant is now known as *Young's modulus* after the British scholar Thomas Young (1773–1829), who developed a similar relation in 1807.

It is now known that the underlying mechanism of elastic deformation involves small atomic motions, such as stretching (without breaking) of the chemical bonds holding the atoms of the material together. Most solid objects will undergo elastic deformation, provided the stress is below a characteristic threshold for the material, called the *elastic limit* (sometimes called yield point or yield strength). Beyond this stress level, the stress–strain curve ceases to be linear, and the sample undergoes *plastic deformation*. Furthermore, the body does not return to its original shape or microstructural condition upon removal of the forces that cause plastic deformation. The mechanism of plastic deformation involves *slip* and is developed later. Some general terms used to describe mechanical properties are defined in Table 9.1.

TABLE 9.1 Some General Mechanical Engineering Terms and Their Definitions

Elasticity	The ability of a material to deform under loads and return to its original dimensions upon removal of the load
Stiffness	A measure of the resistance to elastic deformation
Compliance	The reciprocal of stiffness; a low resistance to elastic deformation
Plasticity	The ability of a material to permanently deform under application of a load without rupture
Strength	A measure of the resistance to plastic deformation
Shear strength	The ability to withstand transverse loads without rupture
Compressive strength	The ability to withstand compressive loads without crushing
Tensile strength	The ability to withstand tensile loads without rupture
Ultimate tensile strength	The point at which fracture occurs
Yield strength	The point at which plastic deformation begins to occur
Ductility	The ability to stretch under tensile loads without rupture. Measured as a percent elongation
Malleability	The ability to deform permanently under compressive loads without rupture
Toughness	The ability to withstand shatter. A measure of how much energy can be absorbed before rupture. Easily shattered materials (small strain to fracture) are *brittle*
Hardness	The ability of a material to withstand permanent (plastic) deformation on its surface (e.g., indentation, abrasion)
Creep	Time-dependent plastic deformation at constant stress and temperature
Fatigue	Damage or failure due to cyclic loads

TABLE 9.2   The Elastic Moduli

Quantity	Symbol	Expression	Comments
Bulk modulus	B	$V(\Delta P/\Delta V)$	The change in pressure divided by the volumetric strain
Young's modulus	E	$\sigma_{xx}/\varepsilon_{xx}$	Normal stress divided by normal strain
Rigidity modulus	G	$\sigma_{xy}/\varepsilon_{xy}$	Shear stress divided by shear strain
Poisson ratio	ν	$|\varepsilon_{yy}|/\varepsilon_x$	Transverse strain divided by normal strain

9.2   ELASTICITY

The manner in which a body responds to small external mechanical forces in the elastic regime is determined by the *elastic constants* (Table 9.2). In the case of elastically isotropic bodies (polycrystals and amorphous substances), these include: the bulk modulus, B; Young's modulus, E; the rigidity modulus (also known as the shear modulus), G; and Poisson's ratio, ν, named after French mathematician Siméon-Denis Poisson (1781–1840).

9.2.1   The Elasticity Tensor

Consider a vertically hanging metal rod, to which we can apply a load (e.g., a steel cable supporting an elevator), as in Figure 9.3. The load exerts a tensile force over

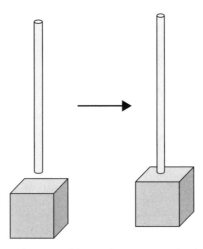

Figure 9.3   In the elastic regime, stretching a wire causes a reduction in its diameter, in order to keep the volume of the wire constant.

the entire cross-sectional area of the rod, which is said to be under *uniaxial stress*, since only the stress along one of the principal axes is nonzero. The stress is equal to the force divided by the cross-sectional area over which it is distributed. In linear elastic theory, according to Hooke's law, the magnitude of the strain produced in the rod by a small uniform applied stress is directly proportional to the magnitude of the applied stress. Hence, we have:

$$\varepsilon = s\sigma \tag{9.1}$$

or, alternatively

$$\sigma = c\varepsilon \tag{9.2}$$

The constants s and c ($\equiv 1/s$) are known as the *elastic compliance constant* and the *elastic stiffness constant*, respectively. The elastic stiffness constant is the elastic modulus, which is seen to be the ratio of stress to strain. In the case of normal stress/normal strain (as in Figure 9.3), the ratio is called the *Young's modulus*, whereas for shear stress/shear strain the ratio is called the *rigidity modulus*. The Young's modulus and rigidity modulus are the slopes of the stress–strain curves, and for non–Hookean bodies they may be defined alternatively as $d\sigma/d\varepsilon$. They are required to be positive quantities. Note that the higher the strain, the lower the modulus.

   If we think about what happens when, say, an elastomer is under tensile stress, we come to realize that the elastic constants, s and c, cannot be scalar quantities, otherwise Eqs. 9.1 and 9.2 do not completely describe the elastic response. This is because, when we change the shape of a body by stretching it, the volume of the body remains constant, necessitating a contraction in length in some other dimension. This is the same behavior exhibited by the metal rod in Figure 9.3. When a uniaxial stress is applied in the x direction to the rod with the tubular axis along the x direction, the rod lengthens along x and slightly narrows in diameter. That is, it deforms in directions *perpendicular* to the stress directions as well. This behavior is quantified by *Poisson's ratio*, defined as $|\varepsilon_{yy}|/\varepsilon_{xx}$. Of course, constant volume is *not* a criterion for elasticity. If the solid were subjected to hydrostatic pressure, the elastic response of the body would be a change in volume, but not shape. This behavior is quantified by the *bulk modulus, B*.

Example 9.1   A wire 3.00 m long with a diameter of 3.19 mm hangs vertically. When a load of 2000 N is applied to the wire, it stretches 3.9 mm. What is the diameter after stretching?

Solution   The volume of the wire before stretching is $V = \pi r^2 h$

$$= \pi(0.00159\,\text{m})^2(3.00\,\text{m}) = 2.39 \times 10^{-5}\,\text{m}^3.$$

The volume of the wire after stretching is the same, so we solve for r using the new length:

$$2.39 \times 10^{-5}\,\mathrm{m}^3 = \pi r^2 (3.0039\,\mathrm{m})$$
$$r = 2.82\,\mathrm{mm}.$$

If stresses and strains were vectors, the elastic constants would be second-rank tensors. However, stresses and strains, themselves, are second-rank tensors:

$$\sigma_{ij} = \begin{pmatrix} \sigma_{xx} & \sigma_{xy} & \sigma_{xz} \\ \sigma_{yx} & \sigma_{yy} & \sigma_{yz} \\ \sigma_{zx} & \sigma_{zy} & \sigma_{zz} \end{pmatrix} \tag{9.3}$$

$$\varepsilon_{ij} = \begin{pmatrix} \varepsilon_{xx} & \varepsilon_{xy} & \varepsilon_{xz} \\ \varepsilon_{yx} & \varepsilon_{yy} & \varepsilon_{yz} \\ \varepsilon_{zx} & \varepsilon_{zy} & \varepsilon_{zz} \end{pmatrix} \tag{9.4}$$

Each component of a second-rank tensor is associated with a pair of axes, i and j. In the case of stress, the convention used is that the component of the force exerted in the $+i$ direction *on* the face whose normal is in the j direction is denoted by σ_{ij}. The diagonal σ_{ij} components are called *normal stresses* and the nondiagonal σ_{ij} components are called *shear stresses*. Similarly, the diagonal components of ε_{ij} are the *tensile strains* and the nondiagonal components are the *shear strains*. Stress and strain tensors are not matter tensors like susceptibility or conductivity, which were covered in earlier chapters. They do not represent a crystal property, but are rather forces imposed on the crystal, which can have any arbitrary direction or orientation. In this respect, they are like electric fields and so are called *field tensors*.

We know from the forms of Eqs. 9.1 and 9.2, and from tensor algebra, that the elastic constants in Eqs. 9.1 and 9.2 must be fourth-rank tensors. In general, an nth-rank tensor property in p-dimensional space requires p^n coefficients. Thus, the elastic stiffness constant is comprised of 81 (3^4) *elastic stiffness coefficients*, c_{ijkl} (n indices are needed for an nth-rank tensor). Hence, Eq. 9.2 must be written as $\sigma_{ij} = c_{ijkl}\varepsilon_{kl}$. A fourth-rank tensor can be written as a $[9 \times 9]$ array, representative of nine equations, each with nine terms on the right-hand side. It follows that each of the nine components of stress, for example, are related to all nine components of the strain. That is, the σ_{xx} component may be written out as: $\sigma_{xx} = c_{xxxx}\varepsilon_{xx} + c_{xxxy}\varepsilon_{xy} + c_{xxxz}\varepsilon_{xz} + c_{xxyx}\varepsilon_{yx}c_{xxyy}\varepsilon_{yy} + c_{xxyz}\varepsilon_{yz} + c_{xxzx}\varepsilon_{zx} + c_{xxzy}\varepsilon_{zy} + c_{xxzz}\varepsilon_{zz}$. An abbreviated way of expressing such relationships will be introduced shortly.

It can be shown by force equilibrium considerations that in the absence of an applied torque the strain and stress tensors are symmetric; $\sigma_{ij} = \sigma_{ji}$ and $\varepsilon_{ij} = \varepsilon_{ji}$. Consequently, there are really only six *independent* stresses for three directions, say, x, y, and z, that can be applied to strain a body and six independent strains in a stressed body. This simplification reduces the size of the elasticity tensor from $[9 \times 9]$ to $[6 \times 6]$. Thus, Eq. 9.2 reduces to 36 elastic-stiffness coefficients. We now introduce the shorthand notation normally used for the elasticity tensor, namely, that the subscripts become: $1 \rightarrow xx$; $2 \rightarrow yy$; $3 \rightarrow zz$; $4 \rightarrow yz$, zy; $5 \rightarrow zx$, xz; and $6 \rightarrow xy$, yx. With this change, the elastic-stiffness tensor can be written in matrix form as:

$$\begin{pmatrix} \sigma_1 \\ \sigma_2 \\ \sigma_3 \\ \sigma_4 \\ \sigma_5 \\ \sigma_6 \end{pmatrix} = \begin{pmatrix} c_{11} & c_{12} & c_{13} & c_{14} & c_{15} & c_{16} \\ c_{21} & c_{22} & c_{23} & c_{24} & c_{25} & c_{26} \\ c_{31} & c_{32} & c_{33} & c_{34} & c_{35} & c_{36} \\ c_{41} & c_{42} & c_{43} & c_{44} & c_{45} & c_{46} \\ c_{51} & c_{52} & c_{53} & c_{54} & c_{55} & c_{56} \\ c_{61} & c_{62} & c_{63} & c_{64} & c_{65} & c_{66} \end{pmatrix} \begin{pmatrix} \varepsilon_1 \\ \varepsilon_2 \\ \varepsilon_3 \\ \varepsilon_4 \\ \varepsilon_5 \\ \varepsilon_6 \end{pmatrix} \quad (9.5)$$

Each of the six independent stresses that can be applied to strain a body are expressed in Eq. 9.5 in terms of six strains. For example, in matrix form, the normal stress $\sigma_1 (\equiv \sigma_{xx})$ and shear stress $\sigma_4 (\equiv \sigma_{yz})$ can be expressed as:

$$\sigma_{xx} = \sigma_1 = \begin{pmatrix} c_{11}\varepsilon_1 + c_{16}\varepsilon_6 + c_{15}\varepsilon_5 \\ c_{16}\varepsilon_6 + c_{12}\varepsilon_2 + c_{14}\varepsilon_4 \\ c_{15}\varepsilon_5 + c_{14}\varepsilon_4 + c_{13}\varepsilon_3 \end{pmatrix}$$

$$\sigma_{yz} = \sigma_4 = \begin{pmatrix} c_{41}\varepsilon_1 + c_{46}\varepsilon_6 + c_{45}\varepsilon_5 \\ c_{46}\varepsilon_6 + c_{42}\varepsilon_2 + c_{44}\varepsilon_4 \\ c_{45}\varepsilon_5 + c_{44}\varepsilon_4 + c_{43}\varepsilon_3 \end{pmatrix}$$

Again, the significance of this type of stress–strain relationship is that all six of the strain components (ε_1–ε_6) may be nonzero with the application of a single component of stress. For example, a sample might shear as well as elongate under a uniaxial tension. Similarly, a sample may bend as well as twist if pure twisting forces are applied to its ends.

The six independent strain components (Eq. 9.1) can likewise be given, as a function of stress, in terms of 36 *elastic-compliance coefficients*, s_{ijkl}:

$$\begin{pmatrix} \varepsilon_1 \\ \varepsilon_2 \\ \varepsilon_3 \\ \varepsilon_4 \\ \varepsilon_5 \\ \varepsilon_6 \end{pmatrix} = \begin{pmatrix} s_{11} & s_{12} & s_{13} & s_{14} & s_{15} & s_{16} \\ s_{21} & s_{22} & s_{23} & s_{24} & s_{25} & s_{26} \\ s_{31} & s_{32} & s_{33} & s_{34} & s_{35} & s_{36} \\ s_{41} & s_{42} & s_{43} & s_{44} & s_{45} & s_{46} \\ s_{51} & s_{52} & s_{53} & s_{54} & s_{55} & s_{56} \\ s_{61} & s_{62} & s_{63} & s_{64} & s_{65} & s_{66} \end{pmatrix} \begin{pmatrix} \sigma_1 \\ \sigma_2 \\ \sigma_3 \\ \sigma_4 \\ \sigma_5 \\ \sigma_6 \end{pmatrix} \quad (9.6)$$

Now consider a single crystal in place of the rod in Figure 9.1. The elasticity of a crystal is the *physical property* relating a homogeneous stress and homogeneous strain in the crystal. Hence, the number of independent elastic coefficients is further

TABLE 9.3 The Independent Elastic-Stiffness Coefficients for Each Crystal Class

Triclinic	$c_{11}, c_{22}, c_{33}, c_{44}, c_{55}, c_{66}, c_{12}, c_{13}, c_{23}, c_{14}, c_{24}, c_{34}, c_{15}, c_{25}, c_{35}, c_{45}, c_{16}, c_{26},$ c_{36}, c_{46}, c_{56}
Monoclinic	$c_{11}, c_{22}, c_{33}, c_{44}, c_{55}, c_{66}, c_{12}, c_{13}, c_{23}, c_{16}, c_{26}, c_{36}, c_{45}$
Orthorhombic	$c_{11}, c_{22}, c_{33}, c_{44}, c_{55}, c_{66}, c_{12}, c_{13}, c_{23}$
Tetragonal	$c_{11} = c_{22};\ c_{12};\ c_{13} = c_{23};\ c_{33};\ c_{44} = c_{55};\ c_{66},$ plus $c_{16} = -c_{26}$ in the 4 and 4-bar, and 4/m point groups
Trigonal	$c_{11} = c_{22};\ c_{12};\ c_{13} = c_{23};\ c_{14} = -c_{24} = c_{56};\ c_{33};\ c_{44} = c_{55};\ c_{16} = 1/2(c_{11} - c_{12})$ plus $c_{25} = -c_{15} = c_{46}$ in the 3 and 3-bar point groups
Hexagonal	$c_{11} = c_{22};\ c_{12}\ ;\ c_{13} = c_{23}\ ;\ c_{33},\ c_{44} = c_{55};\ c_{16} = 1/2(c_{11} - c_{12})$
Cubic	$c_{11} = c_{22} = c_{33};\ c_{12} = c_{13} = c_{23};\ c_{44} = c_{55} = c_{66}$
Polycrystal	$c_{11} = c_{22} = c_{33};\ c_{12} = c_{13} = c_{23};\ c_{44} = c_{55} = c_{66} = 1/2(c_{11} - c_{12})$

Source: Nye, J. F. 1957, *Physical Properties of Crystals: Their Representation by Tensors and Matrices*, Oxford University Press, London.

reduced by the crystal symmetry (Neumann's principle). For example, with triclinic crystals there are only 21 independent elastic-stiffness coefficients:

$$
\begin{pmatrix}
c_{11} & c_{12} & c_{13} & c_{14} & c_{15} & c_{16} \\
\bullet & c_{22} & c_{23} & c_{24} & c_{25} & c_{26} \\
\bullet & \bullet & c_{33} & c_{34} & c_{35} & c_{36} \\
\bullet & \bullet & \bullet & c_{44} & c_{45} & c_{46} \\
\bullet & \bullet & \bullet & \bullet & c_{55} & c_{56} \\
\bullet & \bullet & \bullet & \bullet & \bullet & c_{66}
\end{pmatrix}
\tag{9.7}
$$

With higher symmetry crystals, the number of independent elastic coefficients is even smaller, as shown in Tables 9.3 and 9.4.

TABLE 9.4 The Independent Elastic-Compliance Coefficients for Each Crystal Class

Triclinic	$s_{11}, s_{22}, s_{33}, s_{44}, s_{55}, s_{66}, s_{12}, s_{13}, s_{23}, s_{14}, s_{24}, s_{34}, s_{15}, s_{25}, s_{35}, s_{45}, s_{16},$ $s_{26}, s_{36}, s_{46}, s_{56}$
Monoclinic	$s_{11}, s_{22}, s_{33}, s_{44}, s_{55}, s_{66}, s_{12}, s_{13}, s_{23}, s_{16}, s_{26}, s_{36}, s_{45}$
Orthorhombic	$s_{11}, s_{22}, s_{33}, s_{44}, s_{55}, s_{66}, s_{12}, s_{13}, s_{23}$
Tetragonal	$s_{11} = s_{22};\ s_{12};\ s_{13} = s_{23};\ s_{33};\ s_{44} = s_{55};\ s_{66},$ plus $s_{16} = -s_{26}$ in the 4 and 4-bar, and 4/m point groups
Trigonal	$s_{11} = s_{22};\ s_{12};\ s_{13} = s_{23};\ s_{14} = -s_{24} = 2s_{56};\ s_{33};\ s_{44} = s_{55};\ s_{16} = 2(s_{11} - s_{12})$ plus $s_{25} = -s_{15} = 2s_{46}$ in the 3 and 3-bar point groups
Hexagonal	$s_{11} = s_{22};\ s_{12}\ ;\ s_{13} = s_{23}\ ;\ s_{33},\ s_{44} = s_{55};\ s_{16} = 2(s_{11} - s_{12})$
Cubic	$s_{11} = s_{22} = s_{33};\ s_{12} = s_{13} = s_{23};\ s_{44} = s_{55} = s_{66}$
Polycrystal	$s_{11} = s_{22} = s_{33};\ s_{12} = s_{13} = s_{23};\ s_{44} = s_{55} = s_{66} = 2(s_{11} - s_{12})$

Source: Nye, J. F. 1957 *Physical Properties of Crystals: Their Representation by Tensors and Matrices*, Oxford University Press, London.

It can be seen from Table 9.3 that, in the case of a polycrystalline aggregate with random crystallite orientation (e.g., the metal rod in Figure 9.1), only two independent elastic-stiffness coefficients, c_{44} and c_{12}, need be specified to fully describe the elastic response. These are often referred to as the Lamé constants, μ and λ, named after French mathematician Gabriel Lamé (1795–1870):

$$c_{44} = \mu \qquad c_{12} = \lambda \qquad (9.8)$$

with $c_{11} = c_{12} + 2c_{44}$. The Lamé constants are related to the elastic moduli. To see this, recall that the Young's modulus, E, is defined as the ratio of normal stress to normal strain. Hence, *for a polycrystal with a random orientation distribution* Table 9.3 gives E as $\sigma_1/\varepsilon_1 = c_{11} = \sigma_2/\varepsilon_2 = c_{22} = \sigma_3/\varepsilon_3 = c_{33} = \lambda + 2\mu$, or $c_{12} + 2c_{44}$. Likewise, the rigidity modulus, G, is given by $\sigma_4/\varepsilon_4 = c_{44} = \sigma_5/\varepsilon_5 = c_{55} = \sigma_6/\varepsilon_6 = c_{66} = \frac{1}{2}(c_{11} - c_{12}) = \mu$, or c_{44}.

From the relations in the preceding paragraph, it is seen that a polycrystal with a random distribution of crystal orientations is an elastically isotropic body macroscopically, even though each crystal of the aggregate is elastically anisotropic. This means the Young's modulus and rigidity modulus are independent of the principal direction along which the tensile or shear stress is applied. The elastic moduli can thus be considered as *polycrystalline elastic constants.*

Obtaining an elastic modulus for a polycrystal from the single-crystal elastic coefficients requires reducing the independent coefficients for the crystal class to the appropriate number of Lamé constants. Several methods for obtaining estimates have been proposed. The Voigt approximation, named after the German physicist Woldemar Voigt (1850–1919), assumes a uniform strain (Voigt, 1910). This leads to volume averaging with an orientation distribution function over the elastic stiffness tensor. For the Young's modulus of a material with cubic symmetry, the Voigt approximation is:

$$E_V = (c_{11} - c_{12} + 3c_{44})(c_{11} + 2c_{12})/(2c_{11} + 3c_{12} + c_{44}) \qquad (9.9)$$

If, on the other hand, one assumes uniform stress throughout the polycrystal, a similar averaging procedure can be performed over the elastic-compliance tensor. This is known as the Reuss approximation (Reuss, 1929). In terms of the elastic-stiffness coefficients, the Reuss approximation for the Young's modulus of a cubic crystal is:

$$E_R = 5c_{44}(c_{11} - c_{12}(c_{11} + 2c_{12})/(c_{11}^2 + c_{11}c_{12} - 2c_{12}^2 + 3c_{44}c_{11} + c_{44}c_{12}) \quad (9.10)$$

Of course, a relation also exists between the single crystal elastic coefficients and the rigidity modulus for a polycrystal. The Voigt and Reuss approximations for the rigidity modulus of a cubic material are given by Eqs. 9.11 and 9.12, respectively:

$$G_V = (c_{11} - c_{12} + 3c_{44})/5 \qquad (9.11)$$

$$G_R = 5c_{44}(c_{11} - c_{12})/[4c_{44} + 3(c_{11} - c_{12})] \qquad (9.12)$$

Hill later proposed that the Voigt and Reuss approximations represent the upper and lower bounds, respectively, for the true elastic moduli of a polycrystal. Hill used the Voigt and Reuss approximations to obtain upper and lower bounds for polycrystals with triclinic symmetry (Hill, 1952).

The condition for isotropic elasticity, as we have seen, is $c_{44} = c_{55} = c_{66} = \frac{1}{2}(c_{11} - c_{12})$. Cubic crystals, because of their high symmetry, almost satisfy this condition. Zener introduced the ratio $2c_{44}/(c_{11} - c_{12})$ as an elastic anisotropy factor for cubic crystals (Zener, 1948a). In a cubic crystal, if the Zener ratio is positive the Young's modulus has a minimum along the $\langle 100 \rangle$ direction and a maximum along the $\langle 111 \rangle$ direction:

$$E_{001} = (c_{11} - c_{12})(c_{11} + 2c_{12})/(c_{11} + c_{12}) \tag{9.13}$$

$$E_{111} = \{[(c_{11} + c_{12})/(c_{11} + 2c_{12})(c_{11} - c_{12})] + \frac{1}{3}[(1/c_{44}) - 2/(c_{11} - c_{12})]\}^{-1} \tag{9.14}$$

The reverse condition holds when the Zener ratio is negative. Similar equations for the Young's moduli for the other crystal classes can be derived from the expressions given in Nye's book (Nye, 1957).

In the early days of elasticity theory, atoms were treated as point charges, and in a centrosymmetric structure they interacted via central forces dependent on distance alone. Thus, it was assumed that in certain ionic solids, for example, those with the NaCl structure, only one elastic coefficient was needed to describe elastic properties. This is known as *Cauchy's relation* ($c_{12} = c_{44}$). This relation implies that Poisson's ratio for such a solid is equal to $\frac{1}{4}$. However, it is known that ν normally is closer to $\frac{1}{3}$ (Section 9.2.2).

Example 9.2 In terms of the Lamé constants, write the equations relating the six stresses to the strains for a polycrystalline material.

Solution Using Eqs. 9.8 with Table 9.3, it is easily shown that the stresses can be written in terms of the two Lamé constants as:

$$\sigma_1 = (\lambda + 2\mu)\varepsilon_1 + \lambda\varepsilon_2 + \lambda\varepsilon_3$$
$$\sigma_2 = \lambda\varepsilon_1 + (\lambda + 2\mu)\varepsilon_2 + \lambda\varepsilon_3$$
$$\sigma_3 = \lambda\varepsilon_1 + \lambda\varepsilon_2 + (\lambda + 2\mu)\varepsilon_3$$
$$\sigma_4 = \mu\varepsilon_4$$
$$\sigma_5 = \mu\varepsilon_5$$
$$\sigma_6 = \mu\varepsilon_6$$

where the subscripts are: $1 = xx$; $2 = yy$; $3 = zz$, $4 = yz$; $5 = zx$; $6 = xy$.

Experimental determination of the elastic coefficients is laborious. One method involves exacting wave-velocity measurements (e.g., pulse-echo) of ultrasonic waves propagating in the principal symmetry directions within a relatively large crystal. Elastic moduli of polycrystalline materials are estimated with the assumption of homogenous strain and stress (i.e., that all grains experience the same strain and stress) and are usually reported as upper and lower bounds, obtained from the Voigt and Reuss approximations.

Typically, high modulus inorganic solids contain elements from the top rows of the periodic table (Be, B, C, Si, Al, N, O), since the presence of these elements usually leads to small atomic spacing (stronger covalent bonds). For example, the Young's modulus of silicon nitride, silicon carbide, boron carbide, and diamond are 300, 440, 450, and 600 GPa, respectively. *Ab initio* (Overney et al., 1993) and tight-binding (Hernandez et al., 1998; Xin et al. 2000) simulations have predicted that the smallest diameter (<1 nm) single-wall carbon nanotubes could have a Young's modulus as high as 5 TPa! This value has been substantiated by micro-Raman spectroscopy (Lourie and Wagner, 1998).

In some applications, low-modulus materials are desirable. In high-power semiconductor components, for example, heat is conducted away from a silicon die (which has a coefficient of thermal expansion (CTE) of $2.49 \times 10^{-6} \, \text{K}^{-1}$) to a copper heat sink (CTE $= 16.5 \times 10^{-6} \, \text{K}^{-1}$) via a thermal interface material (TIM), which is often required to bond to both surfaces. Since the die and heat sink have very different CTEs, not only must the TIM have a high thermal conductivity but, in order to minimize thermomechanical fatigue, it should also possess a low modulus. Most often, a lead–tin or indium-based solder is used for this purpose.

9.2.2 Relationships Between the Elastic Moduli

Sometimes the Lamé constants are not reported for a polycrystal and all one knows is the value for a particular elastic modulus. In those cases, it is still possible to derive values for the other moduli. For example, the rigidity modulus and Young's modulus are related via

$$G = E/[2(1 + \nu)] \tag{9.15}$$

where ν is Poisson's ratio. Poisson's ratio is the dimensionless ratio of relative diameter change (lateral contraction per unit breadth) to relative length change (longitudinal extension per unit length) in elastic stretching of a cylindrical specimen. For many materials, Poisson's ratio has a value in the range 0.2–0.5 (e.g., steel $\sim$0.27, rubber $\sim$0.5). Typically, ceramics range from 0.2 to 0.3. (e.g., Al_2O_3 $\sim$0.26, BeO $\sim$0.34).

The bulk modulus, B, is the inverse of the compressibility. The bulk modulus relates a uniform compressive stress (hydrostatic pressure) to the associated isotropic strains that result in volume change, but no shape change. It too is related to the Young's modulus:

$$B = E/[3(1 - 2\nu)] \tag{9.16}$$

9.2.3 The Relation Between Elasticity and the Cohesive Forces in a Solid

We have spent a great deal of time up to now *describing* elastic behavior. It would obviously be of great value to be able to *predict* the elastic constants for a material. First principles calculations are difficult, which is due, in part, to the sizable matrix of coefficients with symmetries other than cubic (Table 9.3). This requires a large number of distortions for calculation of the full set of elastic coefficients (Ravindran et al., 1998). Intuitively, we expect a property like stiffness ("springiness") to be related to the cohesive forces in a sold. Interatomic potential energy functions, in fact, provide the starting point for estimating elastic constants.

The form of these interatomic potentials may be obtained empirically, by fitting parameters of some generic formula to experimental data, or by atomistic scale simulations such as the *ab initio* molecular dynamics techniques known as the Car-Parrinello method (Car and Parrinello, 1985). Interatomic potentials can be roughly separated into two classes: *pair potentials*, also called the *pairwise interaction* (e.g., Lennard-Jones potential (Lennard-Jones, 1931) or Morse potential (Morse, 1930), in which the interaction between a pair of atoms only depends on their (scalar) radial separation; and *many-body potentials*, in which the interaction between a pair of atoms is modified by the surrounding atoms. A simple test to determine if the elastic properties of a solid *might* be adequately described by a pairwise potential is whether the Cauchy relation $(c_{12} = c_{44})$ holds.

In general, the total interatomic potential between any pair of atoms is the sum of the pairwise interaction and the interactions between three atoms (triplets), four atoms (quartets), and so forth. The problem is that pair potentials are by far the easiest to compute; however, their exclusive use gives results that are only semiquantitative (even with ionic solids), accounting for only up to 90% of the total cohesive energy in a solid. The three-body term simply cannot be neglected, although the higher-order terms often can be.

The connection between the interatomic potential and the elastic constants is seen by treating the interatomic bond as a spring. A force constant for the spring can be obtained, which is a measure of the restoring force for small displacements of the ions from their equilibrium positions. The derivative of potential energy is force and the slope of the force/distance curve at the equilibrium separation between the atoms is proportional to the elastic constant. Starting with a pair potential for the simplest case, an ionic solid, we derive the expression for the elastic modulus. The calculation of elastic constants from interatomic potentials for metallic and covalent solids is considerably more complex because of the many-body terms, so we do not discuss it in detail.

Ionic Solids As we learned in Chapter 2, the ionic bond is nondirectional. Hence, ions prefer maximum coordination numbers, and strongly ionic solids normally adopt highly symmetric crystal structures. The rock salt and other cubic structures are frequently observed. For such cases, the Young's modulus is the appropriate elastic constant. We start by writing the potential energy of a *single* ion in the lattice from the Born model (Eq. 2.7):

$$U_{ion} = [1/4\pi\varepsilon_0][-M(q_+q_-)e^2/r] + B/r^n \qquad (9.17)$$

where $4\pi\varepsilon_0$ is $1.11265 \times 10^{-10}\,C^2J^{-1}m^{-1}$, M is the dimensionless Madelung constant characteristic of a given structure type; q is the ion charge; e is the electron charge ($1.6022 \times 10^{-19}\,C$); r is the interatomic distance (meters); B is a constant; and n is the Born exponent, which is a positive integer, found by measuring the compressibility of the compound. Approximate values for n can be obtained from Table 2.5 in Chapter 2. It is typically in the range of 5 to 12. The term in Eq. 9.17 containing the Born exponent represents the short-range repulsion between nearest neighboring ions, of whatever charge, which keeps the lattice from collapsing. The entire first term on the right-hand side of Eq. 9.17 is the long-range Coulombic attraction. Its units are joules per cation.

Now, the force required to pull ions apart, say, by stretching the bond, is related to the potential energy by dU/dr:

$$F = dU/dr = [1/4\pi\varepsilon_0][M(q_+q_-)e^2/r^2] - nB/r^{n+1} \qquad (9.18)$$

The units of Eq. 9.18 are newtons ($1\,N = 1\,J/m$). Note that B can be solved for by setting Eq. 9.18 to zero, that is, when U is a minimum. The force constant of the spring (bond) is given by d^2U/dr^2. The force constant is related to the elastic modulus since the latter is defined as $d\sigma/d\varepsilon$. However, because stress is equal to force per unit area ($\sigma = F/r_0^2$) and strain, ε, is equal to $\Delta r/r_0$, we get:

$$E = d\sigma/d\varepsilon = 1/r(d^2U/dr^2)|_{r=r_0}$$
$$= 1/r_0\{1/4\pi\varepsilon_0[-2M(q_+q_-)e^2/r_0^3] + n(n+1)B/r_0^{n+2}\} \qquad (9.19)$$

The units of E are pascals ($1\,Pa = 1\,N/m^2 = 1\,J/m^3$). But wait! There is still one final correction needed. Equation 9.19 must be divided by m (the coordination number—the number of anions bonded to the cation), because we want the force constant for just one of the m bonds to the ion:

$$E = (1/m)1/r_0\{1/4\pi\varepsilon_0[-2M(q_+q_-)e^2/r_0^3] + n(n+1)B/r_0^{n+2}\} \qquad (9.20)$$

The slope of the dU/dr curve for the ion pair around r_0 is a measure of the restoring force for small displacements from the equilibrium position. The greater the slope, the higher the modulus, and the stiffer the material. As demonstrated in Example 9.3, the Born model is, at best, only semiquantitative. One reason for this is that ions are treated as point charges, and ion polarizability (Table 2.3) is not accounted for.

Example 9.3

(a) Using Eq. 9.20, estimate the Young's modulus for sodium chloride. Assume $M = 1.75$, $r = 2.81$ Å, and $m = 6$. Compare to the experimental value obtained from polycrystalline NaCl (39.96 GPa).

(b) Use Eq. 9.9 with the measured elastic-stiffness coefficients of single-crystal NaCl ($c_{11} = 49.47$ GPa, $c_{12} = 12.88$ GPa, $c_{44} = 12.87$ GPa) to calculate the upper bound to the Young's modulus and compare to the result calculated from part (a). Note that in NaCl, the Cauchy relation $c_{12} = c_{44}$ is obeyed.

Solution

(a) We first need to obtain an approximate value for the Born exponent, n, from Table 2.5. This is found to be $n = 8$.

Next, covert r to meters (1 Å $= 10^{-10}$). We get $r = 2.81 \times 10^{-10}$ m.
Now, solve for B by setting dU/dr in Eq. 9.18 to zero:

$$B = 1/4\pi\varepsilon_0[M(q_+q_-)e^2r^7]/8 = 6.96 \times 10^{-96} \, \text{J m}^8$$

Finally, substitute the numbers into Eq. 9.20 to obtain *75.2 GPa* (1 GPa $= 10^9$ Pa). This result is noticeably larger than the experimental value (% error $= 88.2\%$). The difference implies that a pairwise potential is not entirely adequate for describing NaCl, even though the Cauchy relation holds.

(b) Simply substitute the values for c_{11}, c_{12}, and c_{44} directly into Eq. 9.9 to obtain: *37.6 GPa.*

Metals   The cohesive forces in an ideal close-packed metal are, like those in ionic solids, nondirectional. Nonetheless, pair potentials are generally adequate only for metals in which the cohesion is due to s and p electrons. The d orbitals of transition metals make an angular-dependent covalent contribution to the bond strength, even in close-packed structures, which can only be accounted for by a many-body potential. One of the most common models for transition metals is the embedded-atom method (EAM) (Daw and Baskes, 1983; 1984). The total potential energy is expressed in terms of a sum of pair potential $U(r_{ij})$, functions only of the distance between atoms i and j, and a many-body "embedding energy", $F(\rho)$, which is a superposition of the electron densities, ρ_i, of the surrounding atoms:

$$U_T = \sum_{i>j} U(r_{ij}) + \sum_i^N F(\rho_i) \tag{9.21}$$

The terms appearing in Eq. 9.21 are derived by guessing some functional form and fitting the parameters to *ab initio* calculations or empirical data. In general, the values of the parameters are chosen to reproduce particular properties of interest as closely as possible, and for transferability among a variety of solids. Analytic potentials have been developed for some metals from each of the following classes: monatomic fcc metals (Johnson, 1988 a, b), fcc alloys (Johnson, 1989), monatomic bcc metals (Oh and Johnson, 1989; Yifang et al., 1996), and monatomic hcp metals (Cleri and Rosato, 1993; Oh and Johnson, 1988; Pasianot and Savino, 1992).

The EAM analytical potentials (Eq. 9.21) are multivariable functions. Their second derivatives yield accurate estimates for the elastic-stiffness coefficients. However, calculating the second derivative of a potential with terms beyond the pair interaction is not trivial. The second derivative with respect to atomic displacements is a matrix of partial second derivatives, known as the *Hessian matrix*, the general name for any second derivative of a multivariable function. Molecular dynamics and Monte-Carlo computational techniques that do not require evaluation of the second derivative for obtaining the elastic constants have been developed, but are beyond the scope of this book.

In spite of the success of this model, interatomic potentials are not always available, particularly for alloys. Ballpark estimates of the cohesive forces in an alloy and its derivatives, such as the elastic modulus, can be obtained from those of the pure components by assuming the alloy is a random solid solution and applying the "rule of mixtures." The additive nature of bond properties gives this type of semiempirical approach its power. It is founded in the assumption that the physical property of interest associated with an *A–B* bond can be approximated as the average of the values associated with the *A–A* and *B–B* bonds. We digress briefly to emphasize an important criterion that must be met in order for this approach to yield reasonable results. The rule-of-mixtures is only a valid approximation for a bulk physical property when a well-defined analogous property can be associated with the chemical bonds. This is the reason that electrical conductivity, for example, is not an additive property. An individual bond, localized between two atomic centers, does not constitute a *delocalized* conduction band, in spite of the fact that it has a transfer, or hopping, integral associated with it. Conduction-band formation in a solid requires a large number of atoms (Section 11.2.1), although, interestingly, it has been recently discovered that some aromatic molecules (e.g., benzene) can function as molecular transistors (Section 5.3.1).

Let us now proceed to the problem at hand. We have already established that modeling individual bonds as interatomic springs can yield reasonable estimates for elastic constants. In a binary alloy $xA–yB$ (where x and y are the atomic percentages of A and B, respectively) there are four different types of bonds present: $A–B$, $B–A$, $A–A$, and $B–B$. The *sum* of the fractions of each type of bond must equal one. That is:

$$\sum p_{ij} = 1 \qquad (9.22)$$

where p_{ij} is the fraction of bonds between atoms i and j. In the case of a binary solid solution with random mixing between atoms with the same coordination number, the following relations hold:

$$p_{AA} = x_A^2 \qquad p_{AB} = p_{BA} = x_A x_B \qquad p_{BB} = x_B^2 \tag{9.23}$$

where x_i is the atomic percent of species i.

In order to obtain an approximate value for the elastic constant for the alloy, we assume it is a random solution (i.e., we assume Eq. 9.23 holds) and approximate the force constant of the A–B and B–A "springs" $(d^2\phi^{ab})$ as the average of the known elastic constants of pure A and B. In the language of molecular dynamics, this is equivalent to letting the A–B interatomic potential in the alloy be some linear combination of the potentials of the pure elements: $\phi^{ab} = c_a \phi^{aa} + c_b \phi^{bb}$. Actually, this is the assumption used for the pair potential in the EAM model. Note that due to the constraint imposed by Eq. 9.22, the very simple expression $x_A E_A + x_B E_B$ (where E_i is the elastic constant for pure i) will yield exactly the same result as using the relations in Eq. 9.23 and the average value of the modulus, namely, $x_A^2 + 2x_A x_B E_{\text{avg}} + x_B^2 E_B$. This simple approach can lead to surprisingly good estimates, even for insoluble components, eutectics, and ordered intermetallic phases, as illustrated in Example 9.4.

Example 9.4   Consider the following solder alloys, with the phases present given in brackets:

$$91.6 \text{ at } \% \text{ Pb–8.40 at } \% \text{ Sn} = [\text{Pb–Sn} \text{eutectic} + (\text{Pb, Sn})]$$
$$80.68 \text{ at } \% \text{ Bi–19.32 at } \% \text{ Ag} = [(\text{Bi}) + (\text{Ag})]$$
$$70.67 \text{ at } \% \text{ Au–29.33 at } \% \text{ Sn} = [(\text{Au}_5\text{Sn}) + (\text{AuSn})]$$
$$63.56 \text{ at } \% \text{ Sn–26.90 at } \% \text{ Ag–9.54 at } \% \text{ Sb} = [(\text{Sn}) + (\text{Ag}_3\text{Sn}) + (\text{Sn}_3\text{Sb}_2)]$$

The tabulated shear moduli, G, of the pure metals (in GPa), are as follows:

Pb = 5.6;   Sn = 18;   Bi = 12;   Ag = 30;   Au = 27;   Sb = 20

Calculate the shear modulus for each alloy, assuming it is an additive property, and compare to the experimentally measured values given as follows:

The accepted room-temperature experimental values, obtained from an impulse excitation technique, are (Lalena et al., 2002):

$$\text{Pb–Sn} = 9.34 \text{ GPa}$$
$$\text{Bi–Ag} = 13.28 \text{ GPa}$$
$$\text{Au–Sn} = 21.26 \text{ GPa}$$
$$\text{Sn–Ag–Sb} = 25 \text{ GPa}$$

Solution   Using the elastic moduli of the elements, we obtain for the alloys:

$$Pb\text{--}Sn : (0.916 \times 5.6) + (0.084 \times 18) = 6.64\,GPa$$
$$Bi\text{--}Ag : (0.8068 \times 12) + (0.1932 \times 30) = 15.47\,GPa$$
$$Au\text{--}Sn : (0.7067 \times 27) + (0.2933 \times 18) = 24.36\,GPa$$
$$Sn\text{--}Ag\text{--}Sb : (0.6356 \times 18) + (0.2690 \times 30) + (0.0954 \times 20) = 21.42\,GPa$$

The percent errors are determined via:

$$(|Accepted - calculated|/Accepted) \times 100$$

Alloy	Accepted Value	Calculated Value	% Error
Pb–Sn	9.34	6.64	28.90
Bi–Ag	13.28	15.47	16.49
Au–Sn	21.26	24.36	14.58
Sn–Ag–Sb	25	21.42	14.32

Covalent Solids   Interatomic potentials are the most difficult to derive for covalent solids. The potential must predict the directional nature to the bonding (i.e., the bond angles). Most covalent solids have rather open crystal structures, not close-packed ones. Pair potentials used with *diatomic* molecules, such as the Lennard-Jones and Morse potentials, are simply not adequate for solids, because atoms interacting via only radial forces prefer to have as many neighbors as possible. Hence, qualitatively wrong covalent crystal structures are predicted.

A simple model, which has been quite successful in solids with the diamond or zinc-blende structure, was introduced by Stillinger and Weber (Stillinger and Weber, 1985). The first term in the potential is the product of a Lennard-Jones–like pairwise interaction and a cutoff function smoothly terminating the potential at some distance r_c. The second term is a multivariable three-body potential written as a separable product of two radial functions and an angular function:

$$U = \sum_j \sum_{j>i} (Br_{ij}^{-p} - Ar_{ij}^{-q}) \exp[\mu/(r_{ij} - r_c)]$$
$$+ \sum_i \sum_{j>i} \sum_{k>j} Zg(r_{ij})g(r_{ik})\left[\cos\theta_{jik} + \frac{1}{3}\right]^2 \tag{9.24}$$

The cutoff function also defines the radial functions of the three-body term:

$$g(r_{ij}) = \exp[\mu/(r_{ij} - r_c)]^\alpha$$
$$g(r_{ik}) = \exp[\mu/(r_{ik} - r_c)]^\alpha$$

As can be seen, there are eight parameters in the Stillinger–Weber potential: A, B, p, q, μ, r_c, Z, and α. These parameters are fitted to experimental data, such as lattice constants and cohesive energy, for the diamond structure. The angle θ_{jik} is the angle centered on atom i. If θ_{jik} is 109.47°, $\cos \theta_{jik} = -\frac{1}{3}$, and the angular function has a minimum of zero, which makes sp^3 hybridization of the ith atom, that is, the zinc-blende structure, energetically preferred. Calculation of the elastic-stiffness constants by the method outlined here requires that Eq. 9.24 be a differentiable function. As with the EAM potential, the second derivative is a matrix of partial second derivatives. It is obvious from the form of Eq. 9.24 that a shorter internuclear distance, r, leads to a larger binding force (dU) and modulus, or stiffness (d^2U). The Stillinger–Weber model reproduces the elastic constants for Si reasonably well, but it is not very transferable to other structure types because of the 'built-in' tetrahedral angle.

9.3 PLASTICITY

One might argue that the atoms in any solid should be held very rigidly in place by strong interatomic forces. If this presumption were strictly true, all types of solids would be predicted to fracture under stresses greater than the elastic limit, that is, for atomic displacements greater than elastic, or vibrational motion about their equilibrium positions. This is generally the case for most ionic and covalent substances. However, it will be recalled from Chapter 3 that, for metals, it is not a bad approximation to consider all the valence electrons to be delocalized into extended wave functions, and *not* localized in bonding regions between particular "ions" (consisting of the nuclei plus the filled electron shells). Cohesion results from the electrostatic attraction between the electrons and ions. The ion cores are not involved in bonding of any type, however, and they can be regarded as being displaceable, provided the activation energy is supplied.

   Plasticity is the property that allows materials to undergo permanent deformation without rupture, which is of great utility for fabricating pieces into specific shapes. The three most common plastic-deformation fabrication techniques are extrusion, wiredrawing, and rolling. In an extrusion press, a cylindrical ingot is pushed through a cone-shaped or converging die to reduce its diameter, thus forming wire. In wiredrawing, the ingot is pulled through, rather than being pushed. The material inside the die is subjected to similar (but not identical) stresses as in extrusion. An important attribute of a material, in order that it form into and be useful as a wire, is its *ductility*. This is the ability to withstand tensile loads. Rolling, on the other hand, requires that a material be *malleable*, or resistant to rupture from compressive loads. The plasticity of materials is due to the gliding motion, or *slip*, of planes of atoms. For example, when a tensile force, F, is applied to a cylindrical-shaped single crystal, this places a shear stress, σ_s, on certain crystal planes, whose normal are at an angle, ϕ, to the cylinder axis. If the shear stress is sufficiently strong, movement occurs along these parallel planes in a slip direction, which

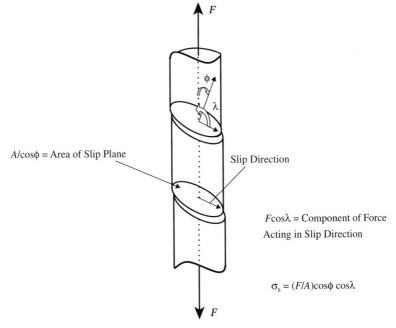

Figure 9.4   Application of a tensile force to a cylindrical single crystal cuases a shear stress on some crystal planes. When the shear stress is equal to the critical resolved shear stress (the yield stress), glide proceeds along the slip direction of the planes.

makes an angle, λ, with the cylindrical axis. This causes sections of the crystal to slide relative to one another, as shown in Figure 9.4.

According to *Schmid's law* (Schmid and Boas, 1935), plastic flow in a pure and perfect single crystal occurs when the shear stress acting along (parallel to) the slip direction on the slip plane reaches some critical value known as the *critical resolved shear stress*, τ_c. From Figure 9.4, it can be seen that the component of the force acting in the slip direction is $F \cos \lambda$, which acts over the plane of area $A / \cos \phi$ (where A is the cross-sectional area). Thus the resolved shear stress is $\sigma_s = (F/A) \cos \phi \cos \lambda$. Schmid's law states that slip occurs at some critical value of σ_s, denoted as τ_c:

$$\tau_c = \sigma \cos \phi \cos \lambda \qquad (9.25)$$

where σ is the applied tensile stress (yield stress) at which slip begins, and ϕ and λ are the slip-plane normal and slip direction, respectively. The critical resolved shear stress is a constant for a substance at a given temperature. This is because one slip system, termed the *primary slip system*, with the greatest τ_c acting upon it dominates the plastic deformation. Slip normally occurs first in slip systems with orientations with the maximum value of the $\cos \phi \cos \lambda$ term (the *Schmid factor*) in Eq. 9.25. However, when the material is subjected to high temperatures or high stresses, other slip systems may also become operative once their τ_c is reached.

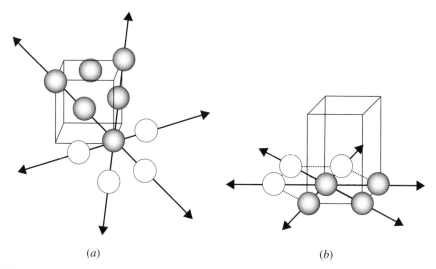

(a) (b)

Figure 9.5 The three close-packed directions in the close-packed planes of the (a) ccp lattice and (b) hcp lattice. The lightly shaded spheres, completing the hexagonal coordination around the sphere at the corner of the cubes, are in neighboring unit cells.

Schmid's law is named after its discover, the Austrian physicist Erich Schmid (1896–1983), who began investigating crystal plasticity in the 1920s.

Experimentally, it is observed that slip most readily occurs on specific crystal-lographic planes—close-packed planes of high atomic density—in close-packed directions (the slip direction must lie in the slip plane). A slip plane together with the slip direction constitutes a slip system. Cubic-close-packed metals contain four close-packed planes. These belong to the set with indices $\{hkl\} = \{111\}$ and include the (111), (11\overline{1}), (1\overline{1}1), and (\overline{1}11) planes. Figure 9.5a shows the three distinct close-packed directions (with $[uvw] = [1\overline{1}0]$, $[10\overline{1}]$, and $[01\overline{1}]$) within a (111) plane. Similarly, the remaining three planes in the $\{111\}$ set also have three directions each (the reader should derive the face-diagonal indices for these other close-packed directions). Thus, there are 12 easy-glide slip systems in the ccp lattice, which are denoted simply as $\{111\}\langle 110 \rangle$. Slip may occur in either direction along the slip vector on a given slip system.

The Austrian mathematician Richard von Mises (1883–1953) recognized that, for general plastic strain of a polycrystalline (isotropic) sample, five independent slip systems must be available for the strain to be accommodated purely by glide. Plastic deformation produces an arbitrary shape change at constant volume. Therefore, of the six components to the strain tensor $(\varepsilon_{xx}, \varepsilon_{yy}, \varepsilon_{zz}, \varepsilon_{xy}, \varepsilon_{xz}, \varepsilon_{zy})$ only five components are independent, due to the condition that $\Delta V = \varepsilon_{xx} + \varepsilon_{yy} + \varepsilon_{zz} = 0$, that is, the trace of Eq. 9.4, $tr(\varepsilon)$, must equal zero. Since the operation of one independent slip system produces one strain component, there must be a minimum of five independent slip systems operating for plastic deformation. This is known as the *von Mises criterion* (von Mises, 1928). A slip system is independent if the shape change it produces cannot be duplicated by combinations of slip on other systems. The material will plastically deform when the distortional energy reaches some critical value, satisfying the von Mises yield condition:

$$[(\sigma_1 - \sigma_2) + (\sigma_2 - \sigma_3) + (\sigma_3 - \sigma_1)]^2 - k^2 = 0 \qquad (9.26)$$

in which k is a constant dependent on the prior strain history of the sample. The term in brackets represents the distortional energy, where σ_1, σ_2, and σ_3 are the three *principal stresses* acting along three mutually perpendicular principal axes that can be derived from any general stress tensor, Eq. 9.3, via coordinate transformation.

Of the 12 slip systems possessed by the cubic-close-packed structure, five are independent, which satisfies the von Mises criterion. For this reason, and because of the multitude of active slip systems in polycrystalline cubic-close-packed metals, they are the most ductile. Hexagonal close-packed metals contain just one close-packed layer, the (0001) *basal* plane, and three distinct close-packed directions in this plane: [1120], [2110], [1210], as shown in Figure 9.5b. Thus, there are only three easy-glide primary slip systems in hcp metals, and only two of these are independent. Hence, hcp metals tend to have low ductility. In fact, they are usually classified as semibrittle metals. The c/a ratio of the ideal hcp lattice is 1.633. It is found experimentally that basal slip is favored for metals with $c/a > 1.633$ (e.g., Zn, Mg, Cd, Co). For $c/a < 1.633$ (e.g., Ti, Cr), slip is favored in the [1210] direction on the prismatic (1010) plane (giving three slip systems), or on the pyramidal (1011) plane in the [1120] direction (six slip systems).

Body-centered cubic metals contain no close-packed planes, but do contain four close-packed directions, the four [111] body diagonals of the cube. The most nearly close-packed planes are those of the {110} set. In bcc crystals, slip has been observed in the [111] directions on the {110}, {112}, and {123} planes, but that attributed to the latter two planes may be considered the resultant of slip on several different (110)-type planes (Weertman and Weertman, 1992). The von Mises criterion is satisfied, but higher shearing stresses than those of ccp metals are normally required to cause slip in bcc metals. As a result, most bcc metals are classified as semibrittle.

Erich Schmid (Courtesy of Erich Schmid Institute. © Österreichische Akademie der Wissenschaften (Austrian Academy of Sciences). Reproduced with permission.)

ERICH SCHMID

(1896–1983) received his doctorate in physics in 1920 from the University of Vienna. In 1922, Schmid, Hermann Mark, and Michael Polanyi investigated plasticity in zinc crystals at the Kaiser-Wilhelm Institute of Fiber Chemistry in Berlin. This work, together with Schmid's subsequent research, culminated in "Schmid's law" for the onset of plasticity. In 1935, Schmid and Walter Boas published the well-known textbook *Kristallplastizität*, which was published in the English edition, *Crystal Plasticity*, in 1950 and again in 1968. From 1920 until 1951, Schmid was at several academic posts in the DACH region, including the Vienna University of Technology (Austria), the University of Freiburg (Switzerland), and the University of Frankfurt (Germany). He returned to University of Vienna in 1951 and stayed there until his retirement in 1967. Schmid served as president of the Austrian Academy of Sciences for 10 years and he was awarded the Austrian Medal of Science and Arts.

(*Source*: The Erich Schmid Institute. © Austrian Academy of Sciences)

9.3.1 The Dislocation-Based Mechanism to Plastic Deformation

Slip relies on chemical bond breaking and bond reformation as two planes of atoms pull apart. It is observed that the critical resolved shear stress required to cause plastic deformation in real materials is much lower (by several orders of magnitude) than the shear stress required to plastically deform a perfect defect-free crystal, the so-called *ideal shear stress*. The latter is equivalent to the stress required for the *simultaneous* gliding motion (bond breaking and reformation) of all the atoms in one plane, over another plane.

Slip occurs much more readily across close-packed planes containing extended defects called *dislocations*. Dislocations can be one of two extreme types, or a *mixture* (of intermediate character). One extreme is the *edge dislocation* (Figure 9.6a), which is an extra half plane of atoms, or unit cells in the case of compounds (dislocations are stoichiometric defects). The existence of this type of extended defect was first hypothesized independently in 1934 by Geoffrey Ingram Taylor (1886–1975) (Taylor, 1934), Egon Orowan (Orowan, 1934), and the Hungarian physical chemist Michael Polanyi (Polanyi, 1934). (Polanyi (1891–1976) is, perhaps, better known to chemists for developing transtion-state theory with Henry Eyring.) The edge of the extra half plane of atoms is the dislocation line passing through the crystal. This line is perpendicular to the direction of slip, which is in the plane of the page. The second extreme type of dislocation is the screw dislocation (Figure 9.6b), which can be thought of as arising from making a half-cut into the solid and twisting the material on either side of the face of the cut in opposite directions parallel to the line of the cut. Screw dislocations were postulated in 1939 by the Dutch mathematician and physicist Johannes Martinus Burgers (1895–1981), across a slip plane by the movement of a dislocation completely through the crystal. Only dislocations whose Burgers vectors lie in a slip plane are mobile. Slip proceeds in the direction of minimum **b** (e.g., in close-packed directions). Referring to Figure 9.6, the Burgers vector is found by performing a circuit around the dislocation core, one unit of translation (i.e., lattice point) in each direction. The distance needed to close the path gives the magnitude of **b**. Note from Figure 9.6 that, if a circuit encloses a dislocation core, there cannot be an equal number of steps traversed in each direction. If a circuit does *not* enclose a dislocation core, there will be an equal number of steps traversed in each direction (the reader may wish to verify this). For an edge dislocation, **b** is perpendicular to the dislocation line (Figure 9.6b). For a screw dislocation, **b** is parallel to the line of the dislocation (Figure 9.6a). In the case of a mixed dislocation, **b** is at some angle to the dislocation line.

Consider the (111) slip plane in the fcc lattice (Figure 9.5a). The smallest Burgers vector possible is simply equal in magnitude to the distance between nearest neighbors (the close-packed atoms). If the lattice parameter is a, simple trigonometry shows that the distance between these nearest neighbors, which lie along the $[110]$, is equal to $a/(2\cos 45) = a/\sqrt{2}$. Because the atoms are shifted into new sites that are crystallographically equivalent to their original positions as the dislocation moves, this particular dislocation is called a *perfect dislocation*. The smallest Burgers vector available to a perfect dislocation in the fcc lattice can thus be written in vector notation as $\mathbf{b} = \frac{1}{2}a[110]$. Similar considerations show the smallest possible Burgers vectors in the hcp and bcc lattices are $?\,a[2110]$ and $\frac{1}{2}a[111]$, respectively.

Dislocations are present in the natural states of crystalline materials and drastically increase in number (expressed as the dislocation density, or dislocation length per unit volume) with plastic deformation as existing dislocations spawn new ones. A mechanism of dislocation multiplication, involving the successive creation of *dislocation loops*, was proposed by F. C. Frank and W. T. Read (Frank and Read, 1950). It can be understood as follows. As a shear stress is applied to a crystal, portions of dislocation lines in active slip planes bow or curve outward, if their two ends, or other segments, are pinned by impurities or by entanglement with other dislocations (perhaps, pinned by low mobility in nonclosed-packed planes). Once the curvature reaches a critical value, these line segments begin to move outward on its slip plane, as illustrated in Figure 9.7. Eventually, the two curved lines at the top of the figure join to form a closed loop, which is free to move through the crystal.

This process can continually repeat itself from the new dislocation line segments left behind between the original fixed endpoints. Because of the increasing number of dislocation loops, there is a reduction in the strength of the material. However, during their movement through the substance the dislocations eventually begin to pile up at obstacles such as inclusions, grain boundaries, and other dislocations. This begins to have the reverse effect, as the mobile dislocations become immobile, thus strengthening the material. This phenomenon is termed *strain hardening* (also called work hardening).

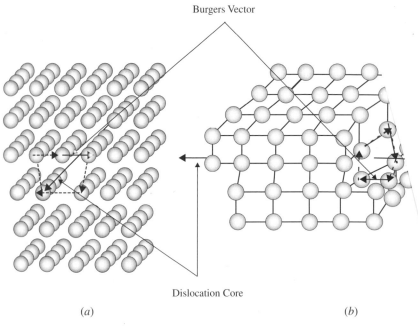

Burgers Vector

Dislocation Core

(a)   (b)

Figure 9.6 The two extreme types of dislocations. (a) In the edge dislocation, the Burgers vector is perpendicular to the dislocation line. (b) In the screw dislocation, the Burgers vector is parallel to the dislocation line.

who later emigrated to the United States (Burgers, 1939). A screw dislocation is parallel to the slip direction. The dislocation line (also called dislocation core) marks the boundary between the slipped and unslipped material.

With dislocations, slip occurs consecutively, a line of atoms at a time. Hence, a smaller shear stress is required to move the dislocation line through the crystal, causing the same atomic displacement that would be obtained with simultaneous motion of all the atoms in the plane. The magnitude of the external force required for moving a dislocation must be greater than the periodic forces exerted on the dislocation by the lattice, the so-called *Peierls stress* (Peierls, 1940). (It is also known as the *Peierls–Nabarro stress*, since Nabarro was the first to attempt an extensive calculation of its magnitude.) In many materials, such as those with the diamond structure, the Peierls stress is believed to be quite large. However, theoretical estimates and experimental measurements of the Peierls stress are elusive, and a discussion of these topics is beyond the scope of this book. The reader is referred to the textbook on elementary dislocation theory by Weertman and Weertman (1992).

A dislocation is characterized by the *Burgers vector*, **b**, which quantifies the atomic disturbance associated with the dislocation, in terms of magnitude and direction. The Burgers vector gives the direction and amount of slip produced

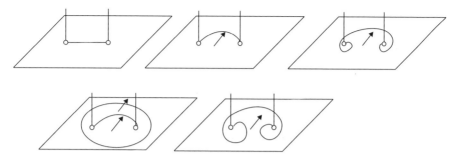

Figure 9.7   The Frank–Read dislocation-loop mechanism of dislocation multiplication. A shear stress causes the portion of a dislocation that is between two pinned segments to bow outward on the slip plane (indicated by arrows). Eventually, the dislocation loop reaches the configuration shown in the middle. When the two curved segments meet, the dislocation loop is freed and a new loop is formed to continue the process.

Egon Orowan (Courtesy of M.I.T. Museum. © Massachusetts Institute of Technology. Reproduced with permission.)

EGON OROWAN

(1902–1989) studied physics, chemistry, mathematics, and astronomy at the University of Vienna. In 1932, he earned his Ph.D. in physics, under Richard Becker, from the Technical University of Berlin. Orowan is most famous for his contributions to the field of crystal plasticity, having postulated, independently of Taylor and Polanyi, the edge-dislocation-based mechanism of plastic deformation in 1934. Orowan was in the Department of Physics at the University of Birmingham from 1937 to 1939, and at Cambridge from 1939 to 1950. From

1950 until his death, he was the George Westinghouse Professor in the Mechanical Engineering Department at M.I.T.. Although Orowan spent much of hs career studying the mechanical behavior of engineering materials, in the 1960s and 1970s he applied the physics of deformation to geological-scale problems, such as continental drift, ocean floor spreading, volcanism, and even the origin of lunar surface features! Orowan also spent time at the Carnegie Institute of Technology and the University of Pittsburgh, where he occupied himself with the evolution of societies and economics. Orowan was elected to the U.S. National Academy of Sciences in 1969.

(*Source*: F. R. N. Nabarro and A. S. Argon, Biographical Memoirs of the U.S. National Academy of Sciences, Vol. 70, **1996**, pp. 261–319.)

Dislocation Reactions   Dislocations can combine or dissociate over their entire lengths, or only partially. Such processes are termed *dislocation reactions.* One reason it is important to study dislocation reactions in ductile materials is that they affect mechanical behavior. For example, the product dislocation(s) of a reaction between two or more parent dislocations can be immobile, or *sessile*, forming an obstacle to the motion of other dislocations. This is the mechanism of strain hardening, first proposed by Taylor (Taylor, 1934). Figure 9.8 shows two separate dislocations (each of mixed character) joining into a single dislocation. The point where the two dislocations join is called a *node*. *Frank's rule* states that the sum of the Burgers vectors for the dislocations whose positive directions point toward the node is equal to the sum of the Burgers vectors for the dislocations whose positive directions point away from the node. This is so because the closure failure for a Burgers circuit enclosing two or more dislocations is equal to the sum of the Burgers vectors for each separate dislocation, regardless of their positions.

As is usually the case with any type of reaction, we can use the criterion of energy minimization to determine whether a dislocation reaction is allowed. The displacement field of a dislocation line represents stored energy, which is proportional to the square of the Burgers vector for the dislocation. The total energy of a

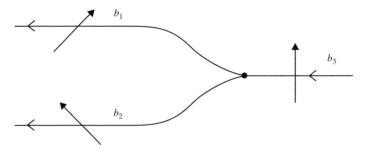

Figure 9.8   Two mixed dislocations can join at a node. The Burgers vector for the combined dislocation is equal to the sum of the Burgers vectors for the uncombined dislocations.

group of dislocations is proportional to the sum of the squares of the Burgers vectors for each dislocation. Hence, the sum of the energies of the resultant Burgers vectors must be less than the sum of the energies of the parent Burgers vectors for a reaction to take place. Each type of crystal lattice has its own permissible Burgers vectors and dislocation reactions, however, that cannot be transferred into different crystal systems.

It may be energetically favorable for a dislocation, $\mathbf{b}_1$, to split into two dislocations if the product dislocation Burgers vectors $\mathbf{b}_2$ and $\mathbf{b}_3$ satisfy the condition: $\mathbf{b}_1^2 > b_2^2 + b_3^2$. Dislocation reactions can even produce stable *imperfect dislocations*, if they result in a greater energy minimization. An imperfect dislocation is one in which the atoms are shifted into new sites, not equivalent to their original positions, as the dislocation moves.

One such imperfect dislocation is a *Frank dislocation*, which is the insertion or removal of a *portion* of a close-packed plane. This produces two edge dislocations of opposite sign, as illustrated in Figure 9.9a. The Frank dislocation also introduces stacking faults (close-packed layer-stacking-sequence error) into the lattice. The Burgers vector of a Frank dislocation is directed normal to the close-packed planes and is equal in length to the spacing between adjacent close-packed planes. Because the Burgers vector is not in the slip plane, these dislocations are stationary and so are sometimes referred to as *Frank sessile dislocations*. As mentioned earlier, sessile dislocations are obstacles to the movement of other dislocations and are responsible for strain hardening.

Example 9.5 Consider the reaction between a [110] edge dislocation on the (111) plane and a [101] edge dislocation on the (111) plane in nickel (c.c.p.). a) Write the dislocation reaction. b) Determine if the reaction is energetically favorable. c) Is the product dislocation mobile or sessile?

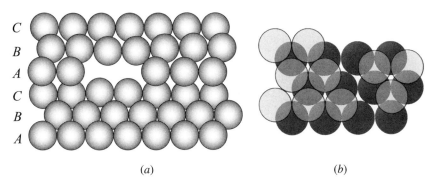

(a)   (b)

Figure 9.9   (a) The removal (or insertion) of a *portion* of a close-packed layer introduces two Frank dislocations of opposite sign and stacking faults, as shown here. (b) The Shockley dislocation is a displacement of a *portion* of atoms in a close-packed layer into adjacent sites of a different type in the same plane.

Solution

(a) The crystal is ccp. Thus, the two parent dislocations may be written as: $\frac{1}{2}$a[110] and $\frac{1}{2}$a[101], where $1/2a\cos 45$ gives the distance between lattice points in the ccp lattice (Figure 9.8).

The parent dislocations are in the same (111) plane. Thus, the product dislocation is also in this plane. Thus the dislocation reaction is:

$$\tfrac{1}{2}a[110] + \tfrac{1}{2}a[101] \rightarrow \tfrac{1}{2}a[011]$$

(b) The sum of the energies of the parent dislocations is proportional to a^2. The energy of the product dislocation is proportional to $a^2/2$. Since there is a net reduction in energy, the reaction is favorable.

(c) A dislocation with a Burgers vector of $\frac{1}{2}$a[011] can slip on any plane containing a [110] direction. This condition is met by the (111) planes. Thus, the product dislocation is favorably oriented for slip. It is mobile.

Example 9.6   Repeat Worked Example 9.5 with a [011] dislocation on the (111) plane and the [101] dislocation on the (111) plane.

Solution   The two parent dislocations are now on different planes. The slip plane of the product dislocation is obtained by taking the cross product of the slip planes of the parent dislocations: $(1)\mathbf{i} + (1)\mathbf{j} + (1)\mathbf{k}$ and $(1)\mathbf{i} + (1)\mathbf{j} + (-1)\mathbf{k}$. This is given by:

$$(a_y b_z - a_z b_y)\mathbf{i} + (a_z b_x - a_x b_z)\mathbf{j} + (a_x b_y - a_y b_x)\mathbf{k}$$

where $\mathbf{a} = (1)\mathbf{i} + (1)\mathbf{j} + (1)\mathbf{k}$ and $\mathbf{b} = (1)\mathbf{i} + (1)\mathbf{j} + (-1)\mathbf{k}$.

Thus, the slip plane of the product dislocation is (001). The dislocation reaction is:

$$\tfrac{1}{2}a[011]_{(111)} + \tfrac{1}{2}a[101]_{(111)} \rightarrow \tfrac{1}{2}a[110]_{(001)}$$

The sum of the energies of the parent dislocations is proportional to a^2. The energy of the product dislocation is proportional to $a^2/2$. Since there is a net reduction in energy, the reaction is favorable.

Since the (001) is not a slip plane, the product dislocation is immobile, or sessile. It provides an obstacle to the movement of other dislocations passing down the (111) and (111) planes. This particular case is known as the *Lomer* lock.

In contrast to the imperfect dislocation discussed in the preceding paragraph, the *Shockley partial dislocation* (Figure 9.9b) is a mobile imperfect dislocation. A Shockley partial dislocation can be considered as a displacement of a portion of the atoms in one close-packed plane into a new set of positions. For example, in the close-packed layer sequence ...*ABCABC*..., a portion of the atoms in, say, the *B* layer, are shifted to *C*-sites.

A very elegant analogy between close-packed metals and two-dimensional bubble-rafts, exhibiting dislocations, was made by L. Bragg and J. F. Nye in 1947 (Bragg and Nye, 1947). Two years later, with the aid of this analogy, Bragg and Lomer (Bragg and Lomer, 1949; Lomer, 1949) showed that plastic deformation proceeds by the motion of dislocations, which gave credence to dislocation theory. The first direct observation of dislocations in solids came with the use of transmission electron microscopy in 1956 (Hirsch et al., 1956). Astonishingly, despite its immense importance to physical metallurgy, no Noble Prizes were ever awarded for the conception or verification of the theories of crystal plasticity. As one observer has commented: " [it seemed] mankind hardly noticed that something big had happened!" Quantum theory understandably took center stage in the physics community from the mid-1920s through the mid-1930s. However, from the mid-1930s through the mid-1950s advances in the theory of plasticity and other fields, regrettably, were overshadowed by the plethora of important discoveries in nuclear and elementary particle physics.

9.3.2 Polycrystalline Metals

In studying the mechanical properties of polycrystalline metals, one must also consider the influence of grain morphology. It is not easy for dislocations to move across grain boundaries because of changes in the direction of slip planes. As discussed earlier, dislocations may actually "pile up" at grain boundaries, after which they become immobile. Polycrystalline materials are therefore stronger (more resistant to deformation) than single crystals. The ease of dislocation motion depends on the relative grain orientation. Dislocations move more readily across small-angle grain boundaries. Reducing the grain size results in more grain boundaries and, hence, more barriers to slip, which further strengthens the metal, while also increasing the ductility.

The empirical Hall–Petch equation (Hall, 1951; Petch, 1953) is well known to express the grain-size dependence to the yield strength:

$$\sigma = \sigma_i + k/d^{-n} \tag{9.27}$$

in which σ_i = friction stress to move individual dislocations, d = average grain diameter, k = strengthening coefficient (a constant), σ = yield strength, and n is normally 1/2, but 1/3 and 1/4 have been reported for some materials.

Equation 9.27 was developed based on the dislocation pileup phenomenon in coarse-grained materials (Section 9.3.1); it is applicable down to a grain size of about 10–30 nm. Materials with grains smaller than this cannot support dislocation activity and may thus deviate from classic behavior (Section 11.2.5). Nanocrystalline materials can plastically deform, but the mechanism is controversial. Some data indicate *inverse* Hall–Petch behavior, in which the material softens once the decrease in grain size reaches a critical limit (10–20 nm), while other results imply that the increase in the total grain boundary volume simply produces a continued increase in yield stress with decreasing grain size. In the latter circumstance, it should be possible to extrapolate the properties from those of the coarse-grained samples. Still, some experiments indicate that the yield stress is independent of grain size in this size regime (Lu and Liaw, 2001).

Other strengthening mechanisms include solid-solution formation and strain hardening. *Solid solution strengthening* involves replacing a small number of atoms in the lattice with substitutional impurities of a slightly different size. This creates strain in the crystal, inhibiting the movement of dislocations and strengthening the material. Note that solid solubility is not reciprocal. Thus, silver will dissolve up to 5% bismuth, but the solubility of silver in bismuth is negligible. Cold working such as rolling, drawing, or extruding at room temperature strain hardens a material by producing elongation of the grains in the principal direction of working. The dislocation density increases with the percent area reduction. This allows dislocations to become entangled and pinned, thus strengthening the material. Unlike decreases in grain size, both strain hardening and solid-solution strengthening lower ductility. Annealing a strain-hardened sample can restore the loss in ductility through a process of recrystallization and grain growth, in which dislocation pileups are removed via, for example, grain boundary migration.

The detrimental effects of cold working can be avoided altogether by hot working. Hot working is the shaping or deformation of a metal just above its recrystallization temperature ($\sim 0.4 T_M$, where T_M is the melting temperature). The material undergoes a grain refinement and there is no strain hardening or loss of ductility in the finished product. However, the surfaces of most metals will oxidize more readily at elevated temperatures, which may be undesirable in some situations.

Another approach to simultaneously increasing the strength and ductility of a metal is by alloying it with a more ductile and stronger metal that does not form a solid solution. It is also necessary to prevent formation of intermetallic phases, which would result in a stronger but more brittle alloy. For example, a hypoeutectic alloy of bismuth and silver will contain large particles of silver, the primary constituent, surrounded by fine bismuth–silver eutectic structure. There are no intermetallic phases in this system and the presence of the silver phase both strengthens the alloy (but not by the solid-solution mechanism) and renders it more ductile than pure bismuth. For example, with a strain rate of 10% per minute, the ultimate tensile strength (UTS) of Bi-11 wt % Ag is 59 MPa (Lalena et al., 2002), while the UTS for pure polycrystalline bismuth is about 25 MPa.

An interesting mechanical behavior exhibited by some fine-grained materials being utilized in many industries today is *superplasticity*, in which elongations > 100% are attained. Superplasticity is important because it allows engineers to fabricate complex shapes out of a material, which might otherwise be unobtainable. It has been observed in metals, intermetallics, and ceramics with grain sizes less than 15 μm. In metals, the mechanism of superplasticty is believed to be grain boundary sliding accompanied by dislocation slip. The phenomenon was first observed in Wootz steel by Hadfield in 1912 (Hadfield, 1912).

9.3.3 Brittle and Semibrittle Solids

The mechanical properties of intermetallic compounds, glasses, and ceramics differ greatly from those of ductile metals. Almost all of these materials are brittle, strong, and hard. Some are semibrittle, exhibiting a very limited plasticity before the onset of fracture. For example, alkali halides (rock-salt structure) plastically deform slightly along the $\{110\}, \langle 110 \rangle$ slip system at room temperature. Other semibrittle solids include glasses, hcp metals, and most bcc metals. The strength and low ductility of brittle and semibrittle substances are due to the presence of ionic and covalent bonding and/or to the lack of a sufficient number of independent slip systems (close-packed layers). Dislocations are present but essentially immobile except at very high temperatures, typically in excess of 1000°C.

In ionic solids, slip is constrained because it requires bringing ions with like charges in contact. Slip systems do not tend to be the "easy-glide" close-packed planes and directions, but rather those in which oppositely charged ions remain close throughout the slip process. In highly ionic cubic systems, the dominant slip system is $\{110\}, \langle 110 \rangle$, but in less ionic cubic crystals, $\{100\}$ or $\{111\}$ planes may dominate (Nabarro, 1967). In covalent solids, there exists strong directional bonding. Although the bonding electrons may be in extended wave functions (Chapter 3), there are definite preferences for certain geometrical configurations of the atoms. Even the *closest*-packed planes tend to be of low atomic density, which impedes dislocation motion. Although many ionic and covalent solids are "almost" close packed (recall the definition of the term *eutactic*), slip appears to be difficult on anything but truly close-packed planes.

This is an important point. A sublattice phase with the face-centered cubic structure should not, generally speaking, be considered cubic closed packed with regard to slip. The atoms or ions on one sublattice may very well be in a ccp-like arrangement, but they can be kept apart by large atoms or ions residing on the other sublattice (the interstitial sites). Slip is easiest along truly close-packed layers of identically sized spheres that are in contact and, preferably, without obstacles such as interstitials. Thus, another reason for low ductility in intermetallics and ceramics is the lack of a sufficient number of *active* slip systems to allow plastic deformation.

Lately, however, some surprising exceptions have been found to the general rule of low plasticity in ceramics. One is the perovskite oxide strontium titanate, $SrTiO_3$. Recent studies on single crystals have revealed a transition from nonductile to ductile behavior in this material, not only at temperatures above 1000°C, but again below 600°C. Even more unexpectedly, it reached strains of 7% at room temperature with flow stresses comparable to those of copper and aluminum alloys. At both the high and low temperatures, the plasticity appears to be due to a dislocation-based mechanism (Gumbsch et al., 2001).

Normally, dislocation-based plastic deformation is irreversible, that is, it is not possible to return the material to its original microstructural state. Remarkably, fully reversible dislocation-based compressive deformation was recently observed at room temperature in the layered ternary carbide Ti_3SiC_2 (Barsoum and El-Raghy, 1996). This compound has a hexagonal structure with a large c/a ratio and it is believed that the dominant deformation mechanism involves dislocation movement in the basal plane.

It is sometimes possible for plastic flow to proceed in nonmetals by modes other than pure dislocation-based mechanisms. For example, at temperatures of about 40–50% of their melting points, grain boundary sliding can become important. Grain boundary sliding accompanied by cation lattice diffusion is believed to be the mechanism of the superplasticty observed in some fine-grained polycrystalline ceramics. Although superplasticity has been known to exist in metals for over 90 years, it has only recently been demonstrated in ceramics, such as fine-grained yttria-stabilized zirconia (Wakai et al., 1986).

In general, however, the low plasticity of intermetallics, glasses, and ceramics hinders their use in many engineering applications and impedes their fabrication by deformation processing. Consequently, materials are typically formed by powder processing (metals and polymers can also be processed from powders). These are multistep manufacturing procedures, which first involve consolidation, or packing, of the particulate to form a "green body." The two basic methods of consolidating powders into a desired shape are: dry-pressing, in which dry powder is compacted in a die; and slip-casting (or filter pressing), where the particles are suspended in a liquid and then filtered against the walls of a porous mold. Either consolidation step is followed by densification (sintering) to improve homogeneity and to reduce moisture content and the number of pores.

We note in passing that the mechanical properties of all polycrystalline materials (including metals), such as strength and toughness, are modified by reinforcement with macroscopic quantities of insoluble components, thereby forming *composites*. The multiphase nature of a composite is clearly discernible in the microstructure. Composites constitute a very important field of materials science, but one outside the theme of this textbook, which, for the most part, is concerned with the properties of single-phase materials.

9.4 FRACTURE

Unlike slip, which relies on chemical-bond breaking *and* reformation, fracture occurs when bonds rupture without reformation. Fracture may result from the application of a single static (sustained) load in excess of the ultimate yield strength or, over time, from cyclic loads of smaller magnitude. The latter phenomenon is called *fatigue damage* and it is due to the build up of substructural or microstructural changes induced by the cyclic straining. Even the highest strength materials, however, fail at a fraction of the stress levels required to break an individual chemical bond. Such behavior is linked to the presence of microscopic cracks that nucleate at defect sites, called *Griffith flaws*, which are virtually always present due to materials processing. This was postulated by British aeronautical engineer Alan Arnold Griffith (1893–1963) in 1921 (Griffith, 1921). At the surface, Griffith flaws may result from impingement of hard dust particles. In the bulk of the material, crack-nucleation sites can be inclusions, grain boundaries, and other defects. Zener proposed that, in ductile and semibrittle materials, crack nucleation involves dislocation pileup at the Griffith flaws, as well as at grain boundaries (Zener, 1948b). This was subsequently substantiated by a number of experiments. Cottrell later described a mechanism for crack nucleation involving intersecting slip planes even in the absence of defects (Cottrell, 1958).

An applied deformation stress concentrates near the tip of a sharp crack, where it can eventually exceed the cohesive strength (the chemical bonding forces between atoms) of the material and cause crack propagation, or fracture. Any material can fracture, but the mechanisms for crack propagation are different in ductile and brittle solids. Modern phenomenological theories of brittle fracture are founded on the seminal 1921 paper by Griffith. The total energy, U, of a crack system with crack area, A, is given by the sum of the potential energy, W_P, and surface energy, W_S. Griffith postulated that the critical condition for crack growth is when $dU/dA = 0$, or when the increase in surface energy due to crack extension is just balanced by an equal decrease in the total potential energy of the system:

$$dW_P/dA = -dW_S/dA \qquad (9.28)$$

The energy balance considerations in Griffith's original concept were later refined by Orowan and Irwin to include the effects of plasticity and elasticity for applicability to metals (Orowan, 1952; Irwin, 1957). Metals fail by ductile fracture, where the crack growth occurs in the direction of the primary slip system. When the slip plane is inclined to the crack, atoms across the slip plane slide past one another, relieving the stress, which results in a zigzag crack path. This is illustrated in Figure 9.10.

Excellent coverage of fracture mechanics in ductile and brittle solids can be found in many textbooks, such as Suresh's (Suresh, 1998). A relatively new approach, aimed at understanding the fundamental causes of fracture, focuses on the role of the topology of the chemical-bond charge density (Eberhart, 1993, 1999a, 2001a, 2001b). We have seen that bond breaking is integral to fracture. Yet, the traditional picture of chemical bonding does not really allow one to define fracture or to predict, a priori, whether a solid will be ductile or brittle. This is

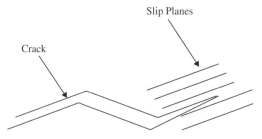

Figure 9.10 The single slip mechanism, in which crack growth occurs in the direction of the primary slip systems, results in a zigzag crack path.

because, as two atoms are pulled apart, the electron density never really goes to zero, but merely begins to flatten out. Thus, it is not possible to unambiguously determine the point at which the bond should be considered broken. By contrast, in the topological view the bond can be considered broken at a well-defined point— when the charge-density *curvature* vanishes. This happens at a special type of relative extremum known as a *saddle point*.

Before we define what a saddle point is, we first have to describe a critical point. A critical point for a function of three variables is defined as a point where all three partial derivatives are equal to zero: $\partial/\partial x = \partial/\partial y = \partial/\partial z = 0$. A critical point can be a local minimum or local maximum. The latter type corresponds to the position of an atom in a molecule or solid. This is the position of maximum electron density, from which the density falls off in all three perpendicular directions of space (Bader, 1990). A saddle point for a function of three variables, by contrast, is a point at which the function is a minimum along two directions and a maximum along the third or, alternatively, a maximum along two directions and a minimum along the third. Again, the latter has special significance for the chemical bonding in an inorganic solid, for that type of saddle point in the electron density between two atoms is the *bond critical point* (Bader, 1990). A saddle point for a function of two variables is illustrated in Figure 9.11.

The advantage of taking this view of the chemical bond, it is hoped, is that it will enable us to predict how alloying can improve fracture resistance and/or ductility. Through quantum mechanical calculations we can model how substituting different elements into a solid transforms the charge-density topology of the chemical bonds and, hence, the failure properties of the material. Eberhart has shown how accounting for contributions from both first and second nearest-neighbor interactions in the bond charge-density topology of a series of transition-metal alumnides with the CsCl structure allows us to do just that. The percentage of the total bond energy of $MA1$ ($M =$ Cr, Mn, Fe, Ni, Co) due to the second nearest-neighbor interactions (of the $dd\pi$ and $dd\delta$ type in the bcc and CsCl lattices) was found to increase in the order CrAl < MnAl < FeAl < NiAl < CoAl. This is identical with

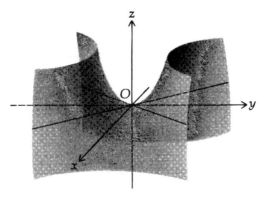

Figure 9.11 A saddle point for a function of two variables.

the trends in the unit cell volume and failure properties, CoAl possessing the smallest unit cell volume and being the most brittle (Eberhart, 2001a). Presumably, this is because $dd\pi$ and $dd\delta$ overlap is less than that due to $dd\sigma$ bonding, which is a component of the first nearest-neighbor interactions. Assuming additivity (the rule of mixtures) applies (Section 9.2.3), replacing a portion of, say, nickel atoms in NiAl with iron atoms would be predicted to improve the failure properties of NiAl. It has been verified experimentally that 10% substitution of nickel by iron does indeed increase the ductility of NiAl.

REFERENCES

Bader, R. F. W. 1990, *Atoms in Molecules—A Quantum Theory*, Oxford University Press, Oxford.

Barsoum, M. W.; El-Raghy, T. 1996, *J. Ceram. Soc.*, *79*, 1953.

Bragg, L.; Nye, J. F. 1947, *Proc. R. Soc. (London)*, *A190*, 474.

Bragg, L.; Lomer, W. M. 1949, *Proc. R. Soc. (London)*, *A196*, 171.

Burgers, J. M. 1939a, *Proc. Kon. Ned. Akad. Wet.*, *42*, 293.

Burgers, J. M. 1939b, *Proc. Kon. Ned. Akad. Wet.*, *42*, 315.

Burgers, J. M. 1939c, *Proc. Kon. Ned. Akad. Wet.*, *42*, 378.

Car, R.; Parrinello, M. 1985, *Phys. Rev. Lett.*, *55*, 2471.

Cleri, F.; Rosato, V. 1993, *Phys. Rev. B*, *48*, 22.

Cottrell, A. H. 1958, *Trans. Met. Soc.*, 192.

Daw, M. S.; Baskes, M. I. 1983, *Phys. Rev. Lett.* *50*, 1285.

Daw, M. S.; Baskes, M. I. 1984, *Phys. Rev. Rev. B*, *29*, 6443.

Eberhart, M. E. 1993, *Philos. Mag. B*, *68*, 455.

Eberhart, M. E. 1999a, *MRS Symp. Proc.*, *539*, 13.

Eberhart, M. E. 1999b, *Sci. Am.*, *66*, 281.

Eberhart, M. E. 2001a, *Philos. Mag. B*, *81*, 721.

Eberhart, M. E. 2001b, *Phys. Rev. Lett.*, *87*, 205503.

Frank, F. C.; Read, W. T. 1950, *Phys. Rev.*, *79*, 722.

Griffith, A. A. 1921, *Philos. Trans. R. Soc. London*, *A221*, 163.

Gumbsch, P.; Taeri-Baghbadrani, S.; Brunner, D.; Sigle, W.; Rühle, M. 2001, *Phys. Rev. Lett.*, *87*, 85505.

Hadfield, R. 1912, *J.Iron Steel Inst.*, *85*, 134–174.

Hall, E. O. 1951, *Proc. Phys. Soc., London, Sect. B*, *64*, 747.

Hernandez, E.; Goze, C.; Bernier, P.; Rubio, A. 1998, *Phys. Rev. Lett.*, *80*, 4502.

Hill, R. 1952, *Proc. Phys. Soc.*, *A65*, 349.

Hirsch, P. B.; Horne, R. W.; Whelan, M. J. 1956, *Philos. Mag.*, *1*, 677.

Irwin, G. R. 1957, *J. Appl. Mech.*, *24*, 361.

Johnson, R. A. 1988a, *Phys. Rev. B*, *37*, 3924.

Johnson, R. A. 1988b, *Phys. Rev. B*, *37*, 6121.

Johnson, R. A. 1989, *Phys. Rev. B*, *39*, 12554.

Lalena, J. N.; Dean, N. F.; Weiser, M. W. 2002, *J. Electron. Mater.*, *31*, 1244.

Lennard-Jones, J. E. 1931, *Proc. Cambridge Philos. Soc.*, *27*, 469.

Lomer, W. M. 1949, *Proc. R. Soc. (London)*, *A196*, 182.

Lourie, O.; Wagner, H. D. 1998, *J. Mater. Res.*, *13*, 2418.

Lu, Y.; Liaw, P. K. 2001, *JOM*, *53*, 31.

Morse, P. M. 1930, *Phys. Rev.*, *34*, 57.

Nabarro, F. R. N. 1967, *Theory of Crystal Dislocations*, Clarendon Press, Oxford.

Nye, J. F. 1957, *Physical Properties of Crystals their Representation by Tensors and Matrices*, Oxford University Press, London.

Oh, D.; Johnson, R. A. 1988, *J. Mater. Res.*, *3*, 471.

Oh, D.; Johnson, R. A. 1989, *J. Mater. Res.*, *4*, 1195.

Orowan, E. 1934, *Z. Physik. 89*, 605, 614, 634.

Orowan, E. *Fatigue and Fracture of Metals*, ed. W. M. Murray, New York, Wiley, 1952.

Overney, G.; Zhong, W.; Tomenek, D. 1993, *Z. Phys. D*, *27*, 93.

Pasianot, R.; Savino, E. 1992, *Phys. Rev. B*, *45*, 12704.

Petch, N. J. 1953, *J. Iron Steel Inst. London*, *74*, 25.

Peierls, R. 1940, *Proc. Phys. Soc. (London)*, *52*, 34.

Polyani, M. 1934, *Z. Physik. 89*, 660.

Ravindran, P.; Fast, L.; Korzhavyi, P. A.; Johansson, B.; Wills, J.; Eriksson, O. 1998, *J. Appl. Phys.*, *84*, 4891.

Reuss, A. 1929, *Z. Angew. Math. Mech.*, *9*, 49.

Schmid, E.; Boas, W. 1935, *Kristallplastizität* (English edition). Springer-Verlag, Berlin. *Plasticity of Crystals*, F. A. Hughes & Co. Limited, 1950.

Stillinger, F. H.; Weber, T. A. 1985, *Phys. Rev.*, *B*, *31*, 5262.

Suresh, S. 1998, *Fatigue of Materials*, 2nd ed., Cambridge University Press, Cambridge, UK.

Taylor, G. I. 1934, *Proc. R. Soc.*, *A145*, 362, 388.

Voigt, W. 1910, *Lehrbuch der Kristallphysik*, Verl. von B. G. Teubner, Leipzig, Germany (reprinted in 1928).

Von Mises, R. V. 1928, *Z. Angew. Math. Mech.*, *8*, 161.

Wakai, F.; Sakaguchi, S.; Matsuno, Y. 1986, *Adv. Ceram. Mater.*, *1*, 259.

Weertman, J.; Weertman, J. R. 1992, *Elementary Dislocation Theory*, Oxford University Press, New York.

Xin, Z.; Jianjun, Z.; Zhong-can, O.-Y. 2000, *Phys. Rev. B*, *62*, 13692.

Yifang, O.; Bangwei, Z.; Shuzhi, L.; Zhanpeng, J. 1996, *Z. Phys. B*, *101*, 161.

Zener, C. 1948a, *Elasticity and Anelasticity of Metals*, University of Chicago Press, Chicago, IL.

Zener, C. 1948b, *Fracturing of Metals*, American Society for Metals, Cleveland, OH.

CHAPTER 10

Phase Equilibria, Phase Diagrams, and Phase Modeling

Phase diagrams, phase equilibria data in graphical form, are a standard tool of materials scientists and engineers. Such compilations of data can be indispensable when, for example, there are questions about the thermodynamic stability of a phase under a given set of working conditions, or in a particular operating environment (e.g., materials compatibility issues). The easiest way for a metallurgical engineer to determine whether an aluminum or tungsten vessel would be the best choice as a container for molten zinc is to consult the Al–Zn and W–Zn phase diagrams. In this particular case, the phase diagrams would show that Al–Zn alloy formation would be expected to occur since zinc exhibits an increasing solid solubility in aluminum with temperature, whereas tungsten is much more resistant to corrosion by molten zinc.

The solid-state chemist interested in preparing new materials also finds phase diagrams valuable. Strictly speaking, phase diagrams describe the phase relationships within single-component or multicomponent systems in *stable thermodynamic equilibrium*. Hence, the phase diagram tells us if a given phase in that system will be accessible under those equilibrium conditions (although metastable phases certainly may be obtainable under appropriate circumstances). Even when a stable phase is accessible, control of the stoichiometry may be difficult to achieve due to complex phase equilibria. For example, many transition metals have a propensity to adjust their oxidation state depending on the temperature and oxygen partial pressure, which can result in a multiphase product (impurities) and/or one with mixed valency.

It is imperative that one be able to properly interpret phase diagrams. In the first part of this chapter, we review some underlying concepts from thermodynamics necessary for understanding phase equilibria. The interpretation of phase diagrams is subsequently taken up. Afterwards, the reader is introduced to the calculation of phase diagrams (CALPHAD) method, in which phase equilibria predictions are made on high-order systems by extrapolation of thermochemical data from the

Principles of Inorganic Materials Design   By John N. Lalena and David A. Cleary
ISBN 0-471-43418-3   Copyright © 2005 John Wiley & Sons, Inc.

lower-order parent systems. For example, a phase diagram for a ternary system *ABC* can be computed from the *AB*, *BC*, and *AC* binary systems. Such calculations are potentially very reliable, depending on the complexity of the system. In the absence of existing experimental data on a high-order system, mathematical modeling is the logical approach for obtaining phase-equilibria information when laboratory work is not feasible or is cost/time prohibitive. The CALPHAD method was pioneered by Hillert, Kaufman, and others over 40 years ago and is still a rapidly growing field.

10.1 THERMODYNAMIC SYSTEMS, PHASES, AND COMPONENTS

Thermodynamics is the branch of physics that enables us to study energy changes accompanying phase transformations and, in general, the equilibrium properties of material systems. The architect of modern equilibrium thermodynamics was the American mathematical physicist Josiah Willard Gibbs (1839–1903). At this point, a few basic definitions are in order. In thermodynamics, the *system* is defined as the macroscopic segment of the world, such as a substance or group of substances, under investigation. Outside the system are the *surroundings*. Taken together, the two constitute the universe. *Open systems* are those that can exchange heat and material with their surroundings. If only heat can be exchanged, the system is *closed*. Systems that can exchange neither heat nor material with their surroundings are *isolated*.

Systems are classified, and named, by the number of *components*, as follows:

Number of components	Type of System
One	Unary
Two	Binary
Three	Ternary
Four	Quaternary
Five	Quinary

The number of components is the number of independent constituents needed to fix the chemical composition of every phase in the system. In metallic systems, the components are usually elements. For systems with covalent and/or ionic bonds, the components may be compounds or stable molecular species.

The definition of a *phase* is more complicated. A phase is often described simply as a homogeneous state of matter. However, this definition is somewhat vague because it gives no indication of the length scale or the degree of homogeneity required. A less ambiguous definition is that a phase is a liquid, gaseous, or solid substance with a physical structure that is, on average, microscopically homogeneous and, as such, has uniform thermodynamic properties (to be described later). The structure of a phase may have long-range 3D translational atomic order (e.g., a crystalline solid) or not (e.g., a liquid or an amorphous solid). Three-dimensional translational order is an important distinction. Some liquid crystals, for example, exhibit 1D translational order.

Two distinct phases may have the same chemical composition (e.g., diamond and graphite, or rutile and anatase), and two substances comprised of the same components, but with different chemical composition, may constitute the same phase (e.g., solid solutions, nonstoichiometric compounds, and intermediate phases). A *polymorphic* substance, such as SiO_2, is one that undergoes a crystal structure, or solid-state phase transformation with changes in temperature or pressure, while maintaining the same chemical composition. The term *allotropic* is used to describe elements with this property.

The *state* or condition of a system is given by a collection of experimentally measurable thermodynamic properties called *state variables* (also known as state functions or state properties), which are independent of the history of the system. State variables can be classified into two categories. *Intensive variables* are those that are independent of the quantity of material present, examples being temperature, T, and pressure, P. By contrast, the values of *extensive variables*, such as volume, V, and internal energy, U, are proportional to the amount of material present. The ratio of two extensive properties is an intensive property. For example, the density of a substance, which is the ratio of its mass to volume, is the same regardless of the size of the sample.

The relationship between the different state variables of a system subjected to no external forces other than a constant hydrostatic pressure can generally be described by an *equation of state* (EOS). For example, in physical chemistry several semiempirical equations ("gas laws") have been derived describing how the density of a gas changes with pressure and temperature. Such equations contain empirical constants characteristic of the particular gas. The density of a solid also changes with temperature or pressure, although to a considerably lesser extent than a gas does. Equations of state describing the pressure–volume–temperature (PVT) behavior of a homogeneous solid utilize thermophysical parameters analogous to the constants used in the various gas laws, such as the *bulk modulus, B* (the inverse of compressibility), and the *volume coefficient of thermal expansion*, β.

The equilibrium state is reached when there is no change with time in any of the system's macroscopic properties. The *phase rule* by Gibbs gives the general conditions for equilibrium between phases in a system. It is assumed that the equilibrium is only influenced by temperature and pressure, that is, surface, magnetic, electrical, and magnetic forces are neglected. In this case, the phase rule can be written as:

$$f = c - p + 2 \qquad (10.1)$$

where f, which must equal zero or a positive integer, gives the *degrees of freedom* (number of independent variables); c is the number of components; p is the number

TABLE 10.1   Invariant Reactions (on cooling)

Monotectic	$L_1 + S \rightarrow L_2$
Eutectic	$L \rightarrow S_1 + S_2$
Metatectic	$S_1 \rightarrow L + S_2$
Monotectoid	$S_1 + S_2 \rightarrow S_2 + S_3$
Euctectoid	$S_1 \rightarrow S_2 + S_3$
Syntectic	$L_1 + L_2 \rightarrow S$
Peritectic	$L + S_1 \rightarrow S_2$
Peritectoid	$S_1 + S_2 \rightarrow S_3$

Abbreviations: $S =$ solid; $L =$ liquid.

of phases in equilibrium; and the factor 2 is for the two variables, temperature and pressure. If the effect of pressure is ignored in condensed systems with negligible vapor pressures, the factor 2 in Eq. 10.1 is replaced by 1, giving the so-called *condensed-phase rule*.

The phase rule(s) can be used to distinguish different kinds of equilibria based on the number of degrees of freedom. For example, in a unary system, an *invariant equilibrium* $(f = 0)$ exists between the liquid, solid, and vapor phases at the triple point, where there can be no changes to temperature or pressure, without reducing the number of phases in equilibrium. Because f must equal zero or a positive integer, the condensed-phase rule $(f = c - p + 1)$ limits the possible number of phases that can coexist in equilibrium within one-component condensed systems to one or two, which means that, other than melting, only allotropic phase transformations are possible.

Similarly, in two-component condensed systems, the condensed-phase rule restricts the maximum number of phases that can coexist to three, which also corresponds to an invariant equilibrium. However, several *invariant reactions* are possible (Table 10.1), each of which maintains the number of equilibrium phases at three, and keeps f equal to zero. The same terms given in Table 10.1 are also applied to the structures of the phase mixtures.

There are other types of equilibria, in addition to the invariant type. For example, from the condensed-phase rule, it can be seen that when n phases of an n-component system are in equilibrium, it is possible to change the value of one variable (temperature *or* composition) without changing the number of phases in equilibrium. This is called *univariant equilibrium* $(f = 1)$. Where a single phase exists, there are two degrees of freedom $(f = 2)$. Both temperature and composition can be changed without changing the kind of equilibrium. This is termed *bivariant equilibrium*.

10.2   THE FIRST AND SECOND LAWS OF THERMODYNAMICS

The term 'thermodynamics' was coined in 1849 by James Thomson (1822–1892). However, basic notions of heat and energy were established in the 1600s. Since that time, a large body of experimental evidence has accumulated supporting the soundness of the basic axioms, or postulates, upon which classic thermodynamics is built. The *first law of thermodynamics* is a corollary of the *law of conservation of energy*, and states that energy (the ability to do work) can neither be created nor destroyed in a thermodynamic system of constant mass, although it may be converted from one form to another. If *external energy*, q, is supplied to a system, it must absorb the energy by increasing its *internal energy*, U (the sum of all the kinetic energies and energies of interactions of the particles in the system, or by doing *work*, W (the effect on the surroundings as a result of changes made to the system). Mathematically, this is written in differential form, as:

$$dU = dq - dW \tag{10.2}$$

For the pressure-volume work done by a gas expanding *reversibly* (through a succession of equilibrium states):

$$dW = PdV \tag{10.3}$$

All of the quantities introduced thus far, U, W, q, P, and V, are thermodynamic state variables.

Another state variable, the entropy, S, is defined by the *second law of thermodynamics*:

$$dS = dq/T \tag{10.4}$$

Combining Eqs. 10.2, 10.3, and 10.4 gives:

$$dU = TdS - PdV \tag{10.5}$$

The *second law of thermodynamics* states that the entropy of an isolated system (a system with constant U and constant V) increases $(dS > 0)$ for a spontaneous processes or stays constant $(dS = 0)$ at equilibrium.

In Eq. 10.5, U is called a *characteristic* state function of two independent extensive variables, S and V, which, in turn, are regarded as the *natural variables* for U. The other variables in Eq. 10.5, T and P, are a special kind of intensive variable that must have the same value at all points in the system at equilibrium. They are called *potentials*. The pair of one potential and one extensive variables in each product on the right-hand side of Eq. 10.5 (i.e., T, S, and $-P, V$) are called *conjugate pairs*.

For systems under constant pressure, it is convenient to define another state variable called the *enthalpy*, or heat content, H:

$$H = U + PV \tag{10.6a}$$

$$dH = d(U + PV) = TdS + VdP \tag{10.6b}$$

From Eq. 10.6b, it is seen that H is a characteristic function with natural variables S and P. Additional relations can be derived from the first and second laws when other experimental conditions are more easily controlled. For example, for a system of constant composition, the Helmholtz free energy, F, has natural variables T and V, and the Gibbs free energy, G, has natural variables T and P, as shown by Eqs. 10.7b and 10.8b, as follows:

$$F = U - TS \tag{10.7a}$$

$$dF = dU - TdS - SdT = -SdT - PdV \tag{10.7b}$$

$$G = H - TS = (U + PV) - TS \tag{10.8a}$$

$$dG = dU - TdS - SdT + PdV + VdP = -SdT + VdP \tag{10.8b}$$

The second law requires that, for a spontaneous change at constant temperature and volume, $dF \leq 0$. Alternatively, under conditions of constant temperature and pressure, $dG \leq 0$. That is, the total free energy of the system decreases spontaneously at constants T and P until it reaches a minimum; at equilibrium, $dG = 0$. The minimization of the G_{total} is one of two criteria defining an equilibrium state.

The Gibbs energy is also an extensive property, that is, it is dependent on the total number of moles, n, of each component, i, present in the system, as well as on the natural variables given in Eq 10.8b. Mathematically, this can be represented by the slope formula for partial derivatives:

$$dG = (\partial G/\partial T)_{P,\,n_i}dT + (\partial G/\partial P)_{T,\,n_i}dP + \sum_i (\partial G/\partial n_i)_{P,\,T,\,n_j}dn_i \tag{10.9}$$

In Eq. 10.9, the first slope, $(\partial G/\partial T)_{P_n i}$, is equal to $-S$, the second slope, $(\partial G/\partial P)_{T,\,n_i}$ is equal to $+V$, and the third slope, $(\partial G/\partial n_i)_{P,\,T,\,n_j}$, defines the *chemical potential* for the ith component, μ_i. The chemical potential is the partial molar Gibbs energy and, being the ratio of two extensive properties, is itself an intensive property. This variable can be included in Eq. 10.8, giving:

$$dG = -SdT + VdP + \sum_i \mu_i dn_i \tag{10.10}$$

The chemical potential gives us another way of looking for phase equilibria: at equilibrium, the value of the chemical potential for each component is the same in every phase at constant T and P. For multiphase equilibria, an equation of the form of Eq. 10.10 holds for each phase.

The second law ensures us of a thermodynamic driving force in the direction of equilibrium. The equilibrium state is finally attained when there is no change with time in any of the system's macroscopic properties. Observing changes, unfortunately, is not a useful test to determine whether a system has reached equilibrium, because there is no standard length scale for time. From a practical standpoint, therefore, three possible types of states can be envisioned: stable, metastable, or unstable.

Stable, or true equilibrium, is the state of lowest free energy. A system is in a *metastable* state (sometimes called metastable equilibrium) when additional energy must be supplied to the system for it to reach true stability. A system is in an *unstable* state (sometimes called unstable equilibrium) when no additional energy need be supplied to the system for it to reach metastability or stability. The latter two states represent situations where the system is not in its lowest energy state or in which chemical potential differences exist within the system. Nonequilibrium condensed phases with extremely long lifetimes under ambient conditions are not uncommon, the archetypal example being diamond. Their existence is simply a manifestation of the extremely slow (normally diffusion-limited) reaction kinetics and phase transformations typical of solids.

10.3 UNDERSTANDING PHASE DIAGRAMS

10.3.1 Unary Systems

Unary phase diagrams are 2D graphs that display the phases of single-component systems (e.g., elements) as a function of both temperature (abscissa) and pressure (ordinate). Since there is only one component, it is not necessary to specify composition. Figure 10.1 shows the phase diagram for sulfur, which exists in two allotropes at 1 atm of pressure, rhombic $(T < 368\ \text{K})$ and monoclinic $(T > 368\ \text{K})$.

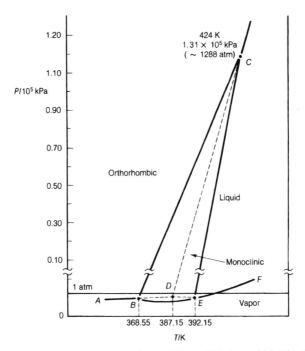

Figure 10.1   Phase diagram for sulfur. (After Laidler and Meiser, 1982, *Physical Chemistry.* Copyright © Benjamin/Cummings Publishing Company. Reproduced with permission.)

10.3.2 Binary Metallurgical Systems

Binary systems would require a 3D graph, since composition, temperature, and pressure are all variable. However, with *condensed* binary systems, the pressure is fixed (normally to 1 atm) and the phase diagram can be reduced to a 2D graph of composition on the abscissa (usually in weight percent or atomic percent) and temperature on the ordinate. The pressure of one atm is usually not the equilibrium value. However, the effect on the behavior of the system is negligible.

The possible types of invariant reactions were illustrated in Table 10.1. These reactions, or their absence, determine the positions and shapes of the areas, known as *phase fields*, in a phase diagram. Three-phase equilibrium is only allowed at a single *point* (an invariant point) in a binary system; that is, three-phase *fields* are not allowed. Binary systems, however, may contain both single-phase and two-phase fields, and when a two-phase field does exist, it must be located between two single-phase fields.

As an example, consider the phase diagram for the Bi–Sb system, shown in Figure 10.2. This diagram is absent of invariant equilibria. A low-temperature single-phase *continuous solid solution* (Bi, Sb) is separated from a high-temperature single-phase liquid solution by a two-phase field (sometimes called the pasty range). The two-phase region is defined by the *solidus* and *liquidus* lines. The solidus shows the temperature at which melting of the solid begins, on heating, or, alternatively, where freezing ends on cooling. The liquidus gives the temperature at which melting finishes (on heating) or freezing begins (on cooling). Thus, for any given alloy composition, the melting-point range is given by the solidus and liquidus lines.

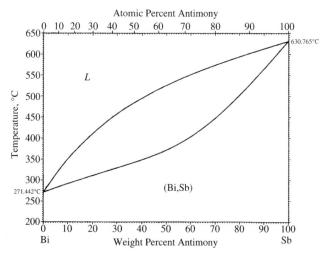

Figure 10.2 Phase diagram for Bi–Sb at one atm pressure. (After Baker, editor, 1992, *ASM Handbook*, Vol. 3, *Alloy Phase Diagrams*. Copyright © ASM International. Reproduced with permission.)

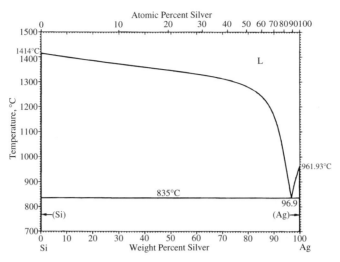

Figure 10.3 Phase diagram for Ag–Si at one atm pressure. (After Baker, editor, 1992, *ASM Handbook, Alloy Phase Diagrams*. Copyright © ASM International. Reproduced with permission.)

Anything less than complete mutual solid solubility between the constituents of a binary alloy system results in a two-phase mixture below the melting point. For example, the Ag–Si system in Figure 10.3, in which there is no solid solubility, shows an invariant equilibrium, the eutectic, occurring at 96.9 wt% silver. An alloy of 96.9 wt % Ag (the eutectic composition) behaves like a pure substance and will melt entirely at a constant temperature of 835°C. Alternatively, on cooling, a liquid with the eutectic composition will precipitate as relatively small *mixed crystals* with the eutectic composition. *Hypoeutectic* compositions consist of a mixture of large crystals of pure Si and small eutectic crystals. *Hyper-eutectic* liquid compositions consist of a mixture of large crystals of pure Ag and small eutectic crystals.

In many cases, there is *partial* solid solubility between the pure components of a binary system, as in the Pb–Sn phase diagram of Figure 10.4. The solubility limits of one component in the other are given by *solvus lines*. Note that the solid solubility limits are not reciprocal. Lead will dissolve up to 18.3% Sn, but Sn will dissolve only up to 2.2% Pb. In Figure 10.4, there are three two-phase fields. Two are bounded by a distinct liquidus line, the solidus line, and the Pb or Sn solid solution boundary. The other two-phase field consists of a mixture of eutectic crystals and, for hypereutectic compositions, crystals of Sn solute dissolved in Pb solvent or, for hypoeutectic compositions, Pb solute dissolved in Sn solvent.

In any two-phase field of a binary phase diagram, an imaginary horizontal (isothermal) line called a *tie line* connects the two points, one on each phase boundary, representing the two phases in equilibrium at the temperature indicated by the line, as shown in Figure 10.5. It is instructive to look at how this originates. A phase diagram may be constructed from a series of Gibbs energy curves for each phase at various temperatures, as a function of composition, as shown for

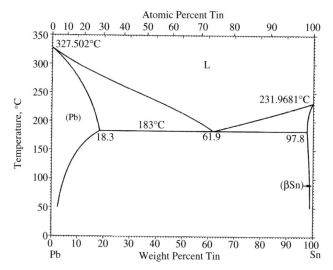

Figure 10.4 Phase diagram for Pb–Sn at one atm pressure. (After Baker, editor, 1992, *ASM Handbook*, Vol. 3, *Alloy Phase Diagrams*. Copyright © ASM International. Reproduced with permission.)

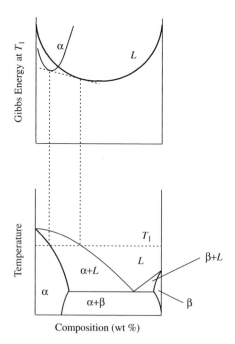

Figure 10.5 A temperature-composition phase diagram (bottom) is generated by a series of Gibbs energy curves for each phase at multiple temperatures. At the top are the Gibbs energy curves for the α solid solution and liquid phases at T_1 (as a function of composition).

temperature T_1 in Figure 10.5. At equilibrium, the chemical potential of each component is the same in every phase, a condition that can only be satisfied with a common tangent line. These two points of tangency are the two points on the phase diagram connected by the tie line.

By employing the *lever rule*, a tie line can be used to determine the fractional amounts of the phases present. For example, a tie line can be drawn in Figure 10.5 below the solidus and between the α and β solvus lines to determine the fractions of those two components in the (α,β) solid solution: f^α and f^β. Since $f^\alpha + f^\beta = 1$, the percentages of the two phases present at any point, x, on the tie line is calculated as follows:

$$\% \ \beta = \frac{\text{Length of line } \alpha x}{\text{Length of line } \alpha \beta} \times 100 \qquad \% \ \alpha = \frac{\text{Length of line } x\beta}{\text{Length of line } \alpha \beta} \times 100 \quad (10.11)$$

The lever rule can also be used in a 2D graph of two molar quantities, for example, molar volume, molar Gibbs energy, to find the average value of those quantities in the system, if f^α and f^β are known. The lever rule only applies, of course, if the molar quantities are plotted with linear scales.

Many alloys contain two-component phases with an ordered crystal structure, as opposed to the random structure of a solid solution. Examples include interstitial solid solutions, line compounds with a very narrow stoichiometric range, or intermediate phases, which exist over wide stoichiometric ranges. Figure 10.6 shows the Ga–La phase diagram. The line compound Ga_2La (mp $= 1450°C$) is a *congruent* phase, meaning that it melts isothermally without undergoing a change in composition. All the other intermetallic phases in this system are *incongruent*;

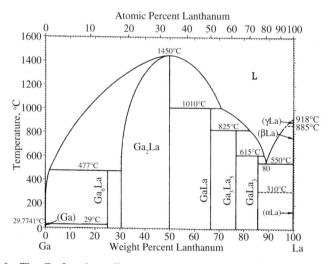

Figure 10.6   The Ga–La phase diagram at one atm pressure showing several line compounds. (After Baker, editor, 1992, *ASM Handbook*, Vol. 3 *Alloy Phase Diagrams*, Copyright © ASM International. Reproduced with permission.)

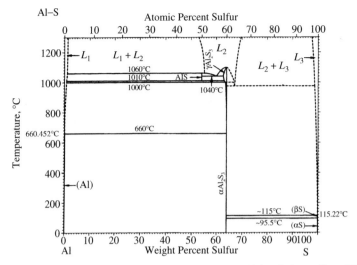

Figure 10.7 The Al–S phase diagram at one atm pressure. (After Baker, editor, 1992, *ASM Handbook*, Vol. 3, *Alloy Phase Diagrams*. Copyright © ASM International. Reproduced with permission.)

that is, two phases are formed from one phase on melting. The phase fields in this system are equilibrium mixtures of *terminal* (end) phases and line compounds, or mixtures of two distinct line compounds. There happens to be no solid solubility between the components in this particular intermetallic system, that is, there are no intermediate phases, but this is not always the case.

10.3.3 Binary Nonmetallic Systems

Consider the Al–S system shown in Figure 10.7. This phase diagram displays several features that are typical of many binary metal chalcogenides. There is only one stoichiometric line compound at room temperature, Al_2S_3. However, at one atm, it may exist as any one of three polymorphs, depending on the temperature. The high-temperature phase, γ-Al_2O_3, melts congruently. There is no solid solubility between any of the phases in this system. Hence, one two-phase field exists to the left of Al_2S_3 (at $T < 660°C$) and another to its right (at $T < 95.5°C$). Another interesting feature are the two dome-shaped regions at temperatures above 1000°C. These are regions of liquid immiscibility, and the lines representing the stability limits are known as *spinodals*, as they fall on a sharp point in property diagrams with potential axes (Hillert, 1998). Within these two-phase fields, there exists liquid aluminum and liquid Al_2O_3 (on the left) and liquid sulfur and liquid Al_2O_3 (on the right). Above the domes, at higher temperatures, the liquids become miscible.

Another nonmetallic binary system is shown in Figure 10.8. Although it contains three elements, the system has just two independently variable constituents, CaO

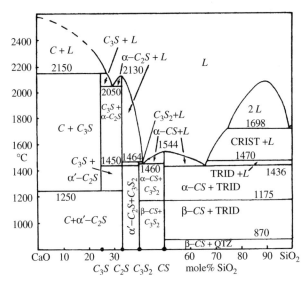

Figure 10.8 The CaO–SiO$_2$ phase diagram at one atm pressure. (After West, 1985, *Solid State Chemistry and its Applications*. Copyright © John Wiley & Sons, Inc. Reproduced with permission.)

and SiO$_2$. There are two congruently melting crystalline phases and two incongruently melting crystalline phases. At the silica (SiO$_2$)-rich end of the phase diagram, there is a liquid immiscibility dome similar to that seen in Figure 10.7 for the Bi–O system. It is worth noting that this liquid immiscibility is responsible for imparting certain optical properties to silica-rich calcium silicate glasses, although the phase diagram does not tell us this. By definition, a glass is a kinetically undercooled amorphous phase that, while lacking long-range order (like solid-solution alloys), has short-range periodicity. Depending on the scale of the immiscibility (two-phase) texture relative to the wavelength of visible light, a silicate glass may be opaque, transparent, or opalescent (West, 1984).

10.3.4 Ternary Condensed Systems

The addition of a third component to a condensed system makes 2D plots of phase equilibria difficult. Since a ternary *ABC* system is comprised of three binary subsystems, *AB*, *AC*, and *BC*, one option is to display the composition for each subsystem on a different edge of an equilateral triangle. When temperature is added, a *solid diagram* is formed with the equilateral triangle as the base and with the binary diagrams as the vertical sides. This figure can be drawn as an isometric projection, as in Figure 10.9.

Unfortunately, reading values from this type of plot is difficult. Therefore, ternary systems are usually represented with a series of 2D vertical or horizontal sections (which remove one degree of freedom) and projections of the solid diagram, like those of Figure 10.10. This method for plotting ternary-phase equilibria was

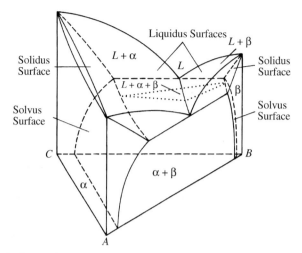

Figure 10.9 An isometric projection showing three-phase equilibria. (After Baker, editor, 1992, *ASM Handbook*, Vol. 3, *Alloy Phase Diagrams*. Copyright © ASM International. Reproduced with permission.)

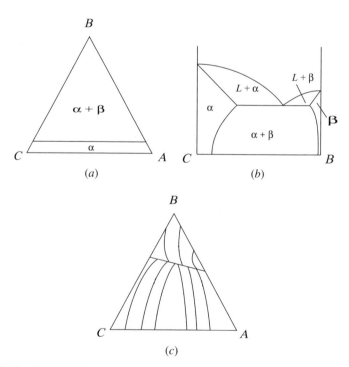

Figure 10.10 (a) An isothermal horizontal section, (b), a vertical section, or isopleth, and (c) the liquidus surface from the isometric projection of Figure 10.9.

introduced by the Dutch physical chemist Hendrik Willem Bakhuis Roozeboom (1845–1907), who succeeded J. H. Van't Hoff at the University of Amsterdam. Between 1901 and 1904, Roozeboom published the first volume and first part of the second volume of the multivolume treatise on heterogeneous equilibria entitled: *Die Heterogenen Gleichgewichte von Standpunkte des Phosenlehre*, or *Heterogeneous Equilibria from the Phase Rule Viewpoint* (Roozeboom, 1901,1904).

Horizontal sections (Figure 10.10a), called *isotherms*, are capable of presenting subsolidus equilibria for any possible composition in a ternary system, but only at a constant temperature. Temperature is the degree of freedom lost. Vertical sections (Figure 10.10b), called *isopleths*, are useful for showing equilibria and the stability ranges of phases over a wide temperature range, but at a constant composition of one component or constant ratio of two components. Sections taken through one corner, or a congruently melting compound on one face, and another congruently melting compound on a different face are (*quasi-*) *binary sections*. When many or most of the ternary compounds are located on the line connecting the primary compositions, the line is called a binary join. In the Ca–Si–O ternary system, for example, the compositions $CaSiO_3$, $Ca_3Si_2O_7$, Ca_2SiO_4, and Ca_3SiO_5 are located between CaO and SiO_2. That system can thus be represented by a binary diagram with components CaO and SiO_2. Liquidus, solidus, and solvus surfaces (projections) of ternary phase diagrams are displayed by adding isothermal contour lines or, as illustrated in Figure 10.10c, temperature troughs with arrows indicating the direction of decreasing temperature to a horizontal section. These diagrams are very useful for locating invariant points, such as ternary eutectics.

The composition of a ternary system can be determined geometrically. From any point within an equilateral triangle, the sum of the distances perpendicular to each side is equal to the height of the triangle. The height is set equal to 100% and divided into 10 equal parts. A network of smaller equilateral triangles is then formed by drawing lines parallel to the three edges through the 10 divisions, although these lines are not always shown in a ternary diagram. Each vertex of the triangle represents one of the three pure components, 100% *A*, 100% *B*, and 100% *C*, while the three edges of the triangle represent the three binary systems with 0% of the third component. Referring to Figure 10.11, the percentage of component *A* in a system with composition at point, *P*, is 40% *A*-40%*B*-20%*C*.

There are some important geometrical constraints for the phase equilibria topology in isobarothermal sections of ternary systems. For example, the Dutch physical chemist Franciscus Antonius Hubertus Schreinemakers (1864–1945) developed rules that determine the arrangement of stable and metastable univariant equilibria where they intersect at an invariant point (Schreinemakers, 1912, 1915). (Incidentally, Schreinemakers also authored the third volume of *Die Heterogenen Gleichgewichte von Standpunkte des Phosenlehre*.) Schreinemakers found that the extrapolations of one-phase field boundaries must either both fall inside a three-phase field or one inside each of the two two-phase fields. There are actually a number of Schreinemakers rules that are helpful for checking the validity of phase diagrams generated by either experimental or computational methods (Hillert, 1998).

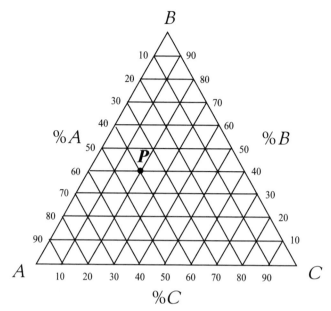

Figure 10.11   The composition at a point P inside a horizontal section is determined by realizing that the sum of the distances perpendicular to each side of an equilateral triangle is equal to the height of the triangle.

Another constraint concerns two-phase equilibria. As in binary systems, tie lines connect two points in equilibrium within a two-phase field. Recall how the points of tangency on the Gibbs energy curves are the points connected by the tie line in the phase diagram. In a ternary system, Gibbs energy curves become surfaces, where it will be found that only specific compositions (termed an *assemblage*) can be in equilibrium in a two-phase field of a ternary diagram. Hence, not all the tie lines in a horizontal section will be parallel to each other and to the edges of the horizontal section. In such a case, the tie line cannot be used to obtain the phase fractions within the two-phase region.

A third constraint relates to three-phase equilibria. Three-phase fields are formed where three two-phase fields converge. The three-phase region is defined by the intersection of three tie lines, one from each of the two-phase fields. The result is a *tie triangle*. Hence, the boundaries of three-phase regions will be triangular, the corners representing compositions that are in equilibrium. Figure 10.12 shows an isothermal section at 750°C from the ternary system Bi_2O_3–CaO–CuO (Tsang et al., 1997), and Table 10.2 lists the three-phase assemblages that are present in this horizontal section.

Any point *within* one of these tie triangles, not representing a distinct single-phase compound, is an equilibrium mixture of the phases at the corners of the tie triangle. For example, heating a mixture comprised of 30% Bi_2O_3, 30% CaO, and 40% CuO to 750°C results in a three-phase equilibrium mixture of CuO, $Bi_6Ca_4O_{13}$, and

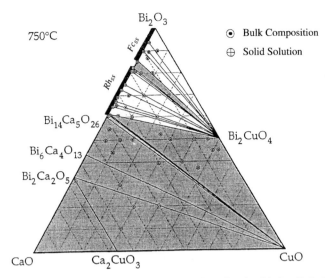

Figure 10.12 The 750°C isotherm (horizontal section) for the Bi_2O_3–CaO–CuO system. (After Tsang et al., 1997. Copyright © American Ceramic Society. Reproduced with permission.)

$Bi_{14}Ca_5O_{26}$. All of the oxides shown in the Bi_2O_3–CaO–CuO phase diagram are binary line compounds. That is, these compounds are the corners of the tie triangle corresponding to the edges of the phase diagram. A stoichiometric *ternary* oxide would appear as a point *inside* a horizontal section of the phase diagram. If, however, an oxide can exist over a narrow compositional range, it will appear as a line segment. This type of behavior is quite common in ternary intermetallic phases, where two components may have a wide stoichiometric range. In such a case, there will be a line extending across the isotherm.

TABLE 10.2   The Three-Phase Fields Shown in Figure 10.13

CaO–Ca_2CuO_3–$Bi_2Ca_2O_5$
CuO–Ca_2CuO_3–$Bi_2Ca_2O_5$
CuO–$Bi_2Ca_2O_5$–$Bi_6Ca_4O_{13}$
CuO–$Bi_6Ca_4O_{13}$–$Bi_{14}Ca_5O_{26}$
CuO–$Bi_{14}Ca_5O_{26}$–Rh_{ss}
CuO–Rh_{ss}–Bi_2CuO_4
Bi_2CuO_4–Rh_{ss}–Fcc_{ss}
CuO–$Bi_{14}Ca_5O_{26}$–Rh_{ss}

Notes: Rh_{ss} = rhombohedral solid solution; *Fcc_{ss}* = face-centered-cubic solid solution.

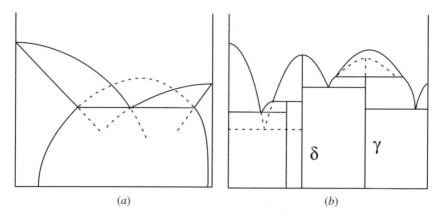

$$(a) \qquad\qquad\qquad (b)$$

Figure 10.13 Metastable extensions of equilibrium phase boundaries. Solvus line extensions usually form a liquid miscibility dome. Extensions of incongruently melting compounds form a congruent melting point and extensions of congruently melting compounds often form eutectics with nonneighboring phases.

10.3.5 Metastable Equilibria

A metastable state is one in which the free energy is higher than the equilibrium value. Metastable states may result from nonequilibrium process conditions. For example, rapid solidification of a molten alloy might produce a bulk metallic glass instead of a crystalline phase. Of course, there is always a thermodynamic driving force to reach equilibrium. Hence, metallic glasses may transform to more stable phases at high temperatures (albeit perhaps very slowly). Partial crystallization of an amorphous alloy on heating may result in nanocrystalline grains within an amorphous matrix.

When the stable boundaries of an equilibrium phase diagram are extended as, for example, in Figure 10.13, regions of metastability are shown. In eutectic systems (Figure 10.13a), metastable equilibria of the solvus lines usually form a liquid miscibility dome. On the other hand, as illustrated in Figure 10.13b, metastable extensions of incongruently melting compounds form a congruent melting point, and metastable equilibria of congruently melting compounds often form eutectics with nonneighboring phases.

10.4 EXPERIMENTAL PHASE-DIAGRAM DETERMINATIONS

As pointed out in Section 10.3.2, a phase diagram can be constructed from a series of Gibbs energy curves. The Gibbs energy, as a function of composition (G/x curves), must be determined for each phase at various temperatures. For example, the top portion of Figure 10.5 shows the G/x curves at temperature $T1$ for an intermediate phase, β, two solid solutions, α and γ, as well as a liquid solution, L. The common points of tangency on the G/x curves are points of equal chemical potential, corresponding to the compositions of the phases in equilibrium. Thus, at $T1$, these points are plotted on an isothermal (horizontal) line. This process is repeated at many other temperatures until the entire phase diagram is generated.

Experimental phase-diagram determination involves the preparation of a large number of samples spanning the entire compositional range, phase identification, and the careful measurement of thermodynamic properties as well as phase transformation temperatures. Accordingly, a variety of experimental techniques is typically utilized in any one case.

Thermal analysis methods, such as differential scanning calorimetry (DSC) and the related differential thermal analysis (DTA), are a type of nonisothermal technique in which phase transformations are signaled by the latent heat evolved or absorbed as a sample is heated and cooled. In DSC, integration of the area of the peak gives a direct measure of the enthalpy of transformation. Another nonisothermal technique involves electromotive-force measurements on electrochemical cells made from the materials under study. Although this method signifies phase transitions by changes in the slope of the electromotive-force versus temperature curves, the relation between electromotive force and temperature is not always linear, which complicates determination of the enthalpy. Measurement of the resistivity of a sample, which changes at a phase transition, and dilatometric techniques (sample expansion and contraction measurements) also fall into the category of nonisothermal methods. The difficulty with determining phase diagrams by nonisothermal methods is that they can easily result in nonequilibrium boundaries, unless sample heating and cooling is carried out slowly enough to allow for equilibration.

Isothermal techniques, by contrast, are inherently better suited for allowing a sample to reach equilibrium. These methods are typically used to identify the phase(s) present and include optical microscopy (metallography), electron-probe microanalysis (EPMA), and X-ray diffraction (XRD). It should be noted, however, that these techniques also could result in inaccurate phase boundaries if an inadequate number of isotherms (measurements at different temperatures) are made.

As can be imagined, the complete experimental determination of an entire phase diagram can be very time-consuming and costly, especially for complex systems. An alternative approach involves the coupling of thermodynamic modeling with experimentation, either to reduce the number of experiments required, or to more efficiently plan them, or both. Thermodynamic modeling is the topic of the remainder of the chapter.

10.5 PHASE-DIAGRAM MODELING

Thermochemical data on the separate phases in equilibrium are needed to construct accurate phase diagrams. The Gibbs energy of formation for a *pure* substance as a function of temperature must be calculated from experimentally determined temperature-dependent thermodynamic properties such as enthalpy, entropy, heat capacity, and equilibrium constants. By a pure substance, one generally means a stoichiometric compound in which the atomic constituents are present in an exact, simple reproducible ratio.

The Gibbs energy of a *solution*, on the other hand, can be computed from those of the pure components plus additional interaction and excess free-energy terms. Liquids and substitutional solid solutions can be described as single-phases, in which all the lattice sites are equivalent. Components randomly mix among these sites, the solute species substituting for the solvent species in the lattice. Substitutional solutions and multiple-phase mixtures are the topic of Section 10.5.1. In other solid solutions, the solute is confined to the interstices between solvent atoms. In this case, the solvent atom sites and interstitial sites constitute two separate sublattices. Sublattice models are discussed in Section 10.5.2. In phase-diagram calculations, the phases in equilibrium can be characterized thermodynamically using different models, each appropriate for a particular phase's type of crystal structure.

10.5.1 Gibbs Energy Expressions for Mixtures and Solid Solutions

At one atm pressure, the Gibbs energy of a two-phase mechanical mixture containing n_A moles of component A and n_B moles of B, in which there is no solubility between components, is:

$$G = x_A \, {}^\circ G_A + x_B \, {}^\circ G_B \qquad (10.12)$$

In this expression, ${}^\circ G_A$ is the standard (${}^\circ$ signifies the value at $P = 1$ atm) molar Gibbs energy of the pure component A, and x_A is the mole fraction of component A. The Gibbs energy of the mechanical mixture serves as a reference state for the properties of a *solution*, in which there is *chemical mixing* between components on an atomic or molecular level.

The total Gibbs energy of a two-component solution containing n_A moles of component A and n_B moles of component B is obtained by integrating the last term of Eq. 10.9 under the condition of constant composition in order that the partial molar Gibbs energies be constant. We can write that expression out explicitly as:

$$dG = (\partial G/\partial n_A)_{P,T,n_B} dn_A + (\partial G/\partial n_B)_{P,T,n_A} dn_B \qquad (10.13)$$

If the solution is at constant temperature and one atm pressure, the result of integration is:

$$(n_A + n_B)G = n_A {}^\circ G_A + n_B {}^\circ G_B \qquad (10.14)$$

where we have used the symbol ${}^\circ G_A$ to represent the partial molar Gibbs energy of the Ath component (which is also the chemical potential, μ_A). Note that this quantity is equal to the molar Gibbs energy when the component is pure. We shall wish to divide each side of Eq. 10.14 by $(n_A + n_B)$ and, in general, use the symbol x_i to represent the mole fraction of the ith component, for example, $x_A = n_A/(n_A + n_B)$. This gives us the following expression for the total Gibbs energy of the solution:

$$G = x_A{}^\circ G_A + x_B{}^\circ G_B \tag{10.15}$$

The Gibbs energy of mixing, ΔG_{mix}, is defined as the difference between Eqs. 10.12 and 10.15.

$$\Delta G_{mix} = (x_A{}^\circ G_A + x_B{}^\circ G_B) - (x_A{}^\circ G_A + x_B{}^\circ G_B) \tag{10.16}$$

We can rewrite Eq. 10.16, using the chemical potential in place of the partial molar Gibbs energy. Doing so, while also allowing for nonstandard state conditions, gives:

$$\Delta G_{mix} = x_A(\mu_A - {}^*\mu_A) + x_B(\mu_B - {}^*\mu_B) \tag{10.17}$$

In Eq. 10.17, μ_A is equivalent to the partial molar Gibbs energy G_A, and ${}^*\mu_A$ is the chemical potential of the pure component A. If both components have low vapor pressures, the chemical potential of the ith component in solution relative to the pure component is approximately equal to:

$$\mu_i - {}^*\mu_i = RT \ln(P_i/{}^*P_i) \tag{10.18}$$

where ${}^*P_i$ is the vapor pressure of the pure component i at the same temperature as the solution. Hence, Eq. 10.17 becomes:

$$\Delta G_{mix} = RT[x_A \ln(P_A/{}^*P_A) + x_B \ln(P_B/{}^*P_B)] \tag{10.19}$$

If the atomic mixing in a solid or liquid solution is a random mingling of similar-size atoms with negligible interactions, the solution obeys Raoult's law (Raoult, 1887) and is said to be *ideal*. In this case:

$$P_i = x_i{}^*P_i \tag{10.20}$$

where x_i is the mole fraction of the ith component. Substituting of Eq. 10.20 into Eq. 10.19, the Gibbs energy of mixing for an ideal two-component solution, $\Delta G_{mix}(\text{ideal})$, may be expressed as:

$$\Delta G_{mix}(\text{ideal}) = RT[x_A \ln x_A) + (x_B \ln x_B)] \tag{10.21}$$

where $-R[x_A \ln x_A + x_B \ln x_B]$ is termed the *ideal entropy of mixing*. Finally, on combining Eqs. 10.16 and 10.21, we obtain:

$$(x_A{}^\circ G_A + x_B{}^\circ G_B) - (x_A{}^\circ G_A + x_B{}^\circ G_B) = RT[(x_A \ln x_A) + (x_B \ln x_B)] \tag{10.22}$$

When ΔG_{mix} has a positive sign, the components are *immiscible*. No solution formation will occur, because the system is in a lower energy state as a two-phase mixture. This is the case with the metallic Ag–Si, for example, as shown in Figure 10.3, where a *two-phase field* expands the entire compositional range at temperatures below the *solidus* (melting point). A polycrystalline Ag–Si alloy of any composition in this system is simply a mechanical mixture of grains of pure Ag, or grains of pure Si, in a matrix of Ag–Si eutectic. Note, however, at temperatures above the *liquidus* a single-phase liquid solution (Ag, Si) does form.

If ΔG_{mix} is negative, a system is able to lower its total Gibbs energy by forming a solution, the extent of which will depend on the magnitude of ΔG_{mix}. In the Bi–Sb system (Figure 10.2), for example, ΔG_{mix} is highly negative below the solidus, corresponding to a high mutual solid solubility. A single-phase *continuous substitutional solid solution*, in which Bi and Sb randomly mix in the lattice, exists over the entire compositional range. By contrast, in the Pb–Sn (Figure 10.4) system, there is only partial solid solubility of one component in another, and a two-phase field exists below the solidus and between the *terminal phases,* which are bordered by *solvus* lines, where the limits of solubility of Pb in Sn and Sn in Pb are located. As in the Ag–Si system, a single-phase liquid solution is present above the liquidus in the Bi–Sb, and Pb–Sn phase diagrams.

Real solutions deviate from ideal behavior. The deviation in the Gibbs energy of mixing is measured with the *excess* Gibbs energy function, G_{EX}, which can be positive or negative, representing positive or negative deviations from ideality:

$$\Delta G_{mix}(\text{real}) = G_{EX} + RT[n_A \ln x_A + n_B \ln x_B] \qquad (10.23)$$

The *regular-solution model*, introduced by the American chemist Joel Hildebrand (1881–1983) is the simplest way to consider these other contributions to the Gibbs energy (Hildebrand, 1929). In this case, G_{EX} is of the form:

$$G_{EX} = x_A x_B \Omega \qquad (10.24)$$

where Ω is an empirical composition-independent interaction energy parameter. Positive values represent repulsive interactions between components and negative values represent attractive interactions. Therefore, for the Gibbs energy of mixing in the binary solution, we have:

$$\Delta G_{mix}(\text{real}) = x_A x_B \Omega + RT[x_A \ln x_A + x_B \ln x_B] \qquad (10.25)$$

Thus, the total Gibbs energy of the binary system, as given by the regular solution model, is:

$$G = x_A{}^\circ G_A + x_B{}^\circ G_B + RT[x_A \ln x_A + x_B \ln x_B] + x_A x_B \Omega \qquad (10.26)$$

Models incorporating linear composition dependencies to Ω (the subregular-solution model), as well as others allowing for complex composition dependencies have been developed. The most commonly used model is the *Redlich–Kister polynomial* (Redlich and Kister, 1948). The total Gibbs energy of a binary system, using the Redlich–Kister model is:

$$G = x_A \,{}^\circ G_A + x_B \,{}^\circ G_B + RT[x_A \ln x_A + x_B \ln x_B] + x_A x_B \sum_v \Omega_i^v (x_A - x_B)^v \quad (10.27)$$

which becomes the regular model if $v = 0$, and the subregular model if $v = 1$. In practice, v does not usually rise above two (Saunders and Miodownik, 1998).

The excess Gibbs energy of a ternary (or higher-order) substitutional solution at a composition point p can be estimated geometrically by extrapolation of the excess Gibbs energies in the binary subsystems at points a, b, and c. There are several possible ways of doing this, but the most commonly used method is by Muggianu, in which the total Gibbs energy becomes equal to (Muggianu et al., 1975):

$$G = x_A \,{}^\circ G_A + x_B \,{}^\circ G_B + x_C \,{}^\circ G_C + RT[x_A \ln x_A + x_B \ln x_B + x_C \ln x_C]$$
$$+ x_A x_B \sum_v \Omega_i^v (x_A - x_B)^v + x_A x_C \sum_v \Omega_i^v (x_A - x_C)^v + x_B x_C \sum_v \Omega_i^v (x_B - x_C)^v$$

$$(10.28)$$

Equation 10.28 is used to model a single-phase liquid in a ternary system, as well as a ternary substitutional solid solution formed by the addition of a soluble third component to a binary solid solution. The solubility of a third component might be predicted, for example, if there is mutual solid solubility in all three binary subsets (*AB*, *BC*, *AC*). Note that Eq. 10.28 does not contain ternary interaction terms, which are usually small in comparison to binary terms. When this assumption cannot, or should not, be made, ternary interaction terms of the form $x_A x_B x_C L_{ABC}$, where L_{ABC} is an excess ternary interaction parameter, can be included. There has been little evidence for the need of terms of any higher order. Phase equilibria calculations are normally based on the assessment of only binary and ternary terms.

10.5.2 Gibbs Energy Expressions for Phases with Long-Range Order

When there are significant differences in the electronegativities, atomic radii, and/ or crystal structures (bonding preferences) between the components, rather than randomly mixing, they assume an atomic arrangement exhibiting long-range order. This class of substance constitutes a very large fraction of the new materials reported in the scientific literature. Examples include interstitial solid solutions, intermediate phases, disordered stoichiometric compounds, and even ionic liquids. Such phases can be modeled thermodynamically as solutions with interlocking sublattices, on which the different species mix.

In general, the sublattice models consist of two sets of positions (e.g., anions and cations), or sublattices, that are distinguishable by different fractional occupancies of each component. If each sublattice is totally occupied by a different species, this corresponds to full long-range order. When detailed crystallographic data are incorporated into the sublattice model, they are often referred to as the *compound energy formalism*. Essentially, this simply corresponds to the use of multiple sublattices, one for each type of site in the crystal structure. The *compound energy*, then, is the Gibbs energy of all the end-member compounds of the solution phase, each with the same crystal structure as the original solvent.

The difference between the compound energy model and the simple two-sublattice model can be illustrated with two ternary intermetallic phases from the Al–Mg–Zn system. One of these two phases is known to contain a constant composition of 54.5 atomic percent magnesium and an extended homogeneity range of aluminum and zinc, corresponding to the formula $Mg_6(Al,Zn)_5$. However, no crystallographic data are available for this phase. Therefore, it is appropriately thermodynamically modeled by two sublattices, in which one sublattice is exclusively occupied by Mg, while Al and Zn are allowed to randomly mix on the second sublattice (Liang et al., 1998).

By comparison, early crystallographic data by Bergman and Pauling indicate that the three components in the bcc phase $Mg_{32}(Zn,Al)_{49}$ (Pearson symbol $cI162$) are distributed with specific fractional occupancies over four distinct lattice sites in the space group $Im3$ (Bergman et al., 1952). This phase was thus modeled with four sublattices (Liang et al., 1998). This compound is a cubic Frank–Kasper phase (Section 2.4.2). Of the 162 atoms in the unit cell, 98 have icosahedral coordination. Metastable icosahedral *quasicrystals* of $Mg_{32}(Zn,Al)_{49}$ with fivefold rotational symmetry (rather than the bcc phase) and, hence, no 3D translational periodicity, can be obtained by rapid solidification (Sastry and Ramachandrarao, 1986).

The Gibbs energy expressions for sublattice solutions are actually quite similar to the regular-solution model. This is because a substitutional solution can be considered as consisting of a single sublattice on which the atoms mix. However, in the compound-energy model, the equations for multiple sublattices can quickly get rather complicated. For the simplest case, in which there is only a single component (z) of fixed stoichiometry on one sublattice (v) and two randomly mixed components (i, j) in a second sublattice (u) (e.g., $Mg_6(Al,Zn)_5$), the Gibbs energy is written as:

$$G = \sum_i y_i {}^\circ G_{i:z} + RT \sum_i y_i \ln y_i + \sum_i \sum_{j>i} y_i y_j \sum_v L_{ij}^v (y_i - y_j)^v \qquad (10.29)$$

In this expression, y_i is the fractional site occupation for component i (the number of atoms of component i on the sublattice, divided by the total number of sites on that sublattice), L_{ij}^v is an interaction energy parameter for mixing between components i and j on the sublattice, and ${}^\circ G_{i:z}$ is the Gibbs energy of the compound when the sublattice u is completely occupied by i.

Interstitial solid solutions are treated similarly. The structure is again approximated with a two-sublattice model, but where one sublattice is occupied by substitutional elements, and one by the smaller interstitial elements (e.g., C, N, H) and vacancies. It follows, from Eq. 10.23, that for a binary interstitial solution, the Gibbs energy is given by:

$$G = y_i \, ^\circ G_{i:z} + y_{Va} \, ^\circ G_{Va:z} + RT(y_i \ln y_i + y_{Va} \ln y_{Va})$$
$$+ y_i y_{Va} \sum L_{iVa}^v (y_i - y_{Va})^v \tag{10.30}$$

where Va denotes a vacancy. All the sites in the second sublattice are vacant in the compound whose Gibbs energy is $G_{Va:z}$, this term is thus the same as that of the pure substance, G_z.

A large number of oxides reported in the literature are actually sublattice solutions. Partial substitution of one species for another in one sublattice is a common approach for tailoring the properties of materials. In cases where the solute ions and solvent ions are of different charge (aliovalent), charge neutrality must be maintained. For example, at temperatures above $1000°C$, lanthanum oxide, La_2O_3, is known to dissolve considerable amounts of SrO, and SrO dissolves a small amount of La_2O_3. It is believed that the dissolution mechanism in the case of strontium dissolution in La_2O_3 involves aliovalent cation (La^{3+}/Sr^{2+}) substitution with the formation of charge-compensating anion (O^{2-}) vacancies (Grundy et al., 2002). For the general case of two randomly distributed cations (i, j) on a cation lattice, with anions (k) and vacancies (Va) randomly distributed on the anion lattice, $(i,j)_n (k, Va)_m$, the Gibbs energy can be expressed as:

$$G = y_i^v y_k^{u\,\circ} G_{i:k} + y_j^v y_k^{u\,\circ} G_{j:k} + y_i^v y_{Va}^{u\,\circ} G_{i:Va} + y_j^v y_{Va}^{u\,\circ} G_{k:Va}$$
$$+ RT[n(y_i^v \ln y_i^v \ln y_j^v) + m(y_k^u \ln y_k^u + y_{Va}^u)] + G_{EX} \tag{10.31}$$

where, in a first approximation, G_{EX} is sometimes set to zero.

Example 10.1 The dissolution of strontium in La_2O_3 has been modeled (Grundy et al., 2002) with a two-sublattice expression, in which Sr^{2+} and La^{3+} randomly mix on the cation sublattice and vacancies are introduced in the anion sublattice to maintain charge neutrality: $(La^{3+}, Sr^{2+})_2(O^{2-}, Va)_3$. Similarly, the dissolution of lanthanum in SrO was modeled by assuming that La^{3+} ions and charge compensating vacancies replace Sr^{2+} in the cation sublattice: $(Sr^{2+}, La^{3+}, Va)_1(O^{2-})_1$. Write the Gibbs energy expression in the compound-energy model for each phase.

Solution Using Eq. 10.31, for the La_2O_3 solid solution, we have:

$$G = y_{La^{3+}}^v y_{O^{2-}}^u \, ^\circ G_{La^{3+}:O^{2-}} + y_{Sr^{2+}}^v + y_O^{2-\,u} \, ^\circ G_{Sr^{2+}:O^{2-}} + y_{La^{3+}}^v y_{Va}^u \, ^\circ G_{La:Va}^{3+}$$
$$+ y_{Sr^{2+}}^v y_{Va}^u \, ^\circ G_{Sr^{2+}:Va} + RT[2(y_{La^{3+}}^v \ln y_{La^{3+}}^v \ln y_{La^{3+}}^v + y_{Sr^{2+}}^v \ln y_{Sr^{2+}}^v)$$
$$+ 3(y_{O^{2-}}^u \ln y_{O^{2-}}^u + y_{Va}^u \ln y_{Va}^u)] + G_{EX}$$

Modifying Eq. 10.31, to account for vacancies on the cation lattice, gives for the SrO solid solution:

$$G = y_{Sr^{2+}}^v y_{O^{2-}}^u {}^\circ G_{Sr^{2+}:O^{2-}} + y_{La^{3+}}^v y_{O^{2-}}^u {}^\circ G_{La^{3+}:O^{2-}} + y_{Va}^v y_{O^{2-}}^u {}^\circ G_{Va:O^{2-}}$$
$$+ RT[(y_{La^{3+}}^v \ln y_{La^{3+}}^v + y_{Sr^{2+}}^v \ln y_{Sr^{2+}}^v + y_{Va}^u \ln y_{Va}^u)] + G_{EX}$$

In the type of solutions described by Equations 10.29 and 10.30, each component is randomly distributed over only one sublattice. In many cases, the constituents are randomly distributed in more than one sublattice. For a random distribution of two components (i, j) on two sublattices (v, u), the Gibbs energy can be expressed (Hillert and Staffansson, 1970) as:

$$G = y_i^v y_i^{u\circ} G_{i:i} + y_j^v y_j^{u\circ} G_{j:j} + y_i^v y_j^{u\circ} G_{i:j} + y_j^v y_i^{u\circ} G_{j:i}$$
$$+ RT(y_i^v \ln y_i^v + y_j^v \ln y_j^v + y_i^u \ln y_i^u + y_j^u \ln y_j^u + G_{EX} \qquad (10.32)$$

Example 10.2   The intermetallic phase β-SbSn has the rock-salt structure with one lattice almost exclusively occupied by antimony and the other by tin. Write the Gibbs energy expression using the compound-energy model for a ternary phase including bismuth, where it is assumed that Bi goes preferentially into the mostly Sb sublattice.

Solution   Using Eq. 10.32, we have:

$$G = y_{Sb}^v y_{Sb}^u {}^\circ G_{Sb:Sb} + y_{Sn}^v y_{Sn}^u {}^\circ G_{Sn:Sn} + y_{Sb}^v y_{Sn}^u {}^\circ G_{Sb:Sn} + y_{Sn}^v y_{Sb}^u {}^\circ G_{Sn;Sb}$$
$$+ y_{Bi}^v y_{Sn}^u {}^\circ G_{Bi:Sn} + y_{Bi}^v y_{Sb}^u {}^\circ G_{Bi:Sb} + RT(y_{Sb}^v \ln y_{Sb}^v + y_{Sn}^v \ln y_{Sn}^v$$
$$+ y_{Bi}^v \ln y_{Bi}^v + y_{Sn}^u \ln y_{Sn}^u + y_{Sb}^u \ln y_{Sb}^u) + G_{EX}$$

Example 10.3   For each of the following oxides, decide if the compound energy model would be appropriate for modeling the Gibbs energy of formation.

(a) Pb_2MnReO_6, prepared under 1-atm pressure at 550°C by direct reaction of the oxides. Crystallographic data: space group $P2_1/n, Z = 2$, cation positions:

Atom	Site	x	y	z	Occ.
Pb	4e	0.9743	0.0126	0.2458	1
Mn1	2c	0.5	0	0.5	0.86
Re1	2c	0.5	0	0.5	0.14
Mn2	2d	0.5	0	0	.86
Re2	2d	0.5	0	0	0.14

(b) $La_2CaB_{10}O_{19}$, prepared under 1-atm pressure at 930°C by direct reaction of the oxides. Crystallographic data: space group $C2, Z = 2$, cation positions:

Atom	Site	x	y	z	Occ.
La	$4c$	0.1624	0	1405	1
Ca	$2b$	0	−0.1855	0.5	1
B1	$4c$	0.4326	0.1919	0.1249	1
B2	$4c$	−0.0397	0.3219	.1608	1
B3	$4c$	0.3272	−0.4884	.2072	1
B4	$4c$	0.1142	−0.4406	.2688	1
B5	$4c$	0.2327	0.0212	0.5289	1

(c) $Ce_2Zr_2O_{7.36}$, prepared under argon at room temperature from reaction of $Ce_2Zr_2O_7$ with NaOBr. Crystallographic data: space group $Fd3m$, $Z = 8$, atom positions:

Atom	Site	x	y	z	Occ.
Ce	$16c$	0	0	0	1
Zr	$16d$	0.5	0.5	0.5	1
O1	$48f$	0.4058	0.125	0.125	1
O2	$8a$	0.125	0.125	0.125	1
O3	$8b$	0.375	0.375	0.375	0.36

Solution

(a) From the crystallographic data, Mn and Re are seen to both occupy the same two sets of lattice sites $(2c, 2d)$. This phase is a sublattice solution and could be described by Eq. 10.32.

(b) This is a stoichiometric compound, not a solid solution. Each type of cation exclusively occupies its own set of lattice sites; the sublattices are not alloyed. The compound energy model does not apply. The free energy of formation must be measured.

(c) This compound was prepared by intercalation of oxygen into the interstices of a host structure. The Gibbs energy of the product should be modeled with Eq. 10.30.

10.5.3   Other Contributions to the Gibbs Energy

The Gibbs energy models for the various types of phases we have described thus far have been based on random distributions of atoms on one or more sublattices. Full long-range order occurs if the solute atoms completely occupy one sublattice and the solvent atoms completely occupy another. However, when the sublattices are not completely occupied, the atoms have a choice of which particular sites on their sublattice they want to reside in. This gives rise to the possibility for short-range ordering. Short-range order means that the atoms do not arrange themselves at random within each sublattice. There is always a tendency for short-range ordering. In order to obtain the highest possible accuracy in a thermodynamic description of a system, the expression for the configurational entropy should account for the occurrence of this ordering. The description of short-range order is usually made by statistical thermodynamic or combinatoric models like the cluster variation method (CVM) and the MC method.

In fact, although the CVM has been extended to the treatment of atoms in a solid solution, it was originally developed to describe cooperative magnetic phenomena, that is, the configurations of spins in a ferromagnet (Kikuchi, 1951). Magnetism is a physical effect that can contribute significantly to the total Gibbs energy for a system. The energy released can often exceed those of ordinary phase transformations (Saunders and Miodownik, 1998). The effects on phase equilibria can include the following: a marked change in solid solubility, distortion of miscibility gaps, and stabilization of metastable phases. In fact, the magnetic Gibbs energy may be large enough to even cause structural changes in unary systems (e.g., pure elements).

Zener was one of the first to examine the influence of magnetism on phase equilibria, although he considered only binary iron alloys (Zener, 1955). He postulated that the effect of alloying on the Gibbs energy is proportional to the effect on the Curie temperature, T_C. Zener assumed that changes to the Curie temperature are linearly proportional to changes in composition, that is, $\Delta T_C / \Delta x$ is constant. With these assumptions, the following expression holds:

$$G_{mag}(\text{alloy}) - G_{mag}(\text{Fe}) = -x(\delta T/\delta x)S_{mag}(\text{Fe}) \qquad (10.33)$$

where S_{mag} is the magnetic entropy. When the end-members exhibit different magnetic states, however, nonlinear behavior in T_C may be observed (Saunders and Miodownik, 1998). It is assumed in Zener's model that, upon alloying, changes in the mean number of unpaired spins per Fe atom, n (see Eq. 7.13), and, hence, the magnetization, will be reflected in changes in T_C. Zener's model was superseded by formulations that attempt to simulate experimental values for the magnetization, in addition to T_C. We refer the reader to more specialized texts, such as Saunders and

Miodownik (1998), for a detailed discussion of magnetic formulations and short-range ordering models.

10.5.4 Phase-Diagram Extrapolations: The CALPHAD Method

The use of computer methods for solving thermodynamic problems is known as computational thermodynamics. When employed specifically for generating phase diagrams, the term CALPHAD is commonly used. The CALPHAD technique, which has made it possible for nonthermodynamic experts to routinely make phase-diagram calculations, is now widely practiced by scientists and engineers from many different disciplines. The journal *CALPHAD*, dedicated to reporting calculated phase diagrams and thermodynamic assessments, has been published since 1977.

An excellent example illustrating the value of the CALPHAD method involves the quest for lead-free solder alloys. It is believed that the major constituents of most lead-free solders will come primarily from a group of 13 elements: Ag, Al, Au, Bi, Cu, Ga, Ge, In, Mg, Sb, Si, Sn, and Zn. One of the most important criteria for any solder alloy is its melting-point range. This information can be obtained directly from the phase diagram. The number of possible alloy systems for the preceding group of 13 elements is given by combinatorics as $n!/[m!(n-m)!]$, where $n = 13$, and $m = 2$ for binary systems, 3 for ternary, and so on. Although all of the phase diagrams for the possible binary combinations have been published, only a fraction of the 286 ternary systems is available. It is obvious that phase-diagram calculations can greatly expedite the acquisition of melting-point data, in comparison to experimentation, in such a situation (Lalena et al., 2002).

The value of the CALPHAD method lies not in the ability to generate phase diagrams for thermodynamically assessed systems (that has already been done experimentally!), but rather in its ability to make phase equilibria *predictions* on multiple-component systems (e.g., ternary) for which phase diagrams are not available. The predictions are based on *extrapolation* from the requisite sub-systems. In fact, the term CALPHAD is normally taken to mean phase-diagram extrapolation. By extrapolation, one is essentially calculating the phase boundaries in a system by modeling the free energies of the substances, as a new component is added, through equations like those given in Sections 10.5.1 and 10.5.2. The substances may be liquids, substitutional solutions, or sublattice phases. For example, a binary oxide with composition $A_x B_y O_z$ can be extrapolated into a ternary phase in which a third metal, C, substitutes on one of the cation sublattices, for example, $(A, C)_x B_y O_z$, $A_x (B, C)_y O_z$, or $(A, C)_x (B, C)_y O_z$. It is very important to understand that extrapolation cannot be use to predict the existence of a hitherto unknown stoichiometric compound or phase with a unique crystal structure. Consequently, it is also impossible to calculate accurate phase boundaries for a system as a new component is added unless thermochemical information for every existing phase is known. For example, Figure 10.14 illustrates how the unaccounted-for existence of a high-melting line compound might result in largely erroneous liquidus and solidus contours in a calculated phase diagram.

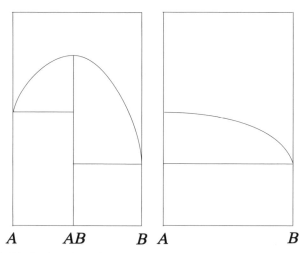

A AB B A B

Figure 10.14 Neglecting the existence of a high-melting line compound (AB) in the phase-equilibria calculations (left) can result in a largely erroneous liquidus or solidus in an extrapolated phase diagram (right).

Extrapolations enable one to find the *stability ranges* for all the known phases of the system as new components are added. For example, extrapolation allows us to predict the solubility extent of A in the lattice of B and the solubility extent of B in the lattice of A. Of course, for the complete phase diagram we must include the Gibbs energy calculations for all the allotropic forms of each element and, indeed, all the structures exhibited by every element (even nonallotropic ones) across the whole system. Of great utility in accomplishing this are *lattice stabilities*, defined as the Gibbs energy differences between all the various crystal structures in which a pure element may exist in the system, as a function of T, P, and V (e.g., $^\circ G_{\text{hcp-Fe}} - {}^\circ G_{\text{fcc-Fe}}$). Equivalently, the lattice stability may be considered as the Gibbs energy of formation of one state of a pure element from another.

It is possible in many cases to predict highly accurate phase equilibria in multicomponent systems by extrapolation. Experience has shown that extrapolation of assessed $(n-1)$ data into an nth order system works well for $n \leq 4$, at least with metallurgical systems. Thus, the assessment of unary and binary systems is especially critical in the CALPAHD method. A thermodynamic assessment involves the optimization of all the parameters in the thermodynamic description of a system, so that it reproduces the most accurate experimental phase diagram available. Even with experimental determinations of phase diagrams, one has to sample compositions at sufficiently small intervals to ensure accurate reflection of the phase boundaries.

Several computer programs for extrapolating phase diagrams are commercially available, including Thermo-calc (Sundam et al., 1985), ChemSage (Eriksson and Hack, 1990), PANDAT (Chen et al., 2001), and MT-DATA (Davies et al., 1995).

Each of these software packages is similarly constructed and consists of no fewer than four integral parts:

1. A user interface allowing one to easily specify the temperature, pressure, and composition ranges over which equilibrium calculations are to be made;
2. A database of *thermodynamically assessed* systems, containing critically checked and internally consistent thermochemical data for the phases taking part in the equilibrium;
3. The algorithms (proprietary!) for making equilibrium calculations, normally based on the minimization of the total Gibbs energy of a system;
4. A graphics routine for plotting or printing the results, that is, the phase diagram.

The general methodology used can be described as follows. The user sets up the problem by specifying the type of calculation to be made (e.g., isotherm, isopleth, or liquidus projection) and then selects the database from which the thermodynamic data will be imported. The system components are defined, as well as the phases that are present and the temperature, pressure, and composition ranges over which to make the equilibrium calculations. All of this is accomplished via the user interface.

Databases with assessed thermodynamic data for hundreds of substances are available, including alloys, semiconductors, geochemical compounds (silicates and other main-group oxides), aqueous solutions, and molten salts. The bulk of the commercially available databases are on metallurgical systems since the CALPHAD method finds ready applicability in the fields of metals processing and alloy development.

Common approaches for the tailoring of nonmetallic (ceramic) materials properties involve topochemical methods (those where the crystal structure remains largely unaffected) and the preparation of phases in which one or more sublattices are alloyed. In principle, such materials are within the realm of CALPHAD. On the other hand, as we have already stated, extrapolation does not really aid the discovery of *new* or novel phases, with unique crystal structures. Furthermore, assessed thermochemical data for the vast majority of ceramic systems, particularly transition-metal compounds, are currently not available in commercial databases for use with phase-diagram software. This does not necessarily preclude the use of the CALPHAD method on these systems. However, it does require the user to carry out their own thermodynamic assessments of the $(n-1)$th-order subsystems and to import those data into a database for extrapolation to nth-order systems, which is not a trivial task.

One normally finds the thermodynamic properties in a database given in the form of parameters to some kind of polynomial function like Eq. 10.34, expressing the temperature dependence to G:

$$G(T) = a + bT + cT\ln(T) + \sum_{2}^{n} d_n T^n \qquad (10.34)$$

Each of the concentration-dependent G terms in the equations presented earlier can have a temperature dependency given by Eq. 10.34. Similar equations can be written for other thermodynamic properties, from which the Gibbs energy can be computed, such as the enthalpy of formation, entropy, and heat capacity. Equation 10.34 is a much more efficient way of incorporating information into a software database than tables containing discrete values, which is important for minimizing computer resource requirements.

Finally, the equilibrium-phase boundaries in a phase diagram are computed by minimizing the total Gibbs energy of the system. Another approach is to minimize the difference between the chemical potentials of the components in all the phases that are present at each value for the composition, temperature, and pressure. The chemical potentials of the components in each phase are the same at equilibrium. The source codes used by the individual programs are highly proprietary, but most of them essentially work by a *local* minimization routine such as the technique of steepest descent using Lagrange's method of undetermined multipliers. The procedure can be found in most undergraduate calculus textbooks. It is a general method, due to Joseph-Louis Lagrange (1736–1813), for finding critical points (minima) of a function subject to constraints. The number of multipliers used is determined by the number of constraints. For thermochemical calculations, example constraints include mass balance or the compositional ranges for each phase.

Once the system and constraints are defined, the user supplies an initial guess (start point) for the equilibrium values. An iterative process begins and convergence is achieved when the difference in the total Gibbs energy between two successive steps is below a predefined limit. It should be pointed out that the closer the start point is to the true value, the fewer iterations required and the less the likelihood that local minima due to metastable equilibria interfere with finding the one true global minimum, or one true equilibrium value for the Gibbs energy. One must always be aware of the possibility of obtaining metastable diagrams from getting trapped in local minima—easy even for skilled and experienced users. Of course, in some cases, the researcher may very well be primarily interested in the metastable equilibria!

Mats Hillert (Courtesy of the Department of Materials Science and Engineering, KTH (Royal Institute of Technology). © Kungl Tekniska Högskolan. Reproduced with permission.)

Larry Kaufman (Courtesy of Lawrence Kaufman. Reproduced with permission.)

MATS H. HILLERT

(b. 1924) and **Larry Kaufman** (b. 1931) studied together at the Massachusetts Institute of Technology. Kaufman (right photo) received his D.Sc. in physical metallurgy in 1955, and Hillert (left photo) his D.Sc. in 1956. Hillert became a research associate with the Swedish Institute of Metals Research in 1948. In 1961, he was appointed Professor of Physical Metallurgy at the Royal Institute of Technology in Stockholm (KTH), becoming emeritus in 1990. Hillert devised Gibbs energy expressions for sublattice phases, time–temperature transformation diagrams for steels, and contributed to our understanding of the energetics of liquid-phase separations, such as spinodal decomposition. He wrote the textbook, *Phase Equilibria, Phase Diagrams, and Phase Transformations*. Hillert was elected a Foreign Associate of the U.S. National Academy of Engineering in 1997.

Kaufman was with the Lincoln Laboratory at M.I.T. from 1955 to 1958, ManLabs from 1958 to 1988, and Alcan Aluminum Corporation from 1988 to 1996. Kaufman derived expressions for thermochemical lattice stabilities for allotropic and nonallotropic elements. He provided an early model for the thermodynamic description of iron, including magnetic contributions, and he was the founding editor of the journal *CALPHAD*. Kaufman is now a consultant and lecturer in the materials science and engineering department at M.I.T.

Through their parallel and independent efforts on both sides of the Atlantic, which began in the 1950s with mathematically modeling known phase diagrams for unary and binary systems, Kaufman and Hillert are considered founding fathers of the CALPHAD method, the field of computational thermochemistry concerned with the extrapolation of phase diagrams for multicomponent systems.

(*Source*: L. P. Kaufman, personal communication, February 08, 2004.)

REFERENCES

Bergman, G.; Waugh, J. L. T.; Pauling, L. 1952, *Nature*, *169*, 1057.

Chen, S.-L.; Zhang, D. F.; Chang, Y. A.; Oates, W. A.; Schmid-Fetzer, R. 2001, *J. Phase Equilibria*, *22*, 373.

Davies, R. H., Dinsdale, A. T.; Gisby, J. A.; Hodson, S. M.; Ball, R. G. J. 1995, in Nash, P., Sundman, B., editors, *Applications of Thermodynamics in the Synthesis and Processing of Materials*, TMS, Warrendale, PA.

Eriksson, G.; Hack, K. 1990, *Metall. Mater. Trans. B21*, 1013.

Grundy, A. N.; Hallstedt, B.; Gaukler, L. J. 2002, *Acta Mater.*, *50*(9), 2209.

Hildebrand, J. H. 1929, *J. Am. Chem. Soc.*, *51*, 66.

Hillert, M. 1998, *Phase Equilibria, Phase Diagrams, and Phase Transformations: Their Thermodynamic Basis*, Cambridge University Press, Cambridge, UK.

Hillert, M.; Staffansson, L.-I. 1970, *Acta. Chem. Scand.*, *24*, 3618.

Kikuchi, R. 1951, *Phys. Rev.*, *81*, 988.

Laidler, K. J.; Meiser, J. H. 1982, *Physical Chemistry*, Benjamin Cummings, Menlo Park, CA.

Lalena, J. N.; Dean, N. F.; Weiser, M. W. 2002, *Adv. Packag.*, *11(2)*, 25.

Liang, P.; Tarfa, T.; Robinson, J. A.; Wagner, S.; Ochin, P.; Harmelin, M. G.; Seifert, H. J.; Lukas, H. L.; Aldinger, F. 1998, *Thermochim. Acta*, *314*, 87.

Muggianu, Y. M.; Gambino, M. Bros, J. P. 1975, *J. Chim. Phys.*, *22*, 83.

Raoult, F. M. 1887, *Compt. Rend.*, *104*, 1430.

Redlich, O.; Kister, A. 1948, *Ind. Eng. Chem.*, *40*, 345.

Roozeboom, H. W. B. 1901, *Die heterogenen Gleichgewichte vom Standpunkte des Phosenlehre*, Vol. 1, Freidrich Vieweg & Sohn, Braunschweig, Germany.

Roozeboom, H. W. B. 1904, *Die heterogenen Gleichgewichte vom Standpunkte des Phosenlehre*, Vol. 2, Freidrich Vieweg & Sohn, Braunschweig, Germany.

Sastry, G. V. S.; Ramachandrarao, P. 1986, *J. Mater. Res.*, *1(2)*, 247.

Saunders, N.; Miodownik, A. P. 1998, *CALPHAD: Calculation of Phase Diagrams, A Comprehensive Gudie*, Pergamon, New York.

Schreinemaker, F. A. H. 1912, *Die Ternäre Gleichgewichte*, Vol. III, Part II, Freidrich Vieweg & Sohn, Braunschweig, Germany.

Schreinemaker, F. A. H. 1915, *Proc. K. Wetensch.*, Amsterdam, *18*, 116.

Sundman, B; Jansson, B,; Andersson, J.-O. 1985, *CALPHAD*, *9*, 153.

Tsang, C.-F.; Meen, J. K.; Elthon, D. 1997, *J. Am. Ceram. Soc.*, *80(6)*, 1501.

West, A. R. 1984, *Solid State Chemistry and Its Applications*, John Wiley & Sons, Chichester, UK.

CHAPTER 11

An Introduction to Nanomaterials

No textbook intended for inorganic materials science and engineering students of the twenty-first century could possibly be considered complete without covering nanomaterials (1 nm = 10Å). Unfortunately, we cannot do full justice to this subject matter with a single chapter. We therefore have chosen only to present a brief history of nanomaterials, explain why their properties differ from those of the macroscopic counterparts, and, in the final chapter of this text, introduce some of the more common preparative techniques. It is hoped that this will be sufficient to motivate the student to pursue further knowledge in this relatively young but rapidly growing field.

The credit for inspiring nanotechnology usually goes to physicist Richard Phillips Feynman (1918–1988), who shared the 1956 Nobel Prize in Physics for his contributions to the field of quantum electrodynamics. In December of 1959, during an after-dinner lecture at the annual meeting of the American Physical Society, Feynman declared that "the principles of physics, as far as I can see, do not speak against the possibility of maneuvering things atom by atom ... a development which I think cannot be avoided." (Feynman, 1960).

Among the many avenues of technology, which Feynman believed would be touched by this research, was information storage. He speculated that one day we would be able to place all the printed information since the Gutenberg Bible in a cube of material 0.1 mm wide. This corresponds to one bit of information per $5 \times 5 \times 5$ cube of 125 atoms. If we are talking about atoms with 1.5Å radius and that this collection of 125 atoms is a discrete particle then, roughly, this corresponds to one bit per 1.7×10^3 ($125 \times (4\pi r^3/3)$) cubic angstroms of volume. How close are we to this prediction today? At the time of this writing, magnetic storage devices (e.g., computer hard disks) with storage densities of at least 30 gigabits per square inch, which corresponds to one bit per 2.1×10^6 square angstroms, are commercially available. Clearly, excitement is high!

Nanotechnology has already begun to bear fruit in other areas as well. Nanosized tungsten carbide grains are now used in machining tool drill bits in order to improve fracture toughness, wear resistance, and hardness. A tiny fuel cell for mobile devices

Principles of Inorganic Materials Design   By John N. Lalena and David A. Cleary
ISBN 0-471-43418-3   Copyright © 2005 John Wiley & Sons, Inc.

using carbon nanohorns as the electrodes has been developed. This fuel cell has about 10 times the energy capacity of a standard lithium battery. Field-effect transistors made from carbon nanotubes have been shown to outperform the most advanced silicon transistors. Such molecular-scale circuits will allow electronic computers to approach the theoretical limits for size and speed, before the anticipated optical computers, with their photonic signal-processing circuits, totally revolutionize the industry.

11.1 HISTORY OF NANOTECHNOLOGY

The study of nanosized particles has its origin in colloid chemistry, which dates back to 1857 when Michael Faraday (1791–1867) set out to systematically investigate the optical properties of thin films of gold. Faraday prepared a suspension of ultrasmall metallic gold particles in water by chemically reducing an aqueous solution of gold chloride with phosphorus (Faraday, 1857). To this day, nanoscale metal particles are still produced by chemical reduction in aqueous solutions.

Faraday called his ruby-colored mixture *colloidal gold*. He showed that, like a solution, the mixture was transparent when looked through, but when a ray of light was shown into the fluid, the particles within the ray created a blue opalescence due to scattering of the light (which subsequently became known as the Tyndall effect). He further showed how the color of the mixture changed from ruby to blue upon addition of a salt, which he reasoned as being due to particle coagulation. Finally, he demonstrated how this effect was not observed when gelatin was added to the mixture, thereby preventing coagulation. Faraday's suspension survives in the collections of the Royal Institution of Great Britain in London and has yet to settle!

In 1861, chemist Thomas Graham (1805–1869) discovered that certain substances, such as starch and gelatin particles, which he called *colloids*, diffuse very slowly through water and that they do not form crystals (Graham, 1861). Based on the slow diffusion and lack of sedimentation, Graham deduced that the particles were about 1–100 nm in size. The size limits of colloids were later extended to the currently accepted range of 1–1000 nm by Wolfgang Ostwald (1883–1943), son of physical chemist and Nobel laureate Friedrich Wilhelm Ostwald.

The term "nanomaterials" now encompasses *clusters* (aggregates containing between 3 and 1000 atoms), *nanostructured* single particles (e.g., nanotubes), *nanocrystalline phases* (e.g., bulk polycrystalline samples with nanosized crystallites, as well as superlattice assemblies of nanocrystals), and *nanometer-thick thin films*. The preparation of the latter can be traced back to the 1930s with the development of the Langmuir–Blodgett method for depositing mono- and multilayer organic thin films by repeatedly dipping a substrate into water covered with a monolayer film (Langmuir, 1917; Blodgett, 1934, 1935). This method is named after Nobel laureate Irving Langmuir (1881–1957) and his longtime colleague Katherine Blodgett (1898–1979), both of whom were GE scientists. The preparation of metal nanopowders by thermal evaporation also dates back to 1930, with the work of American physicist August Hermann Pfund (1879–1949) and the Dutch physicist Hermann Carel Burger (1893–1965) (Pfund, 1930; Burger and van Cittert, 1930).

By 1960, arc discharge, plasma, and flame methods had also been used to produce submicron particles. However, the state of affairs in existence today had to await advancement in numerous techniques for characterizing nanosized particles, including spectroscopy, diffraction, and microscopy. For example, in 1982, IBM Zurich scientists Gerd Binning and Heinrich Rohrer introduced the scanning tunneling microscope (STM), an instrument for imaging the topography of metal surfaces with atomic resolution (Binning and Rohrer, 1982). They were awarded the 1986 Nobel Prize in Physics for this development. In 1990, D. M. Eigler and E. K. Schweizer used an STM to produce the now famous image of the letters "IBM," spelled out of 35 xenon atoms on a nickel surface (Eigler and Schweizer, 1990). This development was a major milestone for nanotechnology, proving it was now possible to manipulate individual atoms.

High-resolution mass spectroscopy enabled the 1985 discovery of the soccer ball-like C60 allotrope of carbon, buckminsterfullerene, from the products of laser vaporization of graphite (Kroto et al., 1985). The C60 molecule was isolated in 1990 (Taylor et al., 1990). Multiwalled carbon nanotubes were reported in 1991 from high-resolution transmission-electron microscopy studies of the material deposited on graphite rod electrodes during the arc-discharge synthesis of fullerenes (Iijima, 1991). In 1993, it was reported that single-walled nanotubes were obtained from the addition of metals such as cobalt to the graphite electrode (Iijima, 1993).

Since the discovery of carbon nanotubes, a large amount of effort has been devoted to understanding their formation conditions and to synthesizing nanotubes of other layered solids, such as MoS_2 (Feldman et al., 1995) and BN (Chopra et al., 1995). Nanoparticles of layered structures with a high fraction of their atoms on the prismatic face, perpendicular to the basal plane (parallel to c), are more likely to contain a larger number of unsaturated or dangling bonds on that face that destabilize the planar topology below some critical size. These particles are more stable as nanotubes and other fullerene-like hollow structures, although the exact folding mechanism is not clear. The analogous driving forces under which hollow macroscopic fibers are formed from naturally occurring minerals with layered polyhedral networks was investigated as early as 1930 by Pauling (Pauling, 1930; Tenne et al., 1998). There are also examples of substances with nonlayered crystal structures that have been synthesized in nanotube form, including: GaN (Goldberger et al., 2003), SiO_2 (Yu et al., 1998), Al_2O_3, (Ajayan et al., 1996), and bismuth (Li et al., 2001). Because they do not form spontaneously, these particles are usually synthesized with some sort of templating process or by thin-film rolling techniques.

Many chalcogenide semiconductors were also at the center of early attention, such as ME, with $M = $ Zn, Cd, Hg, and Pb, and $E = $ S, Se, and Te (Babcock et al., 1998). Initially, research on nanocrystalline ceramics was focused on simple binary compounds. Among the earliest reported nanoceramics were MgO (Utampanya and Klabunde, 1991), TiO_2 (Melendres et al., 1989), and AlN (Chow et al., 1994).

11.2 PROPERTIES OF MATTER AT THE NANOSCALE

The key to understanding why the properties of nanomaterials differ from those of their macroscopic counterparts is the intermediate location of the nanoscale. The nanoscale range is usually taken to be from 1 nm to 100 nm (although some researchers use the range 1–1000 nm). This scale is *between* those traditionally of interest to condensed-matter physicists (e.g., a mole of copper atoms with a volume of 7.1×10^{21} nm^3) and chemists (e.g., the water molecule with a diameter of about 0.3 nm). Because of its intermediate nature, this field of study is sometimes referred to as *mesoscopic physics*.

The smallest nanoparticles do not behave like free atoms, molecules, or extended solids. In fact, if we were to take a crystalline solid and mechanically divide it up into smaller and smaller pieces, until we approached a single unit cell, we would find, perhaps to our surprise, that the properties of these small pieces *do not* correspond to those of the macroscopic solid. Rather, nanomaterials have unusual optical, magnetic, and electronic properties that are size-dependent.

Why is this? As a first step toward answering this question, consider electronic structure. Recall from Section 3.2 how the density and energy levels of molecular orbitals or crystal orbitals depend on the number of atomic orbitals combined. It was pointed out that the energy separation between both types of orbitals decreases with an increase in the number of atomic orbitals. Hence, in a macroscopic solid with $\sim 10^{23}$ atoms, quasi-continuous energy bands with an infinitesimal separation between the crystal orbitals are produced. As illustrated in Figure 11.1, the situation becomes intermediate between a macroscopic sample, with its bands, and a molecule or cluster, with discrete energy levels, as the sample size reaches the nanoscale range. Quite simply stated, the smaller nanoparticles are on the borderline between molecule and bulk, where the band structure begins to disappear and discrete energy levels begin to dominate.

11.2.1 Electrical Properties

Our picture of a transition from bands to discrete energy levels predicts a decrease in metallic conductivity as the number of atoms in a sample decreases. At what point does a metal cease to be a metal? More generally, one might ask: *At what size does a particle no longer exhibit behavior characteristic of the macroscopic material?* A conservative estimate of the range for conduction-band formation in an isotropic monatomic solid, such as a sphere or cube, might be between 10^4 and 10^5 atoms (Elliot, 1998). The distinction of an isotropic 3D solid is an important one. The smallest mesoscopic systems are small clusters of atoms called *quantum dots*, which exhibit 0D (nonconducting) transport properties because the volume of the sample is less than the electron Fermi wave vector, ~ 10 nm for most metals. As the size of a sample is increased beyond this regime in only one or two dimensions, *quantum wires* (1D) or *quantum wells* (2D), respectively, are obtained.

Although we cannot really use band-structure calculations to study small clusters, we can use them to explain the electronic properties observed in reduced

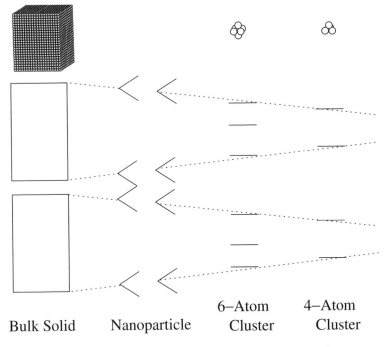

Figure 11.1 The particle-size regime at which a gradual transition occurs from energy bands to discrete energy levels falls somewhere in the nanoscale range.

dimensional systems, such as quantum wires and wells with far fewer than 10^5 atoms. For example, recall from Section 4.2.2 how a nanotube or nanographite ribbon of sufficient length can be regarded as infinitely long and artificial periodic boundary conditions are imposed along the tube or ribbon axis on a macroscopic scale. Most nanographites and nanotubes are 1–100 μm in length, a size regime well above that required for quantum confinement along that axis. Only for extremely short lengths do we need to worry about quantum confinement in all three orthogonal directions, that is, 0D quantum-dot formation.

It is precisely the topological change from an infinite flat sheet (graphene) to a finite-width ribbon or cylinder of nanometer dimensions, and the accompanying boundary conditions along these other directions, that account for the differences between nanographite, nanotubes, and graphene. For example, in a nanotube, electrons are confined to a discrete set of energy levels in the circumferential direction. Only wave vectors satisfying the relation $C \cdot k = 2\pi q$, where C is the chirality vector and q is an integer, are allowed in the corresponding reciprocal space direction (see Eq. 4.37 and let $C = N$, and $q = n$).

Carbon nanotubes can only transport current along parallel 1D channels down the tubular axis. This produces a set of continuous 1D subbands (giving rise to van Hove singularities in the density of states (DOS)), making them *quantum wires*. Nonetheless, the dispersion relations somewhat resemble that of graphene, but with metallic or semiconducting behavior exhibited by the armchair or zigzag nanotubes, respectively. The onset of metallicity has been observed in individual zigzag nanotubes with the vapor phase intercalation of potassium atoms, which donate their electrons to the vacant conduction band, thereby raising the Fermi level (Bockrath, 1999). As the diameters of the semiconducting zigzag nanotubes increase, the band gap closes, corresponding to the semimetallic graphene. Band-structure calculations have thus proven very useful for explaining the electronic properties of these reduced-dimensional systems.

11.2.2 Magnetic Properties

Nanosize particles are also interesting with regard to their magnetic properties. In the size range of about 10 to 100 nm, single magnetic domains are observed, and below this regime, superparamagnetism. At the time of this writing, magnetic hard disk coatings are arrays of cobalt, chromium, and platinum or tantalum particles, between 10 and 20 nm in size, magnetically separated by chromium, which segregates to the grain boundaries, decoupling the magnetic exchange between grains. With storage densities of about 30–35 Gbits/in.^{2}, this corresponds to one bit of data per 10^3 grains. The ultimate goal in magnetic hard disks is to achieve storage densities of one bit per single domain ferromagnetic particle close to, but above, the superparamagnetic limit ($\sim$10 nm in size).

Another interesting aspect, unique to the smallest nanoparticles, is the paramagnetism observed in diamagnetic samples, such as 2–4-nm palladium clusters, at very low temperatures ($<$ 1 K) (Volokitin et al., 1996). A topic of current theoretical interest is the search for a satisfactory explanation for ferromagnetism in ultrasmall nanometals. With discrete electronic energy levels, the validity of the Stoner criterion for itinerant ferromagnetism (Section 7.4.3) comes into question. To be sure, there are other important factors to consider with nanoscale magnetic particles, such as spin frustration on the surface. Nonetheless, the gist of the matter is still: How small can a particle be and still retain its bulklike properties?

11.2.3 Optical Properties

It is believed that nonlinear optical photonic crystals, which can bend and amplify selected photons (wavelengths) of light, will dominate signal-processing technology after the electronics era. This, combined with the intriguing properties of matter at the nanoscale, has made nanooptics an active research field.

The optical properties of solid metallic nanoparticles are accounted for nicely by classic electromagnetic theory (Mie scattering), which is effective for $2\pi r \geq \lambda$, where r is the particle radius, and λ is the wavelength (the wavelength of visible light is in the 400–700-nm range). Mie scattering typically results in an exponential decay profile with decreasing photon energy in the ultraviolet–visual (UV–vis) spectrum. Sometimes superimposed on this is a well-defined *surface plasmon band*. Plasmons are long-wavelength collective oscillations—like a charge-density wave—involving *all* the conduction electrons. Plasmons can be energetically excited, which appears as a pronounced resonance in the optical absorbance spectra, corresponding to a transition from the ground state of zero reflectivity (total transparency, or nonabsorbing) to a reflectivity of unity. This resonance dominates the linear and nonlinear optical response of the material.

Resonance occurs when the wave vectors of the photon and plasmon are equal in magnitude and direction for the same frequency of the waves. The frequency, ν, of the plasmon and, hence, of the photon required for resonance is given by:

$$\nu = [n_0 e^2 / \varepsilon_0 m_e]^{1/2} \qquad (11.1)$$

where n_0 is the electron density, e is the electron charge, m_e is the electron mass, and ε_0 is the vacuum permittivity. The spectral line is of finite width due to excitation lifetime effects. As the particle size decreases, the width of the spectral line is broadened due to a reduction in the lifetime of the excitation.

Line-width broadening may also be caused by other fast relaxation mechanisms in addition to a small particle size. For example, it is well known that, for spherical particles, radiation losses become more pronounced with *increasing* radius. In some metals, these relaxation mechanisms are so strong that a well-defined plasmon resonance is not observed, as in Fe, Pd, and Pt. Nanosized particles are interesting because the optical resonance can be designed in. For example, in a nanoshell consisting of a dielectric core surrounded by a metallic outer layer, the relative dimensions of these components can be varied. This, in turn, varies the optical resonance, possibly over several hundred nanometers in wavelength.

The optical response of a nanocrystalline phase differs from those of an isolated cluster. Linear and nonlinear optical properties in nanocrystalline phases are regulated by particle–particle interactions. This is because the plasmon resonance absorption of a nanoparticle is affected by scattering from the other nanoparticles in the assembly. Generally, the particle–particle interactions lead to shifts in the surface plasmon resonance frequencies and/or the appearance of additional low-energy (long-wavelength) peaks in the optical absorption spectrum, with their location dependent on the geometry of the assembly (Quinten and Kriebig, 1986).

11.2.4 Thermal Properties

We are taught early on that specific heat, thermal conductivity, and melting point are intensive properties, that is, independent of the quantity of material present, and thus are characteristic of a given substance. We might logically argue, then, that these properties would be independent of particle size. However, this intrinsic behavior holds only as long as the fraction of surface atoms is small. For the most part, nanoparticles have higher specific heats, higher thermal conductivities, and lower melting points than their bulk counterparts do.

It must be pointed out that this seemingly anomalous behavior does not represent a breakdown in thermodynamics! In fact, it can be explained by statistical mechanical arguments. All of the aforementioned properties are linked to temperature-dependent atomic vibrations or oscillatory displacements of the atoms about their equilibrium positions. These atomic motions are coupled to give lattice vibrations, called phonons, which are important in describing specific heat and thermal conductivity. Surface atoms can undergo higher-amplitude vibrations at a given temperature than those of the interior, because they are not bound as strongly as atoms in the interior of a particle. The large fraction of surface atoms in a nanocrystal ($\sim$50% for a 3-nm particle of iron) then directly results in an increased specific heat and thermal conductivity, due to the enhanced average atomic displacement. Since melting occurs when atomic displacements exceed a certain fraction of the interatomic distances in the solid, nanocrystals also melt at lower temperatures (sometimes dramatically lower) than bulk crystals.

11.2.5 Mechanical Properties

The mechanical properties of some of the smallest nanoparticles have been investigated. Atomic-force microscopy (Salvetat et al., 1999) and micro-Raman spectroscopy (Lourie and Wagner, 1998) have been used to determine the axial Young's moduli of individual carbon nanotubes. These materials are stiffer than any other known material, with an axial Young's modulus comparable to the in-plane Young's modulus of graphite (about 1 TPa) (Baker and Kelly, 1964). The C—C bonds in graphite and nanotubes are very strong, being intermediate in length between that of a single carbon bond and of a double carbon bond. The smallest diameter nanotubes ($<$1 nm) are predicted to have a Young's modulus perhaps as high as 5 TPa (Overney and Zhong, 1993)! This is due to the presence of σ-bonding, in addition to the extended π system, around the circumference. Of course, the van der Waals attractions between nanotubes in strands or bundles of hexagonally packed nanotubes, like the *inter*planar forces in graphite, are much weaker. Because of their extraordinary strength, nanotubes are being investigated as particle reinforcements for structural materials.

As with nanostructured particles, nanocrystalline phases also possess unique mechanical properties. The elastic moduli of nanocrystalline phases are approximately the same as those for materials with conventional grain sizes until the grain size becomes smaller than about five nanometers. Similarly, the hardness and yield strength increase with decreasing grain size until about 20 nm (Lu and Liaw, 2001). Below this limit, however, things become controversial. Grains in this ultrasmall-size regime cannot support dislocation activity, since the stress required to bow out a dislocation approaches the theoretical shear stress (Legros et al., 2000). Because of the absence of dislocation activity, the empirical Hall–Petch equation (Eq. 9.27) is generally not applicable.

Some data indicate *inverse* Hall–Petch behavior with nanophase metals as the grain size reaches the critical limit of 10–20 nm. In this regime, the material softens with a further decrease in grain size. Other results imply a continued increase in yield stress and ductility with decreasing grain size. Still yet, some experiments indicate that the yield stress is independent of grain size in this size regime. Due to the large surface area of nanoscale grains, a large percentage of the atoms are within

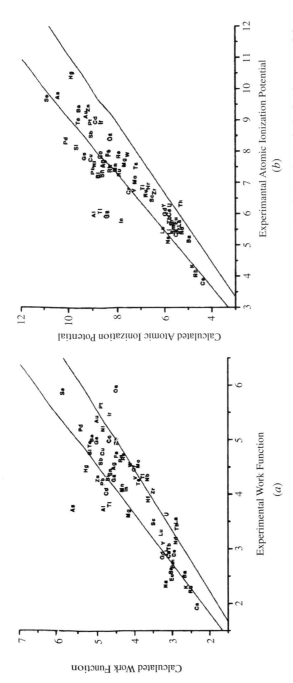

Figure 11.2 (a) A plot of the calculated metal work function versus experimental values and (b) calculated atomic ionization potentials versus experimental values. The solid lines represent regions of ±10% deviation. (After Wong et al., 2003. Copyright © American Physical Society. Reproduced with permission.)

the grain boundary regions. There is also an increase in the
volume because of the very small grain sizes. Molecular dynam
suggested that the plastic deformation mechanism in the sma
metals involves grain boundary sliding and the emission/rea
dislocations (Hemker, 2004).

Nanoceramic powders are more ductile, tougher, and stronge
grained counterparts. They can also be sintered at much lower te
plasticity can generally be observed in ceramics at smaller na
because the superplastic deformation mechanism is thought to
influenced by lattice diffusion than dislocation slip (Nakano et .

11.2.6 Chemical Reactivity

The large surface area/volume ratio of nanoparticles means that m
are on the surface, which allows nanoparticles to react as neai
reagents in chemical reactions, unlike bulk solids. For example, a
in the shape of an octahedron contains 100% of its atoms on the s
low either the ccp or hcp theme, the next smallest close-packed co
we can build has a central atom coordinated to six others in one la
in a layer above, and three in a layer beneath. Hence, $12/13 = 92\%$
this cluster are on the surface. Dense close-packed clusters such a
full-shell clusters. As the number of atoms in a 3D full-shell clust
percentage of surface atoms decreases. It is well known that surface
chemically reactive than atoms in the interior. Thus, nanoparticles w
ed to be very reactive.

Two quantities that can be related to the *intrinsic* chemical reac
clusters, and bulk solids are the electron addition energy, or *electron*
electron removal energy, or *ionization energy*. Note that the ion
in electronvolts is numerically equal to the *ionization potential* in v
all well-understood properties for atoms. There is generally excell
between experimental measurements of these quantities and theoretic
The same cannot be said for solids or even clusters, however. The ion
and electron affinity for a bulk metal are equivalent and given by the v
which is the energy difference between the vacuum level and the higl
electron states. The electron affinity for a semiconductor, however,
difference between the vacuum level and the conduction-band minimu
ionization energies and electron affinities of bulk solids are dependei
cleanliness and the crystal face exposed.

In finite clusters, as with atoms, the properties are no longer equ
furthermore exhibit a particle-size dependency. Figure 11.2a shows the
tion for several metals (Wong et al., 2003). The *atomic* ionization ener
same metals are shown in Figure 11.2b. Note that the atomic ionizati
are normally about twice the magnitude of the work function. Figure
the experimentally measured ionization energies for small-to-large
several of these metals fit by Wong to a model containing only two par

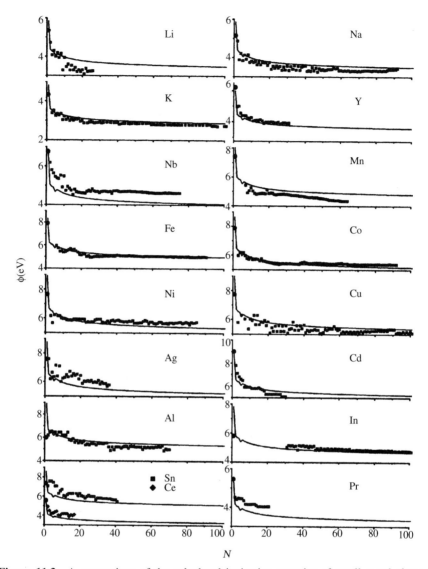

Figure 11.3 A comparison of the calculated ionization energies of small metal clusters versus experimental values. (After Wong et al., 2003. Copyright © American Physical Society. Reproduced with permission.)

circumscribing radius (dependent on the geometry of the cluster), and the number of atoms. It can be seen that even though the actual ionization energies of cluster do not scale *monotonically* (always increasing or always decreasing) with particle size, particularly in the small-cluster regime, the model is successful in yielding values reasonably close to the experimental results, and exceptionally close in some cases.

The electron affinities of clusters behave in a similar manner. This fact undoubtedly has a role to play in the chemistry exhibited by nanometals that has been reported in the literature recently. For example, it has been shown that Au atoms (gold is a noble metal in the bulk state) supported on a TiO_2 substrate show a marked size effect in their ability to oxidize the diatomic gas CO to CO_2 via a mechanism involving O_2 dissociative chemisorption and CO adsorption (Valden et al., 1998). Small Ni particles also have been found to dissociate CO (Doering et al., 1982). Smaller nanoparticles of Ag can dissociate molecular oxygen to atomic oxygen at low temperatures, whereas in the bulk state, the species adsorbed on the Ag surface is O_2^- (Rao et al., 1992).

REFERENCES

Ajayan, P. M.; Stephan, O.; Redlich, P.; Colliex, C. 1995, *Nature, 375*, 564.

Babcock, J. R.; Zehner, R. W.; Sita, L. R. 1998, *Chem. Mater., 10*, 2027, and references therein.

Baker, C.; Kelly, A. 1964, *Philos. Mag., 9*, 927.

Binning, G.; Rohrer, H. 1982, *Helv. Phys. Acta, 55*, 726.

Blodgett, K. B. 1934, *J. Am. Chem. Soc., 56*, 495.

Blodgett, K. B. 1935, *J. Am. Chem. Soc., 57*, 1007.

Bockrath, M. W. 1999. Ph.D Thesis, University of California at Berkeley, Berkeley, CA.

Burger, H. C.; van Cittert, P. H. 1930, *Z. Phys., 66*, 210.

Chopra, N. G.; Luyken, R. J.; Cherrey, K, Crespi, V. H.; Cohen, M. L.;Louie, S. G.; Zettl, A. 1995, *Science, 269*, 966.

Chow, G. M.; Xiao, T. D.; Chen, X.; Gonsalves, K. E. 1994, *J. Mater. Res., 9*, 168.

Doering, D. L.; Dickinson, J. C.; Poppa, H. 1982, *J. Catal., 73*, 91.

Eigler, D. M.; Schweizer, E. K. 1990, *Nature, 344*, 524.

Elliot, S. R. 1998, *The Physics and Chemistry of Solids*, John Wiley & Sons, Chichester, UK.

Faraday, M. 1857, "The Bakerian Lecture: Experimental Relations of Gold (and other metals) to Light," *Philos. Trans. R. Soc., 147*, 145.

Feldman, Y.; Wasserman, E.; Srolovitz, D. J.; Tenne, R. 1995, *Science, 267*, 222.

Feynman, R. 1960, "There's Plenty of Room at the Bottom: An Invitation to Enter a New Field of Physics," *Caltech's Eng. Sci., 23*, 22–36.

Goldberger, J.; He, R.; Zhang, Y.; Lee, S.; Yan, H.; Choi, H.-J.; Yang, P. 2003, *Nature,*

Graham. T. "Liquid Diffusion Applied to Analysis," 1861, *Philos. Trans. R. Soc., 151*, 183.10, 599.

Hemker, J. K. 2004, *Science, 304*, 221.

Iijima, S. 1991, *Nature, 354*, 56.

Iijima, S. 1993, *Mat. Sci. Eng., B19*, 172.

Kroto, H. W.; Heath, J. R.; O'Brien, S. C.; Curl, R. F.; Smalley, R. E. 1985, *Nature, 318*, 162.

Langmuir, I. 1917, *J. Am. Chem. Soc., 39*, 1848.

Legros, M.; Elliot, B. R.; Rittner, M. N.; Weertman, J. R., Hemker, K. J. 2000, *Philos. Mag. A, 80*, 1017.

Li, Y.; Wang, J.; Deng, Z.; Wu, Y.; Kind, H.; Yang, P. 2001, *J. Am. Chem. Soc.*, *123*, 9904.

Lourie, O.; Wagner, H. D. 1998, *J. Mater. Res.*, *13*, 2418.

Lu, Y.; Liaw, P. K. 2001, *JOM*, *53*, 31.

Melendres, C. A.; Narayanasamy, A.; Maroni, V. A.; Siegel, R. W. 1989, *J. Mater. Res.*, *4*, 1246.

Nakano, M.; Nagayama, H.; Sakuma, T. 2001, *JOM*, *53*, 27.

Overney, G.; Zhong, W.; Tomenek, D. 1993, *Z. Phys. D*, *27*, 93.

Pauling, L. 1930, *Proc. Nat. Acad. Sci.*, *16*, 578.

Pfund, A. H. 1930, *Phys. Rev.*, *35*, 1434.

Quinten, M.; Kriebig, U. 1986, *Surf. Sci.*, *172*, 557.

Rao, C. N. R.; Vijayakrishin, V.; Santra, A. K.; Prims, M. W. J. 1992, *Angew. Chem.*, *104*, 1110.; 1992, *Angew. Chem. Int. Ed. Engl.*, *31*, 1062.

Salvetat, J. P.; Briggs, G. A. D.; Bonard, J. M.; Bacsa, R. R.; Kulik, A. J.; Stockli, T.; Burnham, N. A.; Forro, L. 1999, *Phys. Rev. Lett.*, *82*, 944.

Taylor, R.; Hare, J. P.; Abdul-Sada, A. K.; Kroto, H. W. 1990, *J. Chem. Soc. Chem. Commun.*, 1423.

Tenne, R.; Homyonfer, M.; Feldman, Y. 1998, *Chem. Mater.*, *10*, 3225.

Utampanya, S.; Klabunde, K. K. 1991, *Chem. Mater.*, *3*, 175.

Valden, M.; Lai, X.; Goodman, D. W. 1998, *Science*, *281*, 1647.

Volokitin, Y.; Sinzig, J.; de Jongh, L. J.; Schmid, G.; Moiseev, I. I. 1996, *Nature*, *384*, 621.

Wong, K.; Vongehr, S.; Kresin, V. V. 2003, *Phys. Rev. B*, *67*, 35406.

Yu, D. P.; Hang, Q. L.; Ding, Y.; Zhang, H. Z.; Bai, G.; Wang, J. J.; Zou, Y. H.; Qian, W.; Xiong, G. C.; Feng, S. Q. 1998, *Appl. Phys. Lett.*, *73(21)*, 3076.

Zhou, W.; Wiemann, K. J.; Fang, J.; Carpenter, E. E.; O'Connor, C. J. 2001, *J. Solid State Chem.*, *159*, 26.

CHAPTER 12

Synthetic Strategies

In this final chapter we consider some of the synthetic strategies employed in inorganic materials synthesis. Materials synthesis is an intellectually vast and economically important area that cannot be covered in a single text, let alone a single chapter. The purpose of this chapter is to review, and maybe introduce, some of the important techniques being used. The references at the end of the chapter are meant to provide a representative glimpse of this extensive and expanding field as opposed to being an exhaustive list. This final chapter also provides an opportunity to further illustrate some of the fundamental principles discussed in the previous chapters such as phase diagrams, band gaps, and diffusion coefficients.

More importantly, by discussing individual methods, this chapter demonstrates how these stand-alone methods are increasingly being coupled to each other, providing new and exciting opportunities in materials synthesis. One of the difficulties in inorganic solid-state synthesis is the lack of a small set of techniques for the practitioner to master. Unlike organic chemistry, where a few basis skills can lead to a wide variety of syntheses, solid-state inorganic chemistry requires many specialize methods from flux to ion exchange to high temperature, and soon. This is a consequence of the variety of atoms, structures, and phases encountered in solid-state synthesis.

Despite the disjointed nature of solid-state synthesis, there are currently global themes being pursued. One of them, and the subject of part of this chapter, is lower reaction temperatures. As we see in this chapter, this seemingly simple goal is having a profound consequence on the entire field of inorganic materials design. While in the past these innumerable synthetic methods led to isolation and specialization among the users, a growing trend among the synthetic chemists is to combine these well-defined methods. This trend is coupled to the goal of lower reaction temperatures. Is it precisely because more and more synthesis is being done at lower temperatures that hybrid techniques are possible.

Synthetic strategy has at least two meanings. The first involves using a given technique and employing chemical and physical arguments as to what series of compounds should be prepared and why. For example, in 1970, Gamble et al.

Principles of Inorganic Materials Design   By John N. Lalena and David A. Cleary
ISBN 0-471-43418-3   Copyright © 2005 John Wiley & Sons, Inc.

reported that the reaction of TaS_2 with pyridine resulted in an increase in the superconducting temperature of the TaS_2 from 0.7 K to 3.3–3.5 K, a remarkable increase at the time (Gamble et al., 1970). Subsequently, thousands of papers have been published exploring the use of different intercalating agents in a wide variety of transition-metal dichalcogenides and other layered compounds. This work was driven by the search for materials with increasingly large superconducting critical temperatures. The intercalated metal dichalcogenides have been replaced with the metal oxides as the focal point of high-temperature superconductor research, setting off another round of intense systematic synthesis in 1986, although in 1989, the Noble Prize–winning Russian physicist V. L. Ginzburg said, "As I have said in the past, it seems to me that the best candidates [for $T_c \leq 300 - 400$ K] are organic superconductors and inorganic layered compounds, particularly intercalated ones" (Ginzberg, 1989).

The second meaning of *synthetic strategy*, and the subject of this chapter, involves the preparative techniques. Solid-state synthesis has always suffered from the lack of unifying principles upon which to base synthetic attempts. In contrast, organic chemical synthesis can be planned in detail owing to simple concepts such as electronegativity and steric hindrance. Nevertheless, solid-state synthesis has blossomed and continues to be an active area of research. In some sense, the lack of a unifying set of rules has liberated the solid-state synthetic chemist: many an interesting and unexpected result has come from a "let's see what happens when..." approach, or, put more formally by A. W. Sleight: "The main lesson to be learned from the discovery of high temperature superconductors is that we are, in general, not yet clever enough to produce breakthrough compounds by design. Therefore, we must conduct exploratory synthesis of new compounds in order to discover their unexpected properties." (Sleight, 1989).

12.1 SYNTHETIC STRATEGIES

Solid-state synthesis is rich in the variety of techniques used to prepare materials. In this chapter we present and discuss some of the major methods used today. Some of these methods, such as direct combination, have been in use since the beginning of chemistry. Others, such as photonic materials, involved the latest high technology in lithography, patterning, and masking. The methods we discuss include the use of:

- Direct synthesis
- Low temperature
- Defects
- Combinatorial synthesis
- Spinodal decomposition
- Thin films
- Photonic materials
- Nanosynthesis

12.1.1 Direct Synthesis

This classic method of preparation remains the bulwark of solid-state synthesis. The starting materials can be elements as in the preparation TaS_2:

$$Ta(s) + 2S(s) \rightarrow TaS_2(s)$$

In the case of oxides, where use of the elemental oxygen gas presents additional challenges, oxides are often used as reagents:

$$BaO(s) + TiO_2(s) \rightarrow BaTiO_3(s)$$

This type of traditional solid-state chemical synthesis is characterized by high temperatures and long reaction times. If the reagents, such as tantalum and sulfur, react with oxygen at high temperature, then reactions are carried out in vacuum or inert atmospheres. Often, a sealed glass ampule is used as the reaction vessel. This brings additional safety concerns into the synthetic strategy.

The fundamental reason for the use of high reaction temperatures in the direct-combination method is diffusion coefficients. In solution, where organic synthesis typically occurs, the diffusion coefficient for species is typically 10^{-5} cm^2/s. Hence, species bump into each other frequently. In the solid-state, however, if a crystallite of BaO and a crystallite of TiO_2 are in contact with each other (Figure 12.1), the diffusion coefficients for the Ba^{+2} through the $BaTiO_3$ and into the TiO_2, and the Ti^{+4} through the $BaTiO_3$ and into the BaO are orders of magnitude smaller than 10^{-5} cm^2/s, typically 10^{-10} to 10^{-15} cm^2/s. The significance of these numbers is better realized when we considers a 1D diffusion model and ask how far a particle diffuses in a given time with a particular diffusion coefficient. The solution to this problem has been worked out in detail, and according to Fick's law of diffusion, the average displacement in one dimension is (Crank, 1975):

$$\langle x \rangle = \sqrt{2Dt} \tag{11.1}$$

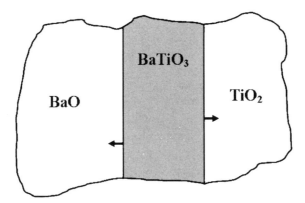

Figure 12.1 BaO and TiO_2 diffusing through $BaTiO_3$.

where D is the diffusion coefficient, and t is the time in seconds. This means that particles will move about 100 microns per hour, with $D = 10^{-12}$ cm^2/s.

Therefore, higher reaction temperatures are used in order to decrease the diffusion coefficients where:

$$D(T) = D_0 e^{E_a/RT} \tag{11.2}$$

where E_a is the activation energy barrier to diffusion. Even with the increase in the diffusion coefficient, long reaction times are still required because the diffusion coefficient does not approach the solution-phase value.

The requirement of high temperature to produce practical diffusion coefficients imposes severe limitations on the type of reactants that can be used. Hence, the two preceding examples involve reagants that are stable at high temperature. Moreover, the products formed at these necessarily high temperatures represent the thermodynamically stable phase. From a practical point of view one is faced with the frustration that the temperature at which most devices operate is significantly lower than the high preparative temperatures. Hence an active goal of synthetic research involves developing low-temperature methods for producing solid-state compounds.

12.1.2   Low Temperature

Low temperature in the solid-state synthesis business covers the range from room temperature to roughly 300°C, the maximum temperature of most oil baths or routine laboratory drying ovens. Low-temperature methods have the advantage of using a wider range of reactants, accessing phases unstable at high temperature, and being less expensive to execute. We discuss three distinct low-temperature methods:

- Sol-gel
- Solvothermal
- Intercalation

and later in the chapter we see how these methods are being combined with other synthetic strategies.

The *sol-gel* method is a variation on the gel method for producing single crystals of ionic materials (Brinker and Scherer, 1990). The gel method has been used for many years to produce single crystals of materials formed from ions where direct combination results in instantaneous product formation, and hence no large single crystals. A good example is the formation of lead iodide from Pb^{+2} and I$^-$. When these two are combined as aqueous solutions,

$$\text{Pb}^{+2}(\text{aq}) + \text{I}^-(\text{aq}) \rightarrow \text{PbI}_2(\text{s})$$

the yellow product forms immediately and has no large crystals, as seen in the photograph in Figure 12.2. When the two ions are held apart from each other by an inert aqueous gel, however, the ions slowly diffuse together and beautiful large

Figure 12.2 Formation of PbI$_2$(s) from Pb^{+2}(aq) and I$^-$(aq).

crystals are produced (Henisch, 1970). The gel is produced by the dehydration of silicic acid (Figure 12.3):

$$H_4SiO_4(aq) \rightarrow SiO_2(s) + 2H_2O(l)$$

The water produced as a result of the dehydration is trapped in the 3D net formed by the SiO$_2$, producing a gel with the consistency of Jello. To produce large single crystals of PbI$_2$, the Pb^{+2} ions are incorporated into the gel and an aqueous solution of I$^-$ is placed on top of the gel once the gel has set. As shown schematically in Figure 12.4, large single crystals of PbI$_2$ grow and are suspended in the gel.

Because the silicic acid spontaneously dehydrates, it is formed when needed by the hydrolysis of sodium metasilicate within an acidic aqueous medium:

$$Na_3SiO_3(aq) + 3H_2O(l) \rightarrow Si(OH)_4(aq) + 3NaOH(aq)$$

It is at this stage that the metal ions are incorporated into the gel. For example, if acetic acid is used to cause the gelling, then lead acetate can be added to the acetic acid and lead ions will become embedded in the gel.

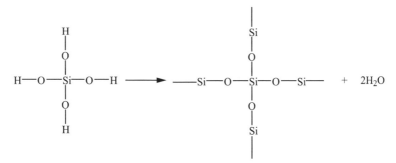

Figure 12.3 Silicic acid condensing to SiO$_2$ and water.

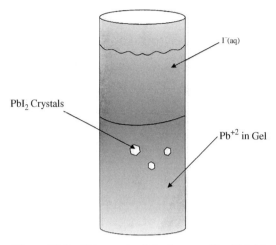

Figure 12.4 Gel growth of single crystalline $PbI_2(s)$.

In the sol-gel synthetic method, the gel graduates from inert background material to product. The underlying chemistry remains the same. The silicon dioxide network is often generated from tetraethyl orthosilicate ($Si(OCH_2CH_3)_4$), commonly abbreviated TEOS. An alcoholic solution of TEOS undergoes hydrolysis and condensation when added to water to produce $SiO_2(s)$ and ethanol:

$$Si(OCH_2CH_3)_4 + 2H_2O \rightarrow SiO_2(s) + 4CH_3CH_2OH$$

By chemically modifiying TEOS, chemically modified $SiO_2(s)$ can be prepared, a feat not easily achieved by attempting to chemically react the robust $SiO_2(s)$ directly. Chaput and co-workers attached a carbazole-9-carbonyl chloride (CB) to 3-amino-propyltriethoxysilane, as shown in Figure 12.5 (Chaput et al., 1996). Similarly, they attached the well-known nonlinear optical dye, Disperse Red 1 (DR1), to a derivative of TEOS, as shown in Figure 12.6. When these chemically modified TEOS are mixed with TEOS and the hydrolysis/condensation occurs, $SiO_2(s)$ is produced that contains organic dyes covalently bonded to the silicon atoms (Figure 12.7), where the spacing between the CB and DR1 groups is determined by the ration of CB to DR1 to TEOS.

The field of sol-gel synthesis has blossomed with this approach of chemically modifying the monomers prior to their polymerization into a covalently bonded 3D

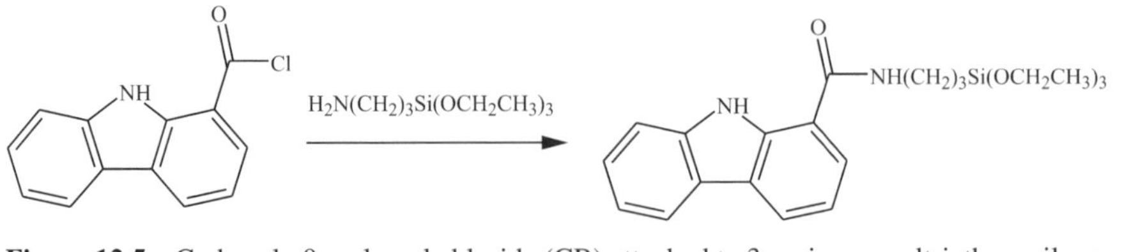

Figure 12.5 Carbazole-9-carbonyl chloride (CB) attached to 3-aminopropyltriethoxysilane.

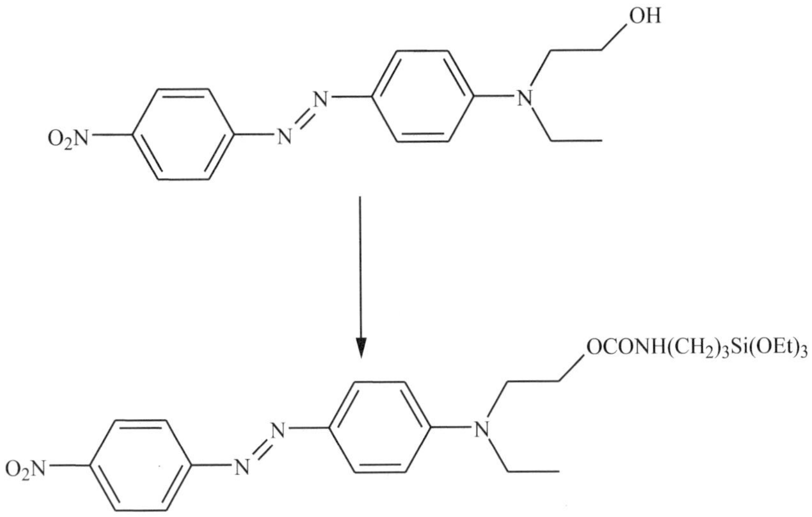

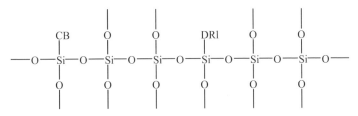

Figure 12.6   Disperse Red 1 (DR1) attached to a derivative of TEOS.

network. This chemistry is reminiscent of traditional organic polymer chemistry where the chemical changes are inflicted on the monomers, which are more labile than the polymer. Hence a polymer can be chemically modified by constructing it with modified monomers. These monomers lend themselves to the usual chemical manipulations of substitution, elimination, extraction, purification, and so forth. The sol-gel method can be performed at room temperature, another highly attractive synthetic feature. The various sol-gel procedures, terminology, and conditions are summarized in Figure 12.8.

The next low-temperature method to discuss, *solvothermal*, requires some heating. Like sol-gel, sovothermal is a derivative method. It comes from the hydrothermal method. Hydrothermal synthesis uses water as a solvent with temperatures and pressures well above 100°C and 1 atm. This method was made famous by Bell laboratories with the synthesis of quartz single crystals (Laudise, 1987). Because of the very high pressures involved in the quartz crystal synthesis, typically over 20,000 psi, expensive equipment is required. The critical temperature and pressure of water are 373.94°C and 3200 psi, respectively. At this point the liquid and vapor phases of water coalesce and form a single phase with a density of 0.322 g cm^{-3}.

Figure 12.7   Spacing of CB and DR1 on a SiO_2 backbone.

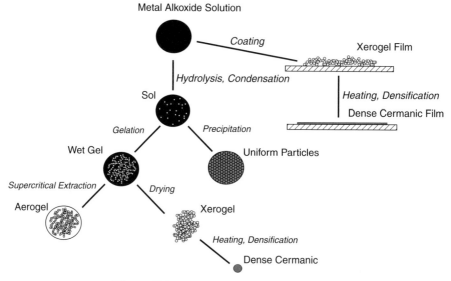

Figure 12.8   Summary of sol-gel processing.

The hydrothermal method for producing single crystals of quartz works by taking advantage of the slight solubility of $SiO_2(s)$ in supercritical water:

$$SiO_2 + 2H_2O \rightleftharpoons Si(OH)_4$$

the reverse reaction from the gel synthesis given earlier. Even at and above the critical point of water, however, this equilibrium lies too far to the left to be of practical use. Hence mineralizers like sodium hydroxide are added to assist in the dissolution, for example,

$$3SiO_2 + 2OH^- \rightleftharpoons Si_3O_7^{-2} + H_2O$$

The dissolution of the $SiO_2(s)$ occurs in the dissolving end of the growth chamber, and the subsequent deposition of $SiO_2(s)$ occurs in the crystal growth end. These ends are defined by their temperatures, with the dissolution end being hotter than the deposition end. This process is used today to grow massive amounts of single-crystal quartz necessary for frequency oscillators in modern electronic devices. The process also occurs naturally, which results in spectacular crystals of quartz in veins and geodes, which can be purchased at most rock shops.

Hydrothermal recrystallization is not confined to quartz. Some II-VI compounds such as ZnSe, ZnTe, CdSe, and CdTe have also been recrystallized into large single crystals using the technique (Kolb et al., 1968), as well as aluminum orthophosphate, $AlPO_4$ (Kolb et al., 1980). The technique can also be used preparatively as in the case of potassium titanyl phosphate (Laudise et al., 1986):

$$KH_2PO_4 + TiO_2 \rightleftharpoons KTiOPO_4 + H_2O$$

Hydrothermal synthesis does not require the water to be above its critical point. Huan and co-workers published a synthesis of $VOC_6H_5PO_3 \cdot H_2O$ prepared from phenylphosphonic acid, $C_6H_5PO(OH)_2$ and vanadium(III) oxide, V_2O_3 (Huan et al., 1990). The two reagents were added to water, sealed in a Teflon acid digestion bomb, and heated to 200°C. Pure water has a vapor pressure of 225 psi at 200°C, well within the bursting pressure of the bomb (1800 psi). Unlike the quartz example, in this case, the solvent became incorporated into the final product.

The solvothermal method uses solvents above their boiling points, but not necessarily above their critical points. The method can be employed using sealed glass tubes, acid digestion bombs, or high-pressure autoclaves. A popular method is to use Teflon digestion bombs and microwave heaters. As in the hydrothermal method, the solvent can become part of the final product. Li and co-workers have reviewed the use of the solvothermal method for preparing metal chalcogenides in an ethylenediamine solvent (Li et al., 1999). Ethylenediamine has a normal boiling point of 117°C, and a critical temperature and pressure of 320°C and 913 psi, respectively. A typical reaction (unbalanced) described by Li et al. includes:

$$CuCl_2 + Sb_2Se_3 + C_2N_2H_8 \rightarrow Cu_2SbSe_3(C_2N_2H_8)$$

where $C_2N_2H_8$ is ethylenediamine, $H_2NCH_2CH_2NH_2$. The reaction temperature would be less than 250°C, the maximum operating temperature for the Teflon-lined digestion bomb. This compound $Cu_2SbSe_3(C_2N_2H_8)$ is an interesing example of a broad class of compounds, known as intercalation compounds, that are discussed next.

The final low-temperature method to be discussed is *intercalation*. This method could be viewed either as a postsynthetic modification of a material or a parallel event occurring during the synthesis of a material. The solvothermal synthesis of $Cu_2SbSe_3(C_2N_2H_8)$ is a good example of the latter. In this section, however, we focus on the former. As a postsynthetic modification, intercalation can be used to chemically alter a material that has been produced using some other technique.

The technique owes its name to the notion of inserting an extra day into the calendar (Whittingham and Jacobson, 1982). The term is used extensively in biochemistry to describe the insertion of one molecular fragment into the space defined by another (Mathews et al., 2000). The most important example is the intercalation of ethidium bromide, a planar molecular species into the space defined by two DNA base pairs, as shown in Figure 12.9. In materials synthesis, the most extensively studies intercalation reactions have involved graphite.

Graphite is a layered material, and under the right conditions, it will accept atomic and molecular species between the adjacent carbon sp^3 sheets. The region between the covalently bonded carbon sheets is known as the van der Waals gap, because the bonding between the neutral sheets is van der Waals in character. If the species being inserted, or intercalated, into the graphite host is too big to fit between the layers, the host will relax and expand as necessary to accommodate the guest. Such intercalation reactions have the virtue of often being carried out at room temperature, or at least under very mild conditions.

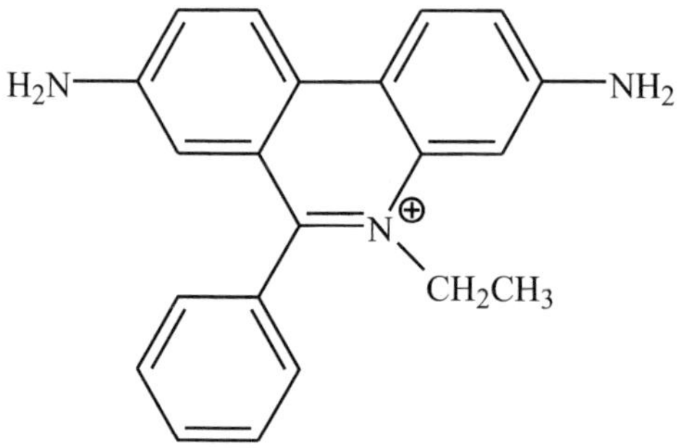

Figure 12.9 Molecular structure of ethidium bromide.

Graphite intercalation is often accompanied by oxidation reduction chemistry (Whittingham and Jacobson, 1982). For example, when graphite is intercalated with potassium, the black graphite converts to a gold, lustrous metallic-looking compound as a result of the graphite host being reduced:

$$8C(s) + K(g) \rightarrow KC_8$$

Intercalated compounds are often written with the host first, so KC_8 is usually written C_8K even though the graphite is reduced. Conversely, when graphite is intercalated with nitric acid the host graphite is oxidized:

$$24C + 4HNO_3 \rightarrow C_{24}NO_3 3HNO_3$$

This ability of graphite to participate in an intercalation reaction as either an oxidizing agent or a reducing agent is a result of the peculiar band structure of graphite.

Graphite is an example of an extended solid with a band gap equal to 0.0 eV at 0 K. Hence it can readily accept electrons into its vacant conduction band or relinquish electrons from its full valence band. Other hosts used in intercalation reactions, such as transition-metal dichalcogendies and transition-metal oxyhalides tend to prefer acting as oxidizing agents only, given the high formal oxidation state of the metal ion.

Intercalation reactions do not always result in electron transfer. The transition metal phosphorus chalcogenide, $M_2P_2S_6$, forms a series of layered compounds with M being a first row divalent transition metal ion (Brec, 1986). Clement and co-workers have intercalated a series of stilbazolium compounds into $Mn_2P_2S_6$, with some of the resulting compounds showing significant second-harmonic non-linear generation efficiencies (Coradin et al., 1996). This work is a classic example of the utility of the intercalation strategy (Gomez-Romero and Sanchez, 2004). The host, $Mn_2P_2S_6$, was prepared by a traditional high-temperature (approxately 700°C) direct-combination reaction, as described earlier:

$$2Mn + 2P + 6S \rightarrow Mn_2P_2S_6$$

A stoichiometric mixture of the elements was heated in a sealed evacuated glass ampule. A near quantitative conversion to the green transparent crystals of $Mn_2P_2S_6$ is often realized. In addition, single crystals of $Mn_2P_2S_6$ will grow in the reaction tube and can be as large as a few cm on edge. Given the layered nature of these compounds, the crystals are always very thin, <0.1 mm. The $Mn_2P_2S_6$ host crystallizes in a layered structure composed of neutral $S(Mn, P_2)S$ sheets. The van der Waals gap $(\cdots)$ occurs between sulfur layers: $S(Mn, P_2)S \cdots S(Mn, P_2)S \cdots S(Mn, P_2)S$.

In Clement's work, the van der Waals gap in the $Mn_2P_2S_6$ was expanded using an aqueous solution, tetramethylammonium chloride, at room (or near room) temperature in an intercalation reaction:

$$Mn_2P_2S_6 + 2x(CH_3)_4N^+ \rightarrow Mn_{2-x}P_2S_6((CH_3)_4N^+)_{2x}H_2O_y + xMn^{+2}$$

In this type of intercalation reaction, where cations are inserted, the host releases metal ions to compensate for the charge. This converts the host to a defect structure, the topic of the next section in this chapter. Finally, the expanded host, $Mn_{2-x}P_2S_6$ $((CH_3)_4N^+)_{2x}H_2O_y$, is converted to the desired product with an ion-exchange reaction:

$$Mn_{2-x}P_2S_6((CH_3)_4N^+)_{2x}H_2O_y + DAMS^+ \rightarrow$$
$$Mn_{1.72}P_2S_6(DAMS^+)_{0.46}H_2O_y + 2x(CH_3)_4N^+$$

where $DAMS^+$, shown in Figure 12.10, is the 4-[4-(dimethylamino)-alpha-styryl]-1-methylpyridinium cation, a known organic species with nonlinear optical properties. It is these low-temperature intercalation reactions that render the final material, $Mn_{1.72}P_2S_6(DAMS^+)_{0.46}H_2O_y$, optically nonlinear. This material could not have been prepared in one step because the reaction temperature required to produce the $Mn_2P_2S_6$ would have destroyed the organic component. In a similar fashion, when ethidium bromide is intercalated into DNA it imparts new properties on the DNA, specifically, rendering the DNA fluorescent, since ethidium bromide itself is fluorescent.

12.1.3 Defects

Here we confine out attention to point defects, such as missing or additional atoms in a crystal lattice, that disrupt the ideal stoichiometry. For example, oxygen

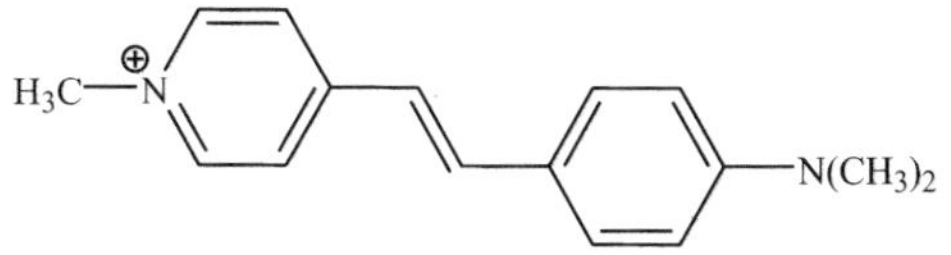

Figure 12.10 Molecular structure of the 4-[4-(dimethylamino)-alpha-styryl]-1-methylpyridinium cation ($DAMS^+$).

atoms missing from TiO_2 to produce TiO_{2-x} make TiO_{2-x} a defect material. In the previous discussion of intercalated materials, $Mn_{2-x}P_2S_6((CH_3)_4N^+)_{2x}H_2O_y$ is a defect structure. Also, Er^{+3} substituting for Y^{+3} in Y_2O_3, producing a material such as $Y_{1.9}Er_{0.1}O_3$, is a defect material. We discuss the utility of both types of strategies.

Solid oxide conductors, such as ZrO_2, which form the basis of chemical sensors rely on the presence of defects to induce ionic conductivity (Madou and Morrison, 1989). If every lattice site was occupied with the correct atom, then it would be impossible for any atoms to move within the material. It would be like an auditorium with no empty seats: in order for one person to move, everyone would have to move. Similarly in a perfect crystal, ionic conduction will be difficult if there are no vacant spots for an ion to move into.

Modifying a stoichiometric oxide into a defect structure can be as simple as heating the material in air or vacuum:

$$ZrO_2(s) \rightarrow ZrO_{2-x}(s) + x/2O_2(g)$$

This type of reaction presents a challenge when trying to prepare an oxide at high temperature. Often, an overpressure of oxygen gas is needed to prevent the type of decomposition shown in the preceding reaction. This proved especially important during the frantic period of solid-state synthesis that followed the first reports of the high-temperature oxide superconductors in 1986 (Bednorz and Muller, 1986). These materials have the formula $YBa_2Cu_3O_x$, where $6 \leq x \leq 7$. When $x = 6$, the material is a semiconductor. When $x = 7$, it is a superconductor, but the optimum composition for the superconducting phase occurs for x at slightly less than 7 (Nelson et al., 1987). In the intercalation example, the defects formed at room temperature as a charge compensation mechanism.

One of the most fruitful areas of research in solid-state synthesis is the synthesis of doped materials. Here the solid-state chemist takes advantage of a powerful concept in chemistry taught from the freshmen chemistry course onward: the periodic law. Elements of a given group display similar chemical properties. Hence, if lanthanum phosphate, $LnPO_4$, can be prepared in a particular crystallographic space group, then all members of the group IIIA should be amenable to the same phosphate structure. Indeed, the lathanide phosphates form such a series (Ushakov, et al., 2001). From a properties perspective, however, one may not want a pure $LnPO_4$ material, but one with a mixture of lanthanide ions.

Heer and co-workers produced materials with a mixture of lanthanide ions with compositions such as $YbPO_4$: 5%Er^{+3} and $LuPO_4$: 49%Yb^{+3}, 1%Tm^{+3} (Heer et al., 2003). Their reason for this formulation has to do with the interplay of the electronic transitions of the dopant ions. The suitability of the hosts $YbPO_4$ and $LuPO_4$ to accommodate a variety of different lanthanide trivalent ions is the key to this type of synthetic strategy. The fundamental principle remains the periodic law. The similarity of the rare-earth trivalent ions provides the synthetic chemist with an overabundance of choices when considering substitutions. In Heer's work, the incorporated ions work together to produce materials known as upconversion materials

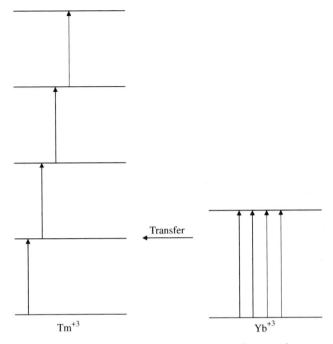

Figure 12.11   Energy transfer from Yb^{+3} to Tm^{+3}.

(Risk et al., 2003). When such compounds are irradiated at long wavelengths of light, they emit light at short wavelengths. In this particular case, four photons of 978-nm radiation are absorbed by the Yb^{+4} ions (Figure 12.11), the energy is transferred to a Tm^{+3}, and emission of one photon of 476-nm radiation results. The radiative, nonradiative, and multiphoton relaxation processes required for this elegant effect are all due to the specific energy levels of the dopant ions (Dieke, 1968).

Many routine, commercially available inorganic materials are the result of this doping for the purpose of achieving particular optical properties. The ruby laser and the neodynimium YAG laser both owe their performance to the use of deliberately doped inorganic materials where the host and the dopant are members of the same class of compounds (Silfvast, 1996). In the ruby laser case, Cr_2O_3 and Al_2O_3 for an almost continuous series (Eliseev et al., 1999). In the neodynium YAG laser, the two relevant end members of the solid solution are $Nd_3Al_5O_{12}$ and $Y_3Al_5O_{12}$. In both cases, the concentration of the substitutional dopant (Cr^{+3} and Nd^{+3}) is about 1%. This level is part of the synthetic strategy to optimize the performance of the laser. For example, at higher concentrations, the linewidth of the laser line broadens.

From the example of $YbPO_4$: 5%Er^{+3} and $LuPO_4$: 49%Yb^{+3}, 1%Tm^{+3}, we see that doping, under favorable conditions allows the synthetic chemist to produce a material with desired properties, in this case optical properties. One of the challenges presented by doping experiments is the infinite number of possibilities for doping levels. While rationale and clever planning may narrow the field considerably, there is still the possibility of needing to prepare a large number of samples, each with a slightly different composition. This leads to our next synthetic strategy, combinatorial synthesis.

12.1.4   Combinatorial Synthesis

The seemingly uncountable number of permutations and the lack of a theoretical basis for selecting which materials to produce has led to an entirely new area of synthetic chemistry called combinatorial chemistry (Xiang and Takeuchi, 2003). Combinatorial synthesis is a fancy term for what previously would have been described as the shotgun approach to synthesis. In other words, lacking a specific stoichiometric target, the synthetic chemists prepares "all" combinations and then sorts through the batch to find the one or ones with interesting or desired properties. The synthesis and characterization of high-temperature superconducting oxides presents a compelling case for such an approach. The original compound had a formula of $YBa_2Cu_3O_{7-x}$. Compositions now look like $Bi_2Sr_2CaCu_2O_8$ (Missori et al., 1994). Attempting to systematically vary the concentration of each of the five elements in this latter compound would be a monumental task. The combinatorial synthetic strategy has matured to the point that there is now an American Chemical Society publication, the *Journal of Combinatorial Chemistry*, devoted to this approach. We say more about this approach, but give a specific example when we consider hybrid strategies at the end of this chapter.

12.1.5   Spinodal Decomposition

Crystallization, whether from the molten state or from a solvent, relies upon slow cooling to produce larger crystals and fast cooling to produce small crystals. In both cases, crystallization results in the segregation of impurities. Impurity atoms usually do not "fit" into the crystal lattice of the solute crystallizing from a solvent. This allows crystallization from a solvent to be used as a purification technique. Likewise, impurity atoms in a polycrystalline metal tend to segregate at the grain boundaries during solidification because of their mismatch with the lattice of the metal atoms. Crystallization is more of a purification strategy than a synthetic strategy. Yet crystallization, as we will show, can be manipulated for synthetic purposes.

A key feature of crystallization is that as tiny crystals form, they are unstable with respect to redissolving. As shown in Figure 12.12, for a tiny crystal to increase its size, it must undergo an increase in its Gibbs free energy, a thermodynamically unfavorable situation. However, upon passing a critical size, further growth is characterized by a decrease in the Gibbs free energy. It is not until the crystals grow beyond this critical size that further growth is thermodynamically favored. This presents the interesting question of how do the big crystals appear if the little crystals are unstable? The answer has been provided by classic nucleation theory and involves fluctuations in particle sizes (Adamson, 1982). The term for separation of one phase from another under these condition is *nucleation-and-growth*, distinguishing it from the mechanism that is the subject of this section, *spinodal*

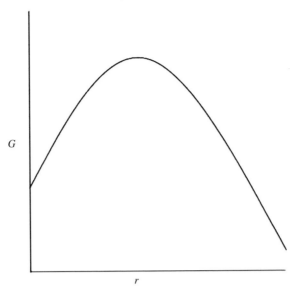

Figure 12.12   Gibbs free energy, *G*, versus crystal radius for small crystals.

decomposition. Nucleation-and-growth occurs in multicomponent systems such as alloys and glasses when the overall Gibbs free energy, *G*, versus composition, *x* (mole fraction), has positive concavity, as shown in Figure 12.13 (Clerc and Cleary, 1995). When *G* versus composition has negative concavity, as shown in the region between compositions *a* and *b*, solidication still occurs, but by the alternate mechanism, spinodal decomposition.

Spinodal decomposition is characterized by spontaneous separation of two or more phases where the formation of the phases does not depend on the growth

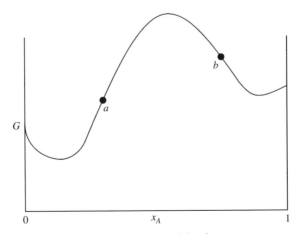

Figure 12.13   Gibbs free energy, *G*, versus composition for a two-component system with a miscibility gap.

of crystallites beyond a critical size in order for the process to be spontaneous. It is rapid and cannot be avoided by dropping the temperature of the sample quickly in order to suppress diffusion. The morphology of systems prepared by spinodal decomposition have characteristic intermingling connections as shown in Figure 12.14a. This spinodal pattern has been exploited in the glass industry by preparing mixtures of the glass components, such as Na_2O, B_2O_3, and SiO_2, that fall in the corresponding three-component spinodal-decomposition range equivalent to region a–b in Figure 12.14 (West, 1984). When this mixture is cooled, an interconnected second phase (minor phase) separates out from the major phase. This minor phase, because of its connectivity, can be chemically etched out of the major phase, Figure 12.14b, leaving a porous form of the major component.

12.1.6 Thin Films

This represents one of the most active and important areas of inorganic materials design, because many of a material's properties depend only on its surface. For

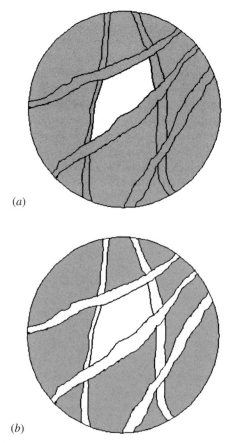

(a)

(b)

Figure 12.14 (a) Spinodal decomposition pattern. (b) With minor component leached out.

example, the hardness of a cutting tool depends on the hardness of the surface of the cutting tool touching the machined item.

In Chapter 8 we examined properties of antireflection coatings. We considered at that time the simplified case of a single film separating air and a substrate. Here we discuss how such a film might be produced. Actual films, such as the anti-reflective coatings on the lenses found in eyeglasses, contain multiple layers plus a protective coating.

Thin films, which we define as ranging from a monolayer to several microns in thickness, are prepared in two ways: physical and chemical deposition. Thin films are deposited on an inert bulk material called the substrate. In physical deposition, the material to be used in the film already exists and is simply being transferred to a substrate. In chemical deposition, the material constituting the film is prepared as part of the film deposition. Both of the methods have a variety of specific variations, and we consider a few here.

The simplest physical deposition involves direct-heating sublimation. This method requires that the material have an appreciable vapor pressure at a temperature where the material is stable. One difficulty with this method is that if the material can be sublimed onto a substrate, it can also be sublimed off! For materials with prohibitively low vapor pressures, three modifications to direct heating have been developed: electron beam heating, sputtering, and laser ablation. In these methods, a sample of the material to be prepared as a thin film is targeted with either an electron beam, an ion (or atomic) beam, or a laser beam. Small bits of the material are dislodged into the gas phase and condense on a cooler substrate. The target material is dislodged either by momentum transfer, in the case of sputtering, or local heating, in the cases of electron beam and laser ablation. In all of these approaches, the stoichiometry of the thin film is an issue. A stoichiometric target, such as MgF_2, does not guarantee a stoichiometric film. Thin-film production requires methods such as X-ray photoelectron spectroscopy or Auger spectroscopy to characterize the film with respect to stoichiometry and composition (Woodruff and Delchar, 1986). The antireflection coatings discussed in Chapter 8 are applied to the lenses in eyeglasses by electron beam heating of SiO_2 and TiO_2 targets to form a four-layer coating, which is effective over a wider range of wavelengths and effective incident angles than can be achieved with a single layer of a single composition.

Chemical vapor methods for thin-film synthesis bring to the substrate surface the chemical reagents, in the gas phase, needed to synthesize the material to be prepared as a thin film. The chemical reaction is allowed to occur at or near the substrate surface, and the resulting material, having a vapor pressure considerably less than the reagent gases, deposits onto the surface. A nice example of this is the synthesis of the classic spinel compound, $MgAl_2O_4$, on a silicon or iron substrate reported by Mathur and co-workers (Mathur et al., 2004). The chemistry behind their approach centers of the following decomposition reaction:

$$MgAl_2(OR)_4(g) \rightarrow MgAl_2O_4(s) + R'(g) + R''(g) + \cdots$$

The challenge for this type of chemistry is to produce a compound with the desired metal and oxygen atoms that has a significant vapor pressure. As a consequence, therefore, thin-film synthesis has provided motivation for continued research and development in molecular organometallic chemistry. The compound $MgAl_2O_4$ is important because of its high melting point, mechanical stability, and chemical inertness, all features that make the processing of $MgAl_2O_4$ into useful forms difficult. Hence a method like the one being developed by Mathur et al. could provide a convenient way to process this important refractory material.

A now famous example of using a decomposition reaction to produce a thin film involves the production of diamond films from the decomposition of methane (May, 2000):

$$CH_4(g) \rightarrow C(s,\ dia) + 2H_2$$

Such a simplified overall reaction is woefully inadequate to describe the details of how this chemistry proceeds. For example, the methane gas is present as a small impurity (1% vol) in hydrogen gas. The methane can be heated at a hot filament or decomposed in a microwave discharge.

A more exotic form of chemical vapor thin-film production uses molecular-beam epitaxy. In this case, individual layers of reagents, sometimes monatomic in thickness, are deposited sequentially. An elegant example of this strategy applied to inorganic materials synthesis is provided by the work of Johnson and co-workers (Harris et al., 2003). $Bi_2Te_3(s)$ and $TiTe_2(s)$ are immiscible. Both are layered structures similar to graphite. By sequentially depositing thin films of tellurium, bismuth, and titanium, Johnson and his co-workers were able to produce a final product with the stoichiometry $(Bi_2Te_3)_x(TiTe_2)_y$ which, although it is metastable, will have a long half-life for the same reasons given at the beginning of this chapter concerning the need for high temperatures in direct combination solid-state reactions: diffusion coefficients in solids are low. Hence the spontaneous thermodynamically favored reaction,

$$(Bi_2Te_3)_x(TiTe_2)_y \rightarrow xBi_2Te_3 + yTiTe_2$$

will be very slow.

This type of reaction represents the ultimate control solid-state synthetic chemists seek: being able to build a compound atom by atom. Johnson et al. took advantage of the 2D nature of Bi_2Te_3 and $TiTe_2$ such that their synthetic challenge was layer by layer. The secret to their success lies in the fact that two important principles control any synthetic efforts: thermodynamics and kinetics. Phase diagrams, which are discussed in Chapter 10, owe allegiance to thermodynamics and have traditionally provided the synthetic chemist with boundaries and targets for planning chemical syntheses. The example of $(Bi_2Te_3)_x(TiTe_2)_y$, however, represents a synthetic achievement that lies outside of traditional phase-diagram constraints. Johnson et al. have prepared materials that are not stable according to the relevant phase diagrams, but have kinetic stability that mimics thermodynamic stability on any relevant time scale.

12.1.7 Photonic Materials

Chapters 3 and 4 emphasize the band gap in materials. This is a critical parameter of a material, determining its utility in transistors, lasers, and detectors. Until recently, altering a material's band gap involved chemical modifications to affect bond lengths, atomic orbital overlaps, and electronegativity. A nice example of this is the series GaP_xAs_{1-x} where the band gap is systematically varied from 2.27 eV to 1.40 eV as x varies from 1 to 0 (Figure 12.15). This represents a traditional chemical approach to achieving a desired property. A more recent approach to preparing materials with specific band gaps involves what have become known as photonic materials.

In photonic materials, the band gap is determined by the geometric arrangement of a dielectric material. The underlying principle of how photonic materials work is best explained using Maxwell's equations (Joannopoulos et al., 1995). Once again, we confront the central importance of Maxwell's equations when optical properties of materials are discussed. In photonic materials, a periodic structure is produced in one, two, or three dimensions. The periodic property is dielectric constant. A trivial macroscopic 1D example would be a collection of individual microscope slides separated by a layers of Saran Wrap. This would produce a 1D modulation in the refractive index, varying between 1.42 and 1.53, on a length scale of millimeters. In order for photonic materials to be of practical interest, the order of the period-

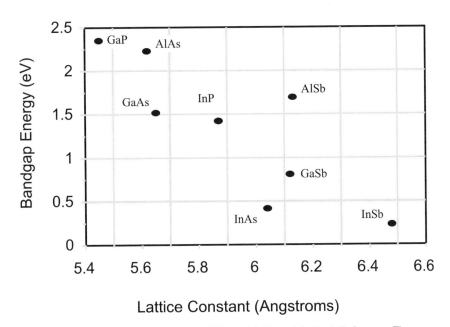

Lattice Constant (Angstroms)

Data taken Ashcroft and Mermin, 1976 and *A Physicist's Desk Reference: The Second Edition of Physics Vade Mecum*, H. L. Anderson, editor-in-chief, 1989, American Physics Institute.

Figure 12.15   Band-gap values for GaP_xAs_{1-x} and other semiconductors.

icity must be similar to the wavelength of visible, ultraviolet, infrared, or micro-wave electromagnetic radiation. Currently, silicon separated by air has been prepared on the length scale of 100 nm (Salib et al., 2004). The refractive index in this case varies between 1.00 and 3.42 at $\lambda = 10$ μm.

This synthetic strategy provides great flexibility to the synthetic chemist. In principle, any band gap in any configuration (loops, lines, pockets, etc.) can be prepared. The drawback to this technique is the high technical demands required for the preparation of the arrays. Lithography, patterning, and masking at X-ray wavelengths limits the number of researchers who can participate in this type of synthesis.

Despite the technical challenges, photonic materials continue to rapidly progress as a area of active research. One of the strongly motivating factors driving this research is the unexpected and unexplored properties accessible in photonic cons-tructions where the limits and restrictions of normal chemical ideas, such as coordination number, valence, and electronegativity, do not apply. For example, photonic materials prepared for the microwave region, where the length scale is micrometers, have been prepared that exhibit a negative refractive index (Pendry and Smith, 2004). In this material, consisting of copper rings separated by either air or Teflon, the modulation of the refractive index at microwave frequencies results in a material that displays a negative bulk refractive index. One of the consequences of this, as shown in Figure 12.16, is that light is bent backwards. This remarkable effect, discussed almost 40 years ago by Veselago, but only recently observed with the advent of photonic materials, provides a glimpse into what might be possible with this newly emerging synthetic strategy (Veselago, 1968). In the field of optical fibers, photon materials (called single-mode photonic-crystal fiber lasers, or PCF lasers) have the potential to revolutionize high power fiber lasers operating in single mode (Limpert et al., 2004).

12.1.8 Nanosynthesis

There are two approaches to the preparation of nanoparticles and nanocrystals. The first is sometimes called the "top-down" approach and the second, the "bottom-up" approach. Top-down methods essentially convert a coarse-grained polycrystalline sample into a nanocrystalline form. For example, the individual grains in a metal will subdivide into domains as small as 20 nm when subjected to large shearing strains while under pressure. This is known as the severe plastic deformation (SPD), technique. Nanocrystals of alloys can also be obtained by the devitrification of a metallic glass, if the alloy can be put into the amorphous state to start with. Unfortunately, these techniques do not necessarily produce well-controlled grain morphologies.

The goal of materials research is really the reverse process, the "bottom-up" method. In this approach, it is hoped that perfect well-controlled nanoparticles, nanostructures, and nanocrystals can be synthesized that may be compacted into macroscopic nanocrystalline samples, or assembled into superlattice arrays, which can, in turn, be used in a variety of applications, such as nanoelectronic or magnetic

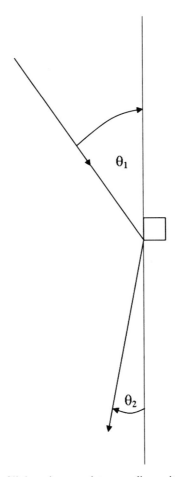

Figure 12.16 Refraction of light as it passes into a medium with a negative refractive index.

devices. Some scientists have even envisioned a time when so-called "molecular assemblers" will be able to mechanically position individual atoms or molecules one at a time in some predefined way (Drexler, 1986). The feasibility of such machines has been hotly debated but, regardless, such systems engineering goals are not really within the scope of this chapter. At present, methods for synthesizing metal and ceramic clusters and nanoparticles fall in one of two broad categories: liquid phase techniques, or vapor/aerosol methods.

Liquid-Phase Techniques   One well-known chemical method for synthesizing nanoscale metal particles is reduction of the metal ions with an aqueous solution of $NaBH_4$ (Dragieva, 1999). Another chemical technique used for preparing nanooxides is the precipitation of a metal hydroxide from a salt solution (e.g., $AlCl_3$) by the addition of a base (e.g., NH_4OH). The product is washed and calcined to obtain a final ultrafine oxide powder (e.g., Al_2O_3).

A soft chemical route known as the sol-gel method has also been employed for the preparation of nanooxides with uniform size and shape. This is a multistep process, usually consisting of hydrolysis of a metal alkoxide in an alcoholic solution to yield a metal hydroxide, followed by polymerization by elimination of water (gel-formation), drying off the solvent, and densification of the product to yield an ultrafine powder (Rao and Raveau, 1998; Khaleel and Richards, 2001)

A relatively new route to nanoparticles, which has been employed for the preparation of colloidal silver nanoparticles (Maillard et al., 2002), as well as gold-coated iron nanoparticles (Zhou et al., 2001), and nanoparticles of compound semiconductors (e.g., CdTe, CdS, and Cdl yZnyS) (Pileni, 1993), is the reverse-micelle method. A reverse micelle is a spherical cluster of surfactant molecules (commonly sodium bis (2-ethylhexyl)-sulfosuccinate (NaAOT) suspended in a nonpolar solvent. In a nonpolar solvent, the hydrocarbon tails of the surfactant molecules become oriented toward the exterior of the aggregate, while the polar sulfonate headgroups are localized in the interior. The combination of the associated surfactant molecules (the reverse micelle) together with the nonpolar solvent constitutes a lyotropic phase and is a type of liquid crystalline ordering. Surfactant in water is another example of a lyotropic phase, but one that results in the formation of micelles instead of "reverse" micelles. Just as micelles can solubilize grease and oil in their nonpolar interior, reverse micelles have the ability to solubilize water, forming a so-called water pool. The water pool can accommodate hydrolysis and precipitation reactions for the preparation of insoluble nanoparticles. Usually, a solvent such as dodecanethiol is added to induce a size-selective synthesis as well as to coat the particles in order to protect them from surface reactions.

There are some major challenges associated with compacting nanoparticles into nanocrystalline phases. Nanoparticles are usually difficult to prepare in monodisperse form and the high surface area to volume ratio imparts enhanced surface reactivity, particularly with metals. For the latter reason, we have seen that nanometals are generally coated with a protective shell of ligand molecules. The nonuniform shapes and sizes of most nanometal particles, along with the more-or-less spherical protective ligand shells surrounding them, makes the assemblage of these clusters into 3D nanocrystals difficult (Schmid, 2001). However, some groups have been able to overcome these obstacles and produce metal nanocrystals with edge lengths of ~100 nm from 4 5-nm-sized clusters (Harfenist et al., 1997; Wang et al., 1998).

Template-directed syntheses are often used to produce nanostructured particles such as mesoporous materials with voids less than 50 nm in diameter. The challenge is to capture the desired size and shape of the particle such that it is preserved once the template is removed, usually by calcination. In the early 1990s, researchers at Mobil Oil Corporation showed that mesoporous silicates could be synthesized by using lyotropic liquid crystalline phases (e.g., surfactant/solvent systems) as the template (Beck et al., 1992; Kreske et al., 1992). With low surfactant concentrations, lyotropic systems are biphasic in nature. That is, the individual micelles are not joined together. Consequently, fine powders of the mesoporous materials are typically obtained from these systems. Attard later showed that with a high surfactant concentration, homogeneous (monophasic) lyotropic systems character-ized by continuous spatially periodic architectures are possible (Attard et al., 1995). The hydrolysis and polycondensation of silica precursors in these monophasic templates were shown to produce monolithic mesoporous silica whose architecture is essentially that of the liquid crystalline phase. For example, silica cylinders with lengths greater than 10 m and diameters of 20–30 Å were prepared from sol-gel precursors.

A very recent novel liquid-phase route to hollow nanocrystals of cobalt oxide and cobalt sulfide takes advantage of the Kirkendall effect (Section 5.4.1). Injection of sulfur or oxygen into a colloidal cobalt nanocrystal dispersion created hollow nanocrystals of the chalcogenide as a direct result of the differing intrinsic diffusion coefficients of cobalt and sulfur or oxygen, the outward transport of cobalt atoms through the chalcogenide, being balanced by an inward flow of vacancies to the metal–chalcogenide interface (Yin et al., 2004).

Vapor/Aerosol Methods   A few vapor/aerosol methods are available for the preparation of nanoparticles or films. One is a PVD process known as the sputtering technique, in which atoms are sputtered (knocked loose) from a source, known as the "target," by bombarding it with high-energy ions in a vacuum chamber filled with an inert gas. The ions are produced by creating a large potential difference between the target and substrate. The ionization of the inert gas forms a plasma (a neutral collection of positively charged ions and negatively charged electrons), and the negative potential on the sputtering target attracts the positive ions, which are accelerated by the electric field. The atoms dislodged from the sputtering target deposit on a substrate.

A related technique is the evaporation method, whereby a bulk metal is heated with an inert gas in a vacuum chamber. The heating may be by various means, say, for example, from an induction current generated by a varying magnetic field. The heating causes the metal to emit atoms, which collide with inert gas molecules, lose kinetic energy, and condense as metal clusters on a cold finger or on a substrate. If metal oxides are desired, oxygen can be introduced into the vacuum chamber along with the inert gas. Alternatively, ammonia or an alkane, respectively, can be intro-duced to produce metal nitrides or carbides.

Rather than produce an atomic vapor by evaporation from a solid surface, an aerosol can be generated from an aqueous salt solution by an "atomization" proce-dure. The aerosol can then be evaporated so that the salt condenses into a particle. This is known as the spray pyrolysis technique. The flame decomposition method is a modification of this technique, in which the aerosol is introduced into a high-temperature flame (1200–3000 K). The precursor is vaporized and oxidized to form metal oxide particles.

Combined Strategies   As we mention at the beginning of this chapter, the well-defined solid-state synthetic methods are themselves being combined to produce new materials. For example, thin-film preparative strategies have been coupled to combinatorial methods. Takahashi et al. reported in the *Journal of*

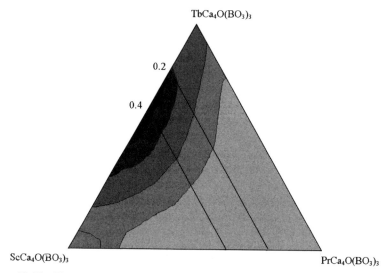

Figure 12.17 Three-component phase diagram for TbCa$_4$O(BO$_3$)$_3$, ScCa$_4$O(BO$_3$)$_3$, and PrCa$_4$O(BO$_3$)$_3$, with resulting luminescence indicated by the contours.

Combinatorial Chemistry a synthesis of Tb$_{1-x-y}$Sc$_x$Pr$_y$Ca$_4$O(BO$_3$)$_3$ deposited on an aluminum oxide substrate (Takahashi et al., 2004). They were interested in the photoemission properties of this series of compounds. Their results are summarized in the three-component phase diagram shown in Figure 12.17, the three components being TbCa$_4$O(BO$_3$)$_3$, ScCa$_4$O(BO$_3$)$_3$, and PrCa$_4$O(BO$_3$)$_3$, with the contours representing white-light emission intensity. From this they conclude that the optimum composition is Tb$_{0.6}$Sc$_{0.4}$Ca$_4$O(BO$_3$)$_3$, and that Pr has no positive effect.

The beauty of this experiment is that the continuous range of x and y can be investigated in a single experiment. Preparing by direct combination enough individual samples to support the conclusion drawn using the combinatorial method would require an enormous investment in time and effort. This illustrates that the combinatorial method's great strength lies in its screening capability. Once the formula Tb$_{0.6}$Sc$_{0.4}$Ca$_4$O(BO$_3$)$_3$ has been established as optimal, more traditional methods can be employed for mass production.

The combinatorial and thin-film strategies make an ideal pairing. Much of the inorganic materials synthesis done using the combinatorial approach results in thin-film products. This should not be interpreted to mean that combinatorial methods are restricted to inorganic thin-film synthesis. Potentially the most important application of combinatorial methods has been in the screening of chemical compounds for their pharmacological utility. This involves the use of organic compounds, which is outside the scope of this text.

Thin-film preparation can be coupled to other synthetic strategies, such as sol-gel synthesis. For example, Qide et al. presented a synthesis of TiO$_2$ thin films prepared by the sol-gel method (Qide et al., 2003). The interest in thin films of TiO_2 lies in the application of these films to photocatalysis. TiO_2 has been shown to be photocatalytic with respect to the oxidation of organic contaminants such as chloroform (Fujishima et al., 1999):

$$CHCl_3(aq) + H_2O(l) + 1/2O_2(aq) \rightarrow CO_2(aq) + 3HCl(aq)$$

This reaction proceeds under ultraviolet light in the presence of TiO_2, presumably at the surface of the insoluble TiO_2 particles.

The approach of Qide et al. to produce thin films of TiO_2 by a sol-gel strategy uses a mixture of titanium isopropoxide, ethanol, water, and triethanolamine. The chemistry is similar to what is used in the silicate gel where a hydrolysis/condensation reaction is exploited (Wright and Sommerdijk, 2001):

$$Ti(OC_3H_7)_4(soln) + 4H_2O \rightarrow TiO_2(s) + 4HOC_3H_7 + \cdots$$

The gel produced from this reaction is coated on an inert substrate, such as glass or stainless steel, and the sample is heated to 500°C for 1 h in air. The films produced are characterized by thickness, grain size, pore size, surface area, and photocatalytic activity. Photocatalytic activity is determined by measuring the rate of discoloring of the aqueous solution of methyl organge, a typical organic dye.

The motivating force for the application of a technique like sol-gel to thin-film synthesis is the simplicity, because in this case simplicity brings low cost. If the preceding example were to become a commercially viable method for removing organic contaminants from water, then the producing method of the TiO_2 thin films would have to be economically viable. The high-vacuum, ultraclean, equipment-intensive approach of methods already discussed, such as molecular-beam epitaxy or laser ablation, would not be able to meet that cost requirement. The sol-gel method however, does bring with it its own limitations. For example, the alkoxides used as starting materials can be expensive and sensitive to moisture, both challenges for large-scale production designs.

A third example of the crossover among these synthetic strategies involves the preparation of doped oxides. We have already discussed the importance of doping as a synthetic strategy used to achieve desired properties. Doping and thin-film preparation come together nicely when chemical vapor deposition is used to prepare the films.

McKittrick and co-workers prepared $(Y_{1-x}Eu_x)_2O_3$ films employing just such a strategy (McKittrick et al., 2000). The objective was to produce a compound with photoluminescent and cathodoluminscent properties suitable for flat-panel displays. Yttrium tris(2,2,6,6-tetramethyl-3,5-heptanedionate) can be decomposed to produce Y_2O_3:

$$Y(C_{11}H_{19}O_2)_3(g) \rightarrow Y_2O_3(s)$$

Owing again to the periodic law, we expect and find a series of compounds $Ln(C_{11}H_{19}O_2)_3$, where Ln is a tripositive lanthanide ion. Hence with careful control of the mixing of $Y(C_{11}H_{19}O_2)_3$ and $Eu(C_{11}H_{19}O_2)_3$ vapors, thin films can be produced with the formula $(Y_{1-x}Eu_x)_2O_3$. In McKittrick's work on this project, x ranged from 0.18 to 0.31. This provides yet another example of how molecular organometallic chemistry supports thin-film synthesis. The reagents, also known as precursors, necessary to produce the lanthanide thin films are commercially available from, for example, Strem Chemicals, Inc.:

Precursor	Melting Point	Form
$Eu(C_{11}H_{19}O_2)_3$	188–189°C	Yellow powder
$Y(C_{11}H_{19}O_2)_3$	170–173°C	White
$Nd(C_{11}H_{19}O_2)_3$	209–212°C	Light purple

An intriguing possibility for preparing a doped material using chemical-vapor deposition involves the diamond thin films mentioned earlier. If p-doped and n-doped diamond could be produced by mixing appropriate gases with the methane in the chemical vapor deposition, it might be possible to produce integrated circuits based on diamond instead of silicon. Because the diffusion coefficient for the dopants in the diamond would be very low (the reason why attempting to dope the diamond directly fails), the doped material would be stable to extremely high temperatures. As an added bonus, control over the final composition would be tighter given that it is easier to purify methane than it is to purify silicon.

12.2 SUMMARY

In this chapter, we have discussed several of the major synthetic strategies used in the preparation of modern inorganic materials. In this discussion it is apparent that inorganic materials synthesis is a diverse field with many specialized pockets of expertise, each suited for a particular synthesis. The cost and technical infrastructure required to participate in this endeavor can be extensive with a method such as molecular-beam epitaxty or photonic-materials synthesis. On the other hand, sol-gel and direct-combination methods can easily be incorporated into undergraduate and even high school curricula. Nonetheless, these low-cost methods make invaluable contributions to the overall field, as evidenced by the discovery of high-temperature superconductors!

In addition to surveying the major strategies, we have also examined the growing trend to combine strategies. This has proven particularly productive in the area of thin-film preparation where combinational and sol-gel techniques are being increasingly adapted.

REFERENCES

Adamson, A. W. 1982, *Physical Chemistry of Surfaces*, John Wiley & Sons, New York.

Attard, G. S.; Glyde, J. C.; Goltner, C. G. 1995, *Nature*, *378*, 366.

Beck, J. S.; Vartuli, J. C.; Roth, W. J.; Leonowicz, M. E.; Kresge, C. T.; Schmitt, K. D.; Brinker, C. J.; Scherer, G. W. 1990, *Sol-Gel Science: The Physics and Chemistry of Sol-Gel Processing*, Academic Press, Boston.

Bednorz, J. G.; Muller, K. A. 1986, *Z. Phys.*, *B64*, 189.

Brec, R. 1986, *Solid State Ionics*, *22*, 3.

Chaput, F.; Darracq, B.; Boilot, J. P.; Riehl, D.; Gacoin, T.; Canva, M.; Levy, Y.; Brun, A. 1996, in Coltrain, B. K., Sanchez, C., Schaefer, D. W., Wilkes, G. L., editors, *Better Cermanics Through Chemisty VII: Organic/Inorganic Hybrid Materials*, MRS Symp. Proc., *435* 583.

Chu, C. T.-W.; Olson, D. H.; Sheppard, E. W.; McCullem, S. B.; Higins, J. B.; Schlenker, J. L. 1992, *J. Am. Chem. Soc.*, *114*, 10834.

Clerc D. G., Cleary D. A., 1995, *J. Chem. Educ.*, *72*, 112.

Coradin, T.; Clement, R.; Lacroix, P. G.; Nakatani, K. 1996, *Chem Mater.*, *8*, 2153.

Crank, J. 1975, *The Mathematics of Diffusion*, 2nd ed., Oxford Univ. Press, Oxford.

Dieke, G. H. 1968, *Spectra and Energy Levels of Rare Earth Ions in Crystals*, John Wiley & Sons, New York.

Dragieva, I. 1999, in Nedkov, I., Ausloos, M., editors, *Nano-Crystalline and Thin Film Magnetic Oxides*, Kluwer Academic Publishing, Dordrecht, The Netherlands.

Drexler, K. E. 1986, *Engines of Creation: The Coming Era of Nanotechnology*, Anchor Books, New York.

Eliseev, A. A.; Lukashin, A. V.; Vertegel, A. A. 1999, *Chem. Mater.*, *11*, 241.

Fujishima, A.; Hashimoto, K.; Watanabe, T. *TiO₂ Photocatalysis: Fundaments and Applications* 1999. BKC, Inc. Tokyo, Japan.

Gamble, F. R.; DiSalvo, F. J.; Klemm, R. A.; Geballe, T. H. 1970, *Science*, *168*, 568.

Ginzburg, V. L. 1989, *Physics Today*, *42*(3), 9–11.

Gomez-Romero, P.; Sanchez, C., editors, 2004, *Functional Hybrid Materials*, Wiley-VCH, Weinheim, Germany.

Harfenist, S. A.; Wang, Z. L.; Alvarez, M. M.; Vezmar, I.; Whetten, R. L. 1997, *Adv. Mater. 9*, 817.

Harris, F. R.; Standridge, S.; Feik, C.; Johnson, D. C. 2003, *Agnew. Chem. Int. Ed.*, *42*, 5296.

Heer, S.; Lehmann, O.; Haase, M.; Gudel, H.-U. 2003, *Angew. Chem. Int. Ed.*, *42*, 3179.

Henisch, H. K., 1970, *Crystal Growth in Gels*, Dover Publications, Inc. New York.

Huan, G.; Jacobson, A. J.; Johnson, J. W.; Corcoran, E. W., Jr. 1990, *Chem. Mater.*, *2*, 91.

Joannopoulos, J. D.; Meade, R. D.; Winn, J. N. 1995, *Photonic Crystals: Molding the Flow of Light*, Princeton University Press, Princeton, NJ.

Khaleel, A.; Richards, R. M. 2001, in Klabunde, K. J., editors, *Nanoscale Materials in Chemistry*, Wiley-Interscience, New York.

Kolb, E. D.; Barns, R. L.; Laudise, R. A.; Grenier, J. C.; 1980, *J. Cryst. Growth*, *50* 404.

Kolb, E. D.; Caporaso, A. J.; Laudise, R. A.; 1968, *J. Cryst. Growth*, *3,4*, 422.

Kresge, C. T.; Leonowicz, M. E.; Roth, W. J; Vartuli, J. C; Beck, J. S. 1992, *Nature*, *359*, 710.

Laudise, R. A. 1987, *Chem. Eng. News*, *74*, 30.

Laudise, R. A.; Cava, R. J.; Caporaso, A. J.; 1986, *J. Cryst. Growth*, *74*, 275.

Li, J.; Chen, Z.; Wang, R.-J.; Proserpio, D. M.; 1999, *Coord. Chem. Rev.*, *190–192*, 707.

Limpert, J.; Liem, A.; Schreiber, T.; Foser, F.; Zellmer, H.; Tunnermann, A. 2004, *Photonics Spectra*,

Madou, M. J.; Morrison, S. R. 1989, *Chemical Sensing with Solid State Devices*, Academic Press, San Diego, CA.

Maillard, M.; Giorgio, S.; Pileni, M. P. 2002, *Adv. Mater.*, *14*, 1084.

Mathews, C. K.; van Holde, K. E.; Ahern Benjamin, K. G. 2000, *Biochemistry*, 3rd ed. Cummings, San Francisco, p. 924.

Mathur, S.; Veith, M.; Ruegamer, T.; Hemmer, E.; Shen, H. 2004, *Chem. Mater.*, *16*, 1304.

May, P. W. 2000, *Philos. Trans. R. Soc., London A*, *358*, 473.

McKittrick, J.; Bacalski, C. F.; Hirata, G. A.; Hubbard, K. M.; Pattillo, S. G.; Salazar, K. V.; Trkula, M. 2000, *J. Am. Ceram. Soc.*, *83(5)*, 1241.

Missori, M.; Bianconi, A.; Saini, N. L.; OYanagi, H. 1994, *Il Nuovo Cimento*, *16 D*, 1815.

Nelson, D. L.; Whittingham, M. S.; George, T. F., editors, 1987, *Chemistry of High-Temperature Superconductors*, ACS Symposium Series, vol. 351, American Chemical Society, Washington, DC.

Pendry, J. G.; Smith, D. R. 2004, *Phys. Today*, *57*(6), 37–43.

Pileni, M. P. 1993, *J. Phys. Chem.*, *97*, 6961.

Qide, W.; Bei, C.; Gaoke, Z. 2003, *Rare Metals*, *22(2)*, 150.

Rao, C. N. R.; Raveau, B. 1998, Transition Metal Oxides: Structure, Properties, and Synthesis of Ceramic Oxides, 2nd ed., Wiley-VCH, New York.

Risk, W. R.; Gosnell, T. R.; Nurmikko, A. V. 2003, *Compact Blue-Green Lasers*, Cambridge University Press, Cambridge, UK.

Salib, M.; Liao, L.; Jones, R.; Morse, M.; Liu, A.; Samara-Rubio, D.; Alduino, D.; Paniccia, M. 2004, *Intel Tech. J.*, *8*, 142.

Schmid, G. 2001, in Klabunde, K. J., editor, *Nanoscale Materials in Chemistry*, Wiley-Interscience, New York.

Silfvast, W. T. 1996, *Laser Fundamentals*, Cambridge University Press, Cambridge, UK.

Sleight, A. W. 1989, *MRS Bull.*

Takahashi, R.; Kubota, H.; Murakami, M.; Yamamoto, Y.; Matsumoto, Y.; Koinuma, H. 2004, *J. Comb. Chem.*, *6*, 50.

Ushakov, S. V.; Helean, K. B.; Navrotsky, A.; Boatner, L. A. 2001, *J. Mater. Res.*, *16* 2623.

Wang, Z. L.; Harfenist, S. A.; Vezmar, I.; Whetten, R. L.; Bentley, J.; Evans, N. D.; Alexander, K. B. 1998, *Adv. Mater.*, *10*, 808.

Veselago, V. G. 1968, *Sov. Phys. Usp.*, *10(4)*, 509.

West, A. R. 1984, *Solid State Chemistry and Its Applications*, John Wiley & Sons, Chichester, UK.

Whittingham, M. S.; Jacobson, A. J., editors. 1982, *Intercalation Chemistry*, Academic Press, New York.

Woodruff, D. P.; Delchar, T. A. 1986, in *Modern Techiques of Surface Science,* Cambridge Solid State Science Series, Cahn, R. W., Davis, E. A., Ward, I. M., editors, Cambridge University Press, Cambridge, UK.

Wright, J. D.; Sommerdijk, N. A. J. M. 2001, *Sol-Gel Materials: Chemistry and Applications*, Advanced Chemistry Texts, Vol 4, Gordon & Breach, Australia.

Xiang, X.-D.; Takeuchi, I., editors, 2003, *Combinatorial Synthesis*, Marcel Dekker, New York.

Yin, Y.; Rioux, R. M.; Erdonmez, C. K.; Hughes, S.; Somorjai, G. A.; Alivisatos, A. P. 2004, *Science*, *304*, 711.

Zhou, W.; Wiemann, K. J.; Fang, J.; Carpenter, E. E.; O'Connor, C. J. J. 2001, *Solid State Chem.*, *159*, 26.

Principles of Inorganic Materials Design   By John N. Lalena and David A. Cleary
ISBN 0-471-43418-3   Copyright © 2005 John Wiley & Sons, Inc.